INDUSTRIAL EFFLUENTS
ORIGIN, CHARACTERISTICS, EFFECTS, ANALYSIS & TREATMENT

N.MANIVASAKAM
Formerly **CHIEF WATER ANALYST**
Principal Public Health Laboratory
Coimbatore - 641018 (T.N.)
INDIA

Chemical Publishing Company, U.S.A.
2016

Industrial Effluents - Origin, Characteristics, Effects, Analysis & Treatment

ISBN: 978-08206-0414-5

eBook ISBN: 978-08206-0415-2

Chemical Publishing Company:
www.chemical-publishing.com
www.chemicalpublishing.net

Printed in the United States of America

DEDICATED
TO THE
LOTUS FEET
OF
LORD MURUGA
AND
TO MY PARENTS
C.K.NATARAJ AND ARUNAGIRI AMMAL

I gratefully acknowledge,

1) Addison - Wesley Publishing Company
Massachusetts,

2) American Public Health Association,
Washington,

3) ELBS and John Wiley & Sons,
Chichester,

4) Thermax Private Ltd., Pune, India, and

5) Oxford & I.B.H. Publishing Co.,
Calcutta, India

for their kind permission to reproduce some of the figures, tables, and paragraphs from their publications,

- N. MANIVASAKAM

PREFACE

It is the intention of the author to offer a comprehensive account on various types of industrial effluents. A vast literature on this subject is available, but this book is a single volume discussion covering all the aspects of effluents. Much effort has been taken to provide all the possible information and the subject is discussed in multiple angles to cater the diverse interests of the reader.

This book is divided into five parts. Part I describes the effects of effluents, types of industries and summary of the origin, characteristics and treatment of effluents. Part II and part III deal with different types of effluents industry wise in detail. They are dealt in the following sequence. Manufacturing processes including raw materials and chemicals, origin and volume of effluents, characteristics and effects of them on surface waters, land and on sewers and sewage treatment systems. It is succeeded by a list of parameters to be determined so that one may not miss the important tests as well minimize the number of tests. Necessary precautions and preliminary treatments to be followed during the determinations are also given. Each chapter is provided with an elaborate list of references so that the reader could peep further for his requirements. Part IV describes in detail the analytical procedures for the examination of effluents. These procedures are applicable to raw effluents, treated effluents, polluted waters, raw and treated waters. In the beginning of the determination of each constituent, a short list of the industries that generate the specific constituent in their effluents is given. The analytical procedures are described step by step in an easy way. Part V illustrates the primary, secondary and tertiary treatments given to an effluent in detail.

Important tables and conversion factors are given in the appendix. An exhaustive index is incorporated at the end.

The book has wide applications and it could be used in many ways. Needless to say that this book would be highly useful to industries and personnel, engineers and chemists engaged in effluent treatment, to water and waste water analysts, to consultants, to pollution control board authorities, to fisheries and to all others who engaged in pollution control programmes. University faculties dealing with Environment, Ecology and Pollution, Public Health Engineering, Industrial Toxicology, Chemistry, Zoology and Botany would find this as an invaluable reference book.

This book is prepared with the assistance of a number of people. Especially I am indebted to Mr. S. Murugesan without whose encouragement this work is not possible. I am delighted to express my acknowledgements to my wife Mrs. M. Sivabagiam, to my son Sakthikumar and my mother Mrs. S.A. Arunagiri Ammal for their patience and encouragements.

I am immensely pleased to express my gratitude and sincere thanks to M/s. Chemical Publishing Company, Los Angeles, U.S.A., for their keen interest shown towards the publication of this book.

- N. MANIVASAKAM

This book is divided into the following parts

CONTENTS

PART - I

BASIC INFORMATION

Chapter - 1

BASIC INFORMATION

Pollution is a serious problem ever since sewage and industrial effluents are disposed of into water courses and on land.

It has increased with the growth of industry as well as of population. Industries can conveniently be classified in to two groups viz. (i) Dry process industries and (ii) Wet process industries. Dry process industries (mostly engineering and assembling industries) do not use water in their process, while wet process industries use water either as raw material or in their process or for both. Only a little quantity of water is absorbed in the process and the rest being discharged as effluent. Practically all the volume of water used is calculated as the volume of effluent. In addition to the process waters, large volumes of cooling water are also discharged as waste. Boiler blow down is also an effluent contributing significant quantities to the bulk. These effluents contain a wider variety of materials of both organic and inorganic nature including toxic substances and are usually discharged with or without treatment into surface waters such as rivers, streams, lakes or into oceans or on land or in to sewers. The solid wastes produced in these industries are not given much importance as they are either recycled, dumped or disposed of to remote places. Liquid industrial wastes are of great concern because of their harmful effects which are described in detail in the following pages.

EFFECTS OF INDUSTRIAL EFFLUENTS

EFFECTS ON SURFACE WATERS

Industries need a wide variety of raw materials and chemicals which are later discharged via their effluents. Acids, alkalis, toxic metals, pesticides and other poisonous substances such as cyanide, dyes, oils, detergents, resins, rubbers are a few to mention. Heated effluents that impart thermal loading on receiving waters and effluents containing radio active materials are also of prime concern. Some of the effluents such as from tanning and meat packing may also contain pathogenic bacteria. The nature and extent of pollution depends on the materials present in the effluent and on the quantity discharged. The constituents present in effluents, their effects on receiving streams etc., are described in detail.

ACIDS

Both organic and inorganic acids are discharged into rivers in the form of industrial effluents. The principal processes which contribute mineral acids are, manufacture of fertilizers, battery manufacture, brewing, textiles, mining, iron and copper pickling. Organic acids are discharged from the manufacture of rayon, fermentation plants, distilleries, dyeing industries, manufacture of leather, chemical industries etc. The main constituent of acid wastes is sulfuric acid. In addition, nitric acid, hydrochloric acid and phosphoric acid are also present.

These acids are considered to cause damage to metals or concrete structures, pumps etc., by their corrosive activity. Moreover, they produce hydrogen sulfide gas when they come into contact with the sludge and mud of the river. The lower the pH value the greater is the amount of hydrogen sulfide liberated leading to a higher degree of atmospheric pollution. Acid pollutants, also destroy microorganisms and thus inhibit self-purification of the stream. Acids are reported to be lethal to fish and other forms of aquatic life.

ALKALIS

Alkalis also cause damage to rivers and streams by altering pH values resulting in a break in the natural buffer system. Chemical manufacture wastes, kier liquors, wool scouring wastes, tannery wastes, cotton mercerizing wastes are the principal contributors of alkalis. The alkali wastes raise the pH values even upto 12. Alkalis behave similarly as acids in destroying bacterial and micro-organisms and thus inhibit the self-purification of a stream. Fish and other aquatic life are also affected by the sudden rise in pH. Strong alkalis are reported to produce asphyxiation by the coagulation of gill secretions in fish. Alums and aluminium salts have a similar action due to the hydrolysis followed by formation of hydrogen ions.

SUSPENDED MATTER

Insoluble matter in suspension is one of the commonest forms of pollution. Suspended matter is present in sewage and in most of the industrial waste waters. Wastes from stone quarrying works, sand-washing, china-clay works, contain primarily inorganic suspended solids. Sewage effluents and many coal washery effluents contain suspended solids which are organic in nature. Suspended matter, though not causing any harm is objectionable as it interferes with the self purification of streams by reducing the photosynthetic activity of water plants and by smothering benthic organisms. Moreover, abrasive materials present in the suspended matter can choke the gills of fish and there by cause damage to fisheries. Deposition of suspended matter on stream bed destroys plant and animal life and thus the food supplies for fish are diminished. The coal washery waste waters which are black and opaque, discolors streams and render it unfit for most uses. Suspended matter containing organic materials, will undergo putrefaction during warm weather. As a result, solids buoyed by gas may rise to the surface leading to the formation of floating masses of evil odor.

TOXIC METALS

Battery manufacturing, paint manufacturing, viscose, rayon manufacturing, electroplating, copper pickling, galvansing and rubber processing are some of the important industries which discharge effluents containing considerable amounts of heavy metals. The primary metals considered to be toxic are lead, arsenic, copper, cadmium, mercury and nickel.

These metallic solutions are toxic and inhibit self-purification of rivers. In addition, aquatic plants and animals are also killed. If the industrial waste is discharged into sewer, metals are precipitated and settle down. Contamination of rivers with toxic metals prevent further use of water for industrial and drinking purposes. Arsenic and lead particularly make water dangerous to drink. These toxic materials are reported to cause chromosome damage and thus interfere with the process of heredity. It is reported that heavy metal ions precipitate the mucous secretion of the gills. These precipitates occupy the interlamellar spaces and the movement of the gill filaments is arrested and respiration is prevented. Copper in about 2 parts per hundred million parts of water is fatal to sticklebacks. The lethal concentration of nickel, lead and zinc is about 1 p.p.m.

In very recent years, a lot of attention has been given to the problems of mercury poisoning. Chlor-alkali industries contribute significant quantities of mercury as it is used as an electrode in the electrolysis of brine. Pulp and paper industries, plastic industries, oil refining and battery manufacture also contribute to mercury pollution. The organic compounds of mercury are toxic while liquid metallic mercury is not toxic. Mercury discharged from industrial processes is mostly in inorganic forms. Anaerobic bacteria in the mud at the bottom of the water course converts inorganic mercury to methyl mercury which is most injurious and concentrate in the food chain, ultimately affecting human beings.

Cadmium ranks next to mercury in its toxicity. The principal contributor of cadmium is electroplating industry.

PESTICIDES

Pesticides if present even in small concentrations in water are fatal to fish, crustaceans and shellfish. They are able to change the metabolic activities of the body. They also cause depression of photosynthesis of planktons.

DYES

Now-a-days thousands of dyes are prepared and they find their way into the streams through the discharge of effluents from dyeing and printing industries, textile industries and to a lesser extent from paper and ink manufacturing industries. These dyes are prepared from coal-tar hydro carbons such as benzene, naphthalene, anthracene, toluene, xylenes etc., Apart from the effluents containing chromium and iron, these dye effluents are the only effluents that impart color to water courses. These organic dyes impart color even when present in micro-quantities. Though the dyes are not harmful, they are undesirable for aesthetic reasons. More over they also add to the B.O.D (Biochemical Oxygen Demand) of receiving waters. In addition to the dye material, salts of chromium and aluminium, and iron which are used as mordant are also found in dye effluents. These mordants also impart color.

OILS

Replacement of coal by oil has increased the problem of pollution. Discharge of water from engineering and metallurgical industries contribute to significant levels of pollutions by oils. Tankers, barges and boats also introduce oils to rivers, canals and lakes.

The presence of oil in a water course is undesirable. It has a tendency to spread over the surface of water and the diffusion of oxygen into water is inhibited and reaeration of water is affected. The absorption of oxygen from air into water depends on the thickness of the film formed by oil. When the thickness of the film increases even above 0.001 mm, a marked effect on the absorption is observed. Oil is reported to coat the gills of fish and thus affect respiration of fish. Oil pollution of sea leads to other problems. Oil is pushed to the shore by water currents and winds. Thus beaches are spoiled and bathing becomes impossible in such places. A report from Britain indicates that sea-birds around Britain are badly affected due to oil pollution.

In addition to the above, 'water insoluble' oils, 'soluble oils' also cause serious pollution. All these soluble oils form milky emulsions with water and the formed emulsions are very stable and cannot be destroyed by easy means.

DISSOLVED SOLIDS

Chlorides, sulfates, nitrates, bicarbonates, and phosphates of sodium, potassium, calcium, magnesium iron and manganese and some of the fluorides are soluble in water. Water softening plants and drainage from salt works contribute many of these salts to river and streams. Fluorides and nitrates are considered to be toxic to man above a certain level.

When waters containing excessive salts are discharged into rivers, the river course becomes brackish which affect fish life and certain vegetation. Many of the freshwater fish varieties die in salt water and this is entirely due to osmotic changes. Not only these soluble salts affect biological organisms but they also cause severe effects to pipelines, pumps and other metallic or concrete structures. Sulfates are converted either into sulfuric acid which aids corrosion or into sulfides which produce odor nuisances. The conversion to sulfides is brought about by sulfate reducing bacteria.

Certain soluble salts of iron and aluminium react with the natural bicarbonate alkalinity to form insoluble hydroxides.

When the river or stream is depleted of dissolved oxygen due to sewage or other contaminants, the water may turn black due to formation of ferrous sulfide. Iron can react with phenols to form various colored compounds and make the streams colored.

SYNTHETIC DETERGENTS

Synthetic detergents have been developed only recently and the use of these compounds increased enormously. Synthetic detergents are used as household cleaning materials and industrial cleaning agents. A point to be noted is that natural detergents exhibit only a little toxicity. It is considered that the toxicity of synthetic detergents may also be due to its non-biodegradability. The intoxicity is completely attributed to the alkyl benzene sulfonate which is the active ingredient.

GASEOUS POLLUTANTS

This category includes ammonia, free chlorine, hydrogen sulfide, ozone and phosphine. Ammonia is liberated during the manufacture of chemicals, fertilizers, and coke. Sewage and other nitrogenous organic matter contribute and increase the ammonia content to dangerous levels. Free chlorine results from the processes of textile bleaching, paper mills and laundries. Hydrogen sulfide occurs by the bacterial action and pH alterations during the discharge of viscose rayon wastes and sulfate and sulfide containing materials. Toxicity of these pollutants are well known and need not be emphasized again.

HEATED EFFLUENTS

Nuclear power plants and thermal power stations are the prime industries the effluents of which contribute to thermal loading which leads to stratification of water course and reduction in dissolved oxygen concentration. Temperature changes affect not only fish life but the entire eco-system. (Also refer chapter 35-"Steam power plants").

RADIO-ACTIVE WASTES

Dealt elaborately in chapter 38 "Radio-active pollution".

EFFECTS ON LAND

1. The excess content of sodium above 60 % and boron (above 2mg/l) are deleterious to crops.
2. The high sodium alkalinity combined with salinity impair the growth of plants.
3. Texture of the soil is affected by sodium and penetration of roots is prevented.
4. Soil permeability is also affected by sodium, and ultimately the soil will lose its productivity.
5. Suspended and colloidal impurities clog the pores and form a mat on the surface of soil preventing the passage of air, water etc.

EFFECTS ON SEWERS AND SEWAGE TREATMENT PLANTS

1. The coarse materials such as rags, feathers cover the whole surface of the bar screens of comminutors and interfere

with operation. Further, they clog nozzles, overload digesters and resist pump operation.

2. Acid/ alkalis in the effluent corrode the treatment structures. They inhibit the biological activity and upset purification.
3. Presence of sulfide leads to 'Crown Corrosion' of sewers.
4. The dissolved solids form incrustation on the surface of sewers.
5. Cyanide, phenols, toxic organic materials, and heavy metals such as copper, chromium(6^+), and zinc suppress the biological activity as they tie up enzymes which carry out oxidation.
6. Fatty materials clog nozzles, pumps, and overload digesters.
7. Synthetic detergents cause foam which affects natural aeration process. The same effect is observed with oil.
8. Solvents if present may cause fire hazards.
9. Workers employed in sewage treatment plants may be exposed to toxic gases.

NATIONAL AND INTERNATIONAL AGENCIES

From the above discussion, it is now clear that it is the prime duty of the industries to treat the effluents to safe levels. The water courses are not merely the receptacles of effluents and city sewage, but they are sources for drinking and industrial water, used for bathing, irrigation, fishing recreation, power and navigation. Also it is the place and medium for numerous living organisms particularly for fish. But unfortunately the water courses become the ultimate points of disposal for effluents in every country. Hence stringent laws came into force to monitor the surface water quality and to check the effluent quality.

There are various national agencies operating in each and every country promulgating laws and ordinances. The agencies include Pollution Control Boards, River Board Authorities, Public Health Departments, Fisheries etc., Various pollution control acts have been promulgated by all of these agencies and tolerance limits for effluents discharged into inland surface waters, into public sewers, on land for irrigation and into marine coastal areas are stipulated by them.

Among the international agencies, United States Environmental Protection Agency needs a special mention. The other well known international agency World Health Organization though not directly involved in pollution control, most of its activities centre around abatement of pollution.

IMPORTANT REFERENCE BOOKS

A select list of reference books dealing about the nature, characteristics, effects, and treatment of effluents is given below. These books not only cater information but are the key sources for various references in this field.

1. N.L. Nemerow - Industrial water pollution, Origin, Characteristics & Treatment, 1978, Addison-Wesley Publishing Co.,

2. J.R. Erichson Jones - Fish and River Pollution-1964 Butterworths, London.

3. Louisklein - River Pollution-Volume-I Chemical Analysis-1969, Butterworths, London.

4. Louisklein - River Pollution-Volume II Causes & Effects - 1967, Butterworths, London.

5. Louisklein - River Pollution-Volume III - Control-1966 Butterworths, London.

6. World Health organization - International Standards for Drinking Water - 3rd ed - 1968.

7. Edmund B.Besselievre - "The Treatment of Industrial Wastes " - 1969, McGraw Hill Book Co., Newyork.

8. H.F.Lund - Industrial Pollution Control Hand book, 1971, McGraw Hill Book Co., Newyork.

9. R.N.Shreeve & Joseph Brink - Chemical Process Industries- 4th ed, - 1978, McGraw Hill International Book Co.,

10. C.N.Sawyer & P.L.McCarty - Chemistry for Environmental Engineering -1978, McGraw Hill Book Co, Newyork.

11. American Public Health Association - Standard Methods for the Examination of Water, & Waste Water, 1982.

12. N. Manivasakam - Environmental Pollution -1984, National Book Trust, New Delhi.

13. N. Manivasakam - Physico -Chemical Examination of Water, Sewage and Industrial Effluents -1984, Pragati prakashan, Meerut.

14. M.N. Rao & A.K. Datta - "Waste water Treatment" - 1979, Oxford & IBH Publishing Co, Calcutta.

Chapter - 2

SUMMARY OF INDUSTRIAL EFFLUENTS

As human needs increase and civilization changes, more and more finished products of different types are required. Accordingly, a large number of industries born and grown in every country. It is not possible to describe all the industries in a single volume, but the author tries to give information of the most common industries. The following table is a brief summary of the type of industrial effluents, their origin, important characteristics and common treatment methods adopted, with some remarks. This summary is given with an intention to provide the readers a quick reference. For more information and details, readers should refer appropriate chapters.

S.No.	Industry	Origin of Effluents	Important Characteristics	Treatment	Remarks
1.	Acid Manufacture	Dilute acid and wash waters	Low pH	Neutralization	
2.	Asbestos Manufacture	Cleaning and Crushing	Suspended asbestos	Detention, Neutralization Land filling.	
3.	Bakery	Washing of pans, floor washings.	High BOD and oil & grease content, sugars and detergents.	Biological oxidation.	

S.No.	Industry	Origin of Effluents	Important Characteristics	Treatment	Remarks
4.	Beet Sugar	a) Process water, screening, juicing and draining from lime sludge. b) Transport water (flume water & wash water) c) Condenser water.	High BOD due to dissolved and suspended organic matter such as sugars and proteins.	Screened, chlorinated and recycled. Screened coagulated and settled. Reused.	Biological treatment of undiluted wastes is not very successful.
5.	Brewery	a) Steeping and Pressing of grain b) Spent liquor from alcohol distillation	High BOD, High content of dissolved organic solids.	Anaerobic digestion, Trickling filtration, Recovery and use in feeds.	
6.	Cane Sugar	a) Spillage from juice extraction and clarification. b) Filter press wash water. c) Spray pond overflow d) Spillage from pan boiling	High BOD (mostly dissolved carbonaceous matter) Variable pH	Neutralization Aerobic oxidation	

S.No.	Industry	Origin of Effluents	Important Characteristics	Treatment	Remarks
		e) Floor washes.			
7.	Cannery	Trimming, culling, juicing, blanching, lye treatment.	High suspended solids, colloidal and dissolved organic matter.	Lagooning or spray irrigation.	
8.	Cement manufacture	Wet scrubbing of gases.	High suspended solids hot effluent.	Neutralization followed by sedimentation	
9.	Coal washery	Cleaning	High suspended solids, low pH, H_2SO_4 and $FeSO_4$ present.	Settling	
10.	Coke manufacture	Coke breeze cooling waters	High suspended solids, tar, ammonia, cyanide, H_2S, phenol, benzole, pyridine, oils are present.	Recirculation, settling, neutralization recovery, reuse, chemical-coagulation.	
11.	Coffee pulping	Pulping and fermenting of coffee bean.	High BOD and suspended solids	Screening. Settling Trickling-filtration	
12.	Corn starch manufacture	Evaporator condensate, bottom sediments, syrup from final washes, wastes from final washes, wastes from 'bottling up'	High BOD (mainly due to starch and allied materials)	Equalization, biological filtration and anaerobic digestion.	

S.No.	Industry	Origin of Effluents	Important Characteristics	Treatment	Remarks
13.	Dairy	Washing of cans, tankers, equipments and floors	High BOD, easily, putrescible, containing dissolved organic matters such as fat, protein and lactose	Equalization, skimming, Biological treatment. Oxidation ditches are more suitable with aeration and sludge return arrangements as in activated sludge process.	Spray irrigation is the most economical method of disposal of treated effluent.
14.	Detergents manufacture	Spill overs and floor washes.	Contains alkyl benzene sulfonate or other active agents with builders such as phosphates, borax, variable pH (usually alkaline)	Floatation, skimming, precipitation with $CaCl_2$	Not amenable for biological treatment
15.	Distillery (alcohol production from molasses)	Spent wash water, waste water from heat exchangers.	Abnormally high BOD and suspended solids and dissolved solids, Characteristic brown color and disagreeable odor.	Anaerobic digestion and biogas production. Evaporation (and recovery of waste solid for use as animal food stuff).	
16.	Edible oil refinery.	Neutralization of excess fatty acids. Filter-washing, spill overs.	High BOD and COD, Oils	Equalization, neutralization, coagulation, settling, aerobic or anaerobic digestion.	

S.No.	Industry	Origin of Effluents	Important Characteristics	Treatment	Remarks
17.	Electroplating	Alkaline degreasing, Acid pickling, plating bath wastes, rinse waters.	Highly acidic, high COD, contain heavy metals and toxic substances.	Neutralization to pH 8.5 and precipitation	
				Chromium wastes: a) Reduction to Cr^{3+} and precipitation with lime	Much sludge produced (Cr and Fe hydroxides)
				b) Reduction with SO_2 at pH 3.0-5.5	Less sludge is produced but expensive.
				Cyanide wastes: a) Alkaline chlorination at pH 10-11	Suitable for weaker rinse waters containing cyanides of alkali metals and Zn, Cd and Cu (but not Ni) Complete removal of CN is possible

S.No.	Industry	Origin of Effluents	Important Characteristics	Treatment	Remarks
				b) Electrolytic oxidation	Suitable for stronger spent wastes
				c) Copperas + lime (pH 7.5-9.0) to give blue insoluble complex ferrocyanide, then sedimentation	Cheapest method. Removes only 86-96% of CN. Suitable when effluent is discharged into sewer.
18.	Explosives manufacture	Acid manufacture wastes, washing TNT and guncotton, washing and pickling of cartridge	Low pH, contain acids, oils, soaps etc.,	Floatation, precipitation, biological treatment, chlorination of TNT, neutralization, absorption	
19.	Fertilizer manufacture	Spill overs from component acids manufacture, and final fertilizer production. Boiler blow down and cooling waters.	Low pH, High phosphate and fluoride content Presence of NH_3 and nitrate	Reuse after sedimentation. Neutralization, Defluoridation and dephosphatisation by lime treatment.	

S.No.	Industry	Origin of Effluents	Important Characteristics	Treatment	Remarks
20.	Food Processing (See Cannery, Coffee pulping, Edible oil refining and meat packing.)	Waste is generated from almost all the wet processes.	High BOD and suspended solids, amenable for biological treatment.	Screening, chemical precipitation and settlement. Lagooning, spray irrigation.	
				Activated sludge process.	Used when fairly complete treatment is required.
21.	Gelatine manufacture	Acid soaking, lime treatment.	Near neutral, High dissolved solids content Moderate BOD.	Chemical coagulation Aerobic biological treatment.	
22.	Glass manufacture	Spill overs	Highly alkaline	Neutralization, settlement.	
	Glass etching	Floor washes	Fluorides	Defluoridation by lime treatment.	
	Optical Glass polishing.		Finely divided iron and glass particles. Alkaline in nature.	Resistance to alum coagulation.	May be coagulated with $CaCl_2$ and lime
23.	Glue (animal) manufacture.	Acid wash, Lime wash, Degreasing.	High BOD and COD, mineral acids.	Chemical coagulation Trickling filter or activated sludge process.	

S.No.	Industry	Origin of Effluents	Important Characteristics	Treatment	Remarks
24.	Hospitals	Washing	Bacteria and other chemicals	Aerobic biological treatment after holding for a considerable period.	
25.	Iron Foundry	Hydraulic discharge of waste sand.	High suspended solids.	Settling and reclamation of sand.	
26.	Laundering	Washing of fabrics	Usually alkaline containing syndets, soaps and other surface active agents.	Chemical coagulation. For complete treatment biological treatment may be employed.	
27.	Metal container	Solvent degreasing, Alkaline degreasing, Washing	Solvents, alkalis, oils, lubricants and dissolved metals, phosphates etc.	Chemical precipitation and activated carbon adsorption.	
28.	Naval stores manufacture (manufacture of wood rosin, wood turpentine, pine oil, dipentene etc.,)	Washing of stumps, spill overs of drop solution, solvent recovery.	Usually acidic, High BOD	Equalization, recirculation and reuse. Final treatment by trickling filtration.	
29.	Nuclear power plants.	Processing of ores and fuels, Cooling waters.	Hot effluents Radio-active materials present.	Segregation, concentration and containing.	

S.No.	Industry	Origin of Effluents	Important Characteristics	Treatment	Remarks
30.	Oil field and oil refinery	Pumping and desalting, various refining processes.	Drilling muds, saline water and oil, odoriferous, high BOD, phenol and sulfur compounds.	Discharged into sea, injection into subterranean formation. Removal of oils, phenols and other substances. Secondary biological treatment. Tertiary activated carbon treatment or ozonation.	
31.	Paint manufacture	Washings Rejected materials.	Containing synthetic resins, solvents, pigments, lacquers, varnishes etc., Also contain heavy metals such as lead, chromium, aluminium etc., High COD.	Settling.	
32.	Pesticide manufacture	Washing and purification	Highly acidic containing residues of pesticides.	Neutralization Dilution Activated-carbon adsorption Alkaline Chlorination	As the pesticides are highly poisonous even traces should not be let out in streams or sewers.

S.No.	Industry	Origin of Effluents	Important Characteristics	Treatment	Remarks
					They must be removed completely the process is more expensive.
33.	Petroleum refinery.	By product water during chemical conversions, Water used as a medium to carry out reactions, Bottom sediments	Hydrocarbons, alcohols, aldehydes, ketones, acids, phenols, oils and metals. High TDS High COD/BOD ratio.	Recovery, equalization, neutralization, coagulation and flocculation, primary settling, oil separation, secondary-biological treatment.	
34.	Pharmaceu-ticals manufacture Synthetic drugs	Process and wash waters	Acidic in nature, various chemical compounds	Segregation, neutralization, chlorination, Final biological oxidation.	
	Antibiotic	Mycelium, Spent filtrate and wash waters	Variable pH, High BOD, COD and suspended solids	Equalization, Neutralization, Clarification followed by biological (aerobic) treatment.	
35.	Phosphorous mining	Mining, processing of phosphate rock to elemental phosphorus.	Low pH, high suspended solids, clays, slimes, phosphorus, fluoride, and silica.	Clarification, coagulation and settling	Volume of waste is more and treatment plants should be designed suitably

S.No.	Industry	Origin of Effluents	Important Characteristics	Treatment	Remarks
36.	Photographic processing	Spent hypo fixing solution	Contain silver (upto 1%). Alkaline and contains various organic reducing agents	Recovery of silver, discharge into municipal sewer.	Treated effluent however contains 2-10 ppm of silver
		Wash waters	Contain 3 to 6 ppm silver	Neutralization Activated carbon adsorption, Ion-exchange and discharge into sewer.	Costlier treatment
37.	Pickle manufacture	Cleaning and wash water, spill overs, brine waters and syrup wastes	Variable pH, high dissolved solids content, acetic acid etc.,	Screening, equalization and neutralization	
38.	Metal Pickling Brass and copper pickling	Spent pickle liquors, wash waters	Highly acidic containing mineral acids and copper.	Copper recovery	
				i)by electrolysis	Suitable for spent pickle liquor.
				ii) by scrap iron treatment	Suitable for wash waters (Copper deposits as sludge but equivalent amount of iron is dissolved)

S.No.	Industry	Origin of Effluents	Important Characteristics	Treatment	Remarks
				After recovery neutralization with lime.	Large capacity tanks needed to avoid carryover of insoluble hydroxides. Copper cannot be recovered from the sludge
	Iron & steel pickling	Spent pickle liquors and wash waters	Mineral acids, Iron.	Either recovery or neutralization. Recovery of $FeSO_4$ by evaporation or used as a coagulant for sewage sludge conditioning.	Lime is the cheapest chemical employed for neutralization but soda ash and caustic sods are more effective and react quickly.
39.	Plastics manufacture	Spill overs, Equipment wash waters.	Acids, phenols, formaldehyde etc., are present.	Reuse and controlled discharge.	
40.	Plywood manufacture	Glue washings	High BOD and COD, phenols.	Settling ponds or incineration	

S.No.	Industry	Origin of Effluents	Important Characteristics	Treatment	Remarks
41.	Poultry processing	Cleaning and washings from killing and evisceration sections, spent scald water.	Highly organic in nature, high BOD and suspended solids.	Preliminary screening, Biological treatment with septic tanks and lagoons, oxidation ponds or percolating filters with provision for recirculation.	
42.	Pulp, paper and board manufacture	Digestion of wood, refining, washing of fibres, screening of pulp, paper making, recovery of chemicals.	Usually low pH, intense brown color, characteristic odor, High suspended, dissolved and colloidal solids content.Resistant to biological oxidation.	Recovery of by-products, preliminary screening, coagulation, flocculation, clarification. Secondary biological treatment.	Activated sludge process is useful when a high quality effluent is required.
43.	Rice processing	Soaking, cooking, washing.	Color and high BOD, suspended solids.	Treatment with lime and settlement.	
44.	Rubber Industry	Coagulation of rubber and latex washings.	Characteristic odor, High BOD and suspended solids, low BOD/PV ratio, variable pH.	Aeration, Activated - carbon treatment, Chlorination, Trickling filter or activated sludge process.	Removes odor but expensive.

S.No.	Industry	Origin of Effluents	Important Characteristics	Treatment	Remarks
45.	Slaughter house and meat packing.	Killing of animals, washing, rendering of bones and fats, residues in condensates.	High BOD due to blood and proteins, high dissolved and suspended solids, oils & grease.	Screening, grease removal (especially by air floatation). Sedimentation and chemical precipitation. Activated sludge process.	The waste is similar to domestic sewage but more strong. This treatment purifies only partially.
				Anaerobic digestion followed by aerobic biological process	Longer aeration period, high amounts of return sludge must be used.
46.	Soaps and detergents manufacture	Soap: spil overs, waste waters from processing.	Unsaponified oil, alkali, high pH and suspended solids.	Skimming of floating fatty acids. Recovery of glycerine etc.,	
	Detergents: Spill overs.	Residual detergents.	Percolation through sand-silt type soils.		
47.	Soft drink manufacture	Bottle and equipment cleaning. Floor washing, Syrup tank drains.	Sugars, high BOD, high pH amenable for biological treatment.	Screening, biological treatment.	

S.No.	Industry	Origin of Effluents	Important Characteristics	Treatment	Remarks
48.	Steam power plants	Cooling water, boiler blow down	High dissolved solids content, (both organic & inorganic) Hot effluents	Cooling. (storage of ash solid waste)	
		Scrubbing effluent	Particulate materials	Settling, pH adjustment and reuse	
49.	Steel industry (See iron pickling also)	Carboni-zation of coal, Blast-furnace operation (Iron production) Steel making, Rolling, Pickling etc.	Low pH, presence of cyanide and phenols, coke, limestone, oils, mill scale and suspended solids.	Recovery and reuse, Neutralization, Chemical-coagulation.	
50.	Synthetic resin manufacture	Synthesis of monomers, polymer preparation, washing of equipments, spill overs.	Variable pH, acids, alkalis, varied type of organic materials including phenols and formaldehyde	Reuse, Solvent extraction	Useful for high phenol concen-trations.
				a) Counter current extraction	Final effluent contains about 50 mg/l of phenos.
				b) Centrifugal counter current extraction	Suitable for treatment of ammoniacal liquors.

S.No.	Industry	Origin of Effluents	Important Characteristics	Treatment	Remarks
					Removes phenols from 2000 mg/l to 5 mg/l
				Chemical oxidation (using chlorine, chlorine-dioxide, ozone etc.,)	Suitable for dilute liquors. For strong liquors it is expensive.
				Biological treatment (activated sludge or percolating filters)	Better treatment can yield a final effluent with less than 1 mg/l phenol
51.	Tannery	Soaking, liming, deliming, spent tan liquors	Characteristic obnoxious odor, high suspended and dissolved solids, specially sodium chloride & sulfide, high BOD and COD, oils and grease, chromium from chrome tanning.	Equalization, settling, chemical coagulation, activated sludge process. Oxidation pond and anaerobic lagoon are recommended for small and isolated tanneries.	No treatment removes sodium chloride present in abnormal concent-ration and disposal of treated effluent is again a problem.

S.No.	Industry	Origin of Effluents	Important Characteristics	Treatment	Remarks
52.	Textile Processing				
	Cotton	Sizing	Starch, generally colored, high BOD and suspended solids and dissolved solids, polyvinyl-alcohols softeners.		
		Desizing, Kier boiling Mercerizing.	Starch, acids, enzymes, alkalis silicates.	Equalization, neutralization, chemical coagulation. Aeration or biological treatment.	
		Bleaching	Chlorine, hypo-chlorites, peroxides		
		Dyeing, Printing, Finishing.	Dyes, alkalis, chromium, phenolic substances, oil, natural and synthetic waxes.		
	Synthetic		Colored, high BOD and COD values, contain alkalis, organic solvents etc.,		
	Wool	Scouring	Hot, intensely colored, High	Screening, grease	

S.No.	Industry	Origin of Effluents	Important Characteristics	Treatment	Remarks
			BOD, COD and suspended solids, grease, soaps and detergents, alkalis etc.,	removal, equalization, coagulation and settlement, Aerobic biological treatments such as trickling filtration or activated sludge process.	
		Oiling	Vegetable oil		
		Slashing	Starch		
		Fulling	Soap, soda ash and detergents		
		Dyeing	Dyes, alkalis, chromium, phenolic substances.	Anaerobic treatment is under study.	
53.	Viscose rayon manufacture.	Wood pulping, Steeping, Xanthation, Filter pressing, Deaeration.	Alkaline waste		
		Spinning, Washing, Sulfur removal, Chlorination	Highly acidic waste. Color of combined effluent is orange yellow to deep yellow brown which is	Screening, Equalization, Neutralization Coagulation with alum.	

S.No.	Industry	Origin of Effluents	Important Characteristics	Treatment	Remarks
			very difficult to bleach. Low pH, high BOD, COD and suspended solids, zinc-sulfate, sulfides and hemicelluloses.	Trickling filtration or Activated sludge process Aerated lagoon or oxidation ponds.	Expensive Compara-tively economi-cal but more space is required.

PART - II

MAJOR INDUSTRIAL EFFLUENTS

Chapter - 3

PULP, PAPER & BOARD INDUSTRY

Paper industry is the most common industry situated all over the world. Production of paper involves the following two major processes.

(i) The wood and allied materials are digested with chemicals to convert them into pulp, and

(ii) The pulp is refined and passed over a fine screen to form paper.

Most of the paper industries have integrated pulp and paper production units.

MANUFACTURING PROCESSES

The manufacturing processes can be divided into three sections and studied in detail

(i) Raw material preparation

(ii) Pulping process

(iii) Paper Making

I. RAW MATERIAL PREPARATION

The cellulosic raw materials commonly used are bamboo, wood, cotton linters, rags, straw mainly of rice and wheat, bagasse, hemp, flax, jute and waste paper.

After debarking, the wooden logs are slashed and conveyed to chippers where they are cut into small chips and the chips are placed in digesters for pulping.

II. PULPING

Pulping is the process in which wood or other cellulosic raw materials are digested with chemicals under high temperature and pressure so that cellulosic fibers of wood are relieved from its binders such as lignin, resin etc., There are three major types of pulping processes employed according to the raw materials and finished products. They are:

(i) Sulfate or Kraft Pulping

(ii) Sulfite Pulping

(iii) Semichemical Pulping

The other pulping processes of less importance are Soda pulping, Mechanical pulping and Chemi-mechanical pulping.

All the pulping processes include washing and bleaching as final steps which contribute to the bulk of effluents. The quantity and quality of the effluents largely depend on the pulping process.

(a) SULFATE (or) KRAFT PULPING

This is the most widely employed method of pulping. Almost any kind of wood, soft or hard can be pulped by this process and the pulp produced is brown in colour and difficult to bleach. As it contains strong and longest fibres, the end products obtained with this pulp are strong-brown wrapping papers, bag making papers, multiwall papers, gumming papers, paper boards used for cartons, corrugated boards and containers, strong white papers from bleached kraft.

The chemicals used for digestion are sodium sulfide, sodium hydroxide and sodiumcarbonate. The digestion is carried out for about 2 to 5 hours at a temp. of 170° to 177°C under 100-135 psi pressure. During digestion most of the lignins are hydrolyzed to alcohol and acids. Toxic materials such as dimethyl sulfide and some mercaptans are also formed. After digestion, the pressure is allowed to drop to 80 psi and the charge or brown stock is blown into a pit (or heat recovery furnace to use the steam) where the fibres are separated from the cooking liquor. The waste liquor thus obtained on separation is called 'black liquor' due to its intense dark color and is sent to recovery unit for recovery of chemicals (described in later pages).

Washing

The fibres are then washed and the concentrated wash water is sent for chemical recovery. The washed pulp is then passed on to screen room where the knots, small slivers of uncooked wood and other non-disintegrated matter are removed by sieving through knotters, rifflers and screens. The pulp is partially dewatered by decker-a cylindrical screen revolving across the path of the pulp. The dilute wash water and other waste water from this section termed as 'brown stock wash' or 'unbleached decker waste' forms major portion of the pulp mill effluent.

Bleaching

The bleaching of pulp involves three successive stages (i.e.) treatment with (i) Chlorine or Chlorine dioxide (ii) Caustic alkali and (iii) Calcium hypochlorite solution. The effluents from chlorination and hypochlorite treatment are light yellow in color and discharged together, while that of the effluent from caustic extraction (or neutralization) is highly colored and discharged separately. After bleaching, the pulp is washed, rethickened and sent to paper mill.

Recovery of chemicals

The black liquor obtained on separation of the pulp after digestion contains about 95 to 98% of the total chemicals charged into the digester. Usually it contains organic sulfur compounds in combination with sodium sulfide. The recovery process involves the concentration of black liquor by burning, and liming. Before incineration, sodium sulfate is added and in the smelting furnace, organic compounds such as lignin and resins are broken down completely and the carbon produced is burnt along with molten inorganic chemicals resulting in the following reaction.

$$Na_2SO_4 + 2C \longrightarrow Na_2S + 2CO_2$$

The molten smelt is dissolved in a weak solution coming from the lime mud washer and the resulting solution termed as 'green liquor' is settled to remove the insoluble impurities and allowed to react with slaked lime where the following reaction takes place immediately.

$$Na_2CO_3 + Ca(OH)_2 \longrightarrow CaCO_3 + 2NaOH$$

The resulting slurry is a mixture of 'white liquor' and 'lime mud' and are separated by passing through settlers, and rotary continuous filters. The filtrate white liquor is reused in cooking of fibres as it contains sodium hydroxide and sodium sulfide and small quantities of sodium carbonate, sulfate, sulfite and thiosulfate. The separated sludge (lime mud) is sent for calcination for conversion to calcium oxide for reuse.

An important by-product obtained from black liquor recovery is 'tall oil' (a black, sticky viscous liquid) used in the manufacture of soaps, greases and emulsions. The digester relief gases yield turpentine amounting to 10 to 40 litres/ton of pulp produced.

A flow chart of sulfate pulping indicative of all the processes including recovery is given in Fig 3-1.

b) ACID SULFITE PULPING

This type of pulping is intended for the manufacture of fine grade papers such as bond paper, other white grades, tissue paper, book paper and bread wrap papers etc., Mostly coniferous woods are used. The pulp produced is dull white in color and easy to bleach. However fibres are weaker than that obtained by kraft pulping. The digesting liquor consists of either calcium bisulfite or magnesium bisulfite and free sulfur. With calcium bisulfite, chemicals are not recovered and heavy pollution is caused whereas use of magnesium bisulfite includes the recovery of chemicals and heat, and pollution load is decreased considerably. However availability of calcium carbonate (to produce calcium bisulfite) is more and is cheaper. The digestion is carried out in special digesters and heated with

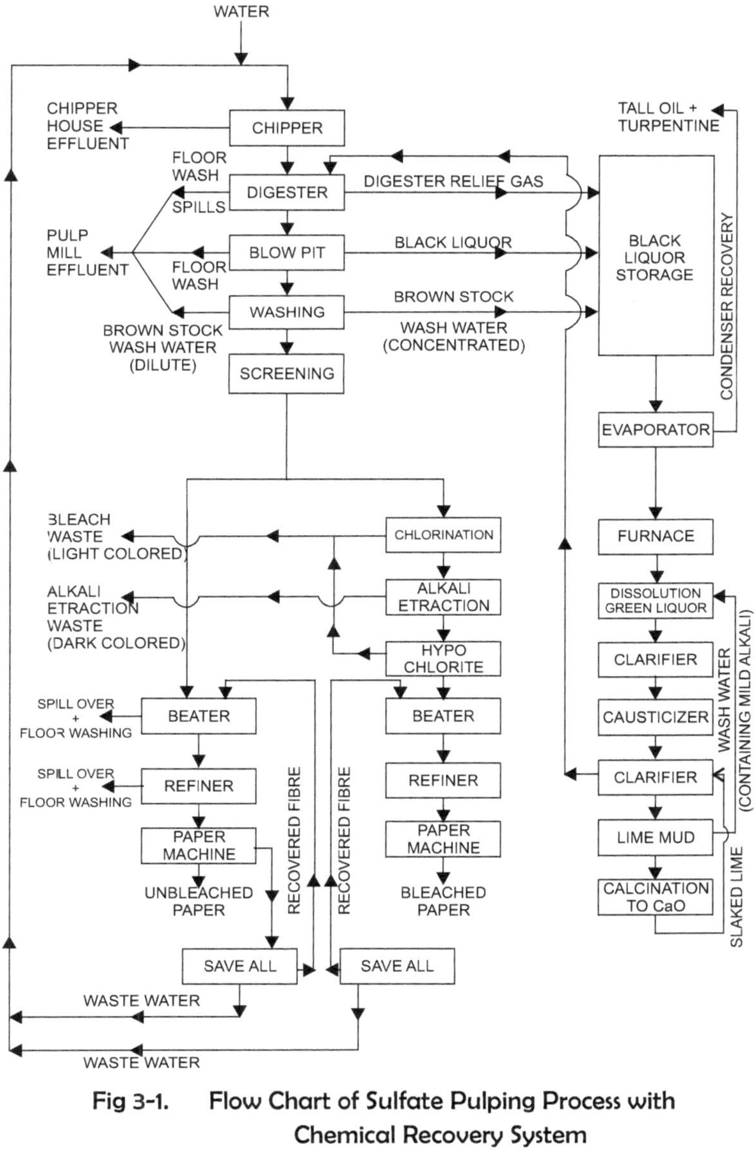

Fig 3-1. Flow Chart of Sulfate Pulping Process with Chemical Recovery System

steam either internally or externally at about 125° to 160°C for a period of about 6 to 12 hours. (The pressure varies from 90 to 110 psi depending on the construction of the digester). During digestion, the cellulose lignin complex is split up and the hemicelluloses are hydrolyzed to simpler compounds. After digestion, the digestate is blown into blow pit or blow tank. The processes after blow pit are similar to that of kraft pulping. A few mills use ammonium or sodium bisulfite in place of magnesium bisulfite. Neutral sulfite pulping is another process employing sodium sulfite and sodium carbonate and is practiced in some mills. Recovery of chemicals is possible with this type of pulping.

Recovery of chemicals

The spent liquor from storage is concentrated and incinerated to yield magnesium oxide which is slaked to magnesium hydroxide and treated with sulfurdioxide to form magnesium bisulfite and is used for digestion.

c) SEMI CHEMICAL PULPING

For high quality bleached pulp, this process is employed with lesser amounts of caustic soda and sodium sulfide as the active pulping chemicals and sodium sulfate as the make-up chemical. Full recovery of chemicals is possible with this process. The pulp is used to manufacture writing, bond, offset, newsprint, and tissue papers, and corrugated and specialty boards.

III. PAPER MAKING

Based on the type of paper required, the pulps are chosen and disintegrated in a beater, while various fillers (to fill the pores of the paper) and sizes and dyes are added to improve quality of the paper. The commonly used fillers are talc, certain clays, calcium carbonate, blancfixe (barium sulfate) and titanium dioxide. (Fillers are not added to absorbent papers such as blotting paper). Usually rosin or a wax emulsion and aluminum sulfate [($Al_2(SO)_4 14H_2O$] also known as paper makers' alum) are added to produce a gelatinous film on the fiber and hardened surface. After beating, the pulp is refined. (Some times the pulps are washed in breaker beater before the addition of chemicals and this produces rather a strong waste. But as the washing continues the waste gets diluted).

The refining is carried out in a machine, 'Jordan' where the fibers are cut to the desired size and the pulp is then passed on to stuffing boxes at which pulp is made to proper consistency and then passed over a traveling belt of fine wire known as the fourdrinier wire. Major portion of the water passes through the screen and the fibers form a mat on the surface and is passed on to rolls and calendaring machine for final surface finishing. The waste water drained usually referred as 'white water' contains water along with some fine fibers and filler and size materials. The fibers separated from the white water are sent to beater, while the liquid is sent to chipper house and used along with fresh water.

SOURCES OF EFFLUENTS

(i) **Raw material preparation (Chipper House)**
Arise by washing, cleaning and chipping.

(ii) **Pulp Mill Effluent**

Consists of the following contributory effluents.

a) Spent liquor or black liquor from blow pit
b) Effluents from brown stock washers (major portion of the effluent)
c) Effluents from chlorination
d) Effluents from caustic extraction
e) Chemical recovery effluents
f) Spills and leakages, floor washes etc.,

(iii) **Paper Mill Effluent**

Generated during stock preparation and excess white water. If the industry possesses its own chlorine (or other bleaching agent) production unit that also contributes to effluent.

VOLUME OF EFFLUENTS

The volume of effluents discharged varies from mill to mill and depends on the production capacity, types of paper manufactured, pulping process, recovery of chemicals and water economy adopted. It is reported that an integrated pulp and paper mill employing kraft process generates 220 to to 320 m^3 ($1m^3$=1000 litres) of effluents per ton of paper manufactured. Mills manufacturing specialty papers generate more volume of effluents as they use more water for washing and bleaching. It is to be mentioned that small mills discharge higher volumes of waste per ton than larger mills as they do not posses recovery units. It is estimated that about 330 m^3 of effluent is produced per ton of paper manufactured in a small mill of capacity 20 ton/day where as 220 m^3/ton is generated in a large mill manufacturing 2000 tons/day.

CHARACTERISTICS OF EFFLUENTS

Raw Material Preparation (Chipper House Effluent)

The quantity of effluents from this section is much less. The BOD and COD values are also comparatively low. BOD value varies from 50 to 200 mg/l. They contain readily settleable solids such as dust, grit and sand.

Pulp Mill Effluents

As the different pulping methods generate different types of wastes, each should be considered separately.

a) Sulfate or Kraft Pulping

The major part of this effluent is contributed by pulp washing and bleaching operations. Pulp washing effluent, also known as brown stock washing effluent or unbleached decker waste is strong and dark brown in color (which is not easily destroyed), alkaline with a high pH and possesses an appreciable BOD (about 100 to 200 mg/l) and a high COD and suspended solids. It also contains significant quantities of sodium, as the digesting chemicals used are sodium compounds. It may also contain dimethyl sulfide and methyl mercaptan. The bleach plant discharges two separate wastes, one is a combined effluent from chlorination and hypochlorite treatment and contains chlorine compounds. It is also brown in color and the pH is near neutral. The BOD and COD values of this effluent are usually less. The other effluent from caustic extraction (the process that follows chlorination) is segregated and discharged separately and is dark brown or black in color and highly alkaline.

b) Sulfite Pulping

Effluent from this pulping is dark colored, highly corrosive and acidic with a very high BOD value. It contains calcium or magnesium salts of lignosulfonic acid, reducing sugars (pentoses and hexoses which contribute to above 70% BOD), sugar, sulfur-dioxide derivatives and calcium.

c) Semi-chemical Pulping

Characteristics similar to that of kraft pulping.

d) Soda Pulping

Similar to that of kraft pulping except high BOD (about 800 mg/l)

Recovery Plant Effluent

This is a combination of three individual effluents viz., surface condenser, barometric condenser, and floor washes. The effluent is slightly colored (some times brown or dark colored) and have a characteristic mercaptan odor. BOD, COD, and suspended solids values are comparatively less. The major components present in it are sodium and calcium hydroxides, calcium carbonate, sulfides and mercaptans.

Paper Mill Effluent

This effluent is white in color with a low pH value

(about 5 to 6) and high suspended solids. BOD and COD values are not so high. It also contains filler and size materials and pigments. In integrated unit, the liquor after separation is sent to chipper house and used along with fresh water. If the mill is having only paper production unit, it is discharged as waste.

The characteristics of an integrated pulp and paper mill (kraft pulping) are given in Table 3-1.

Table 3-1. Characteristics of Combined Effluent of an Integrated Pulp & Paper Mill with Recovery Sysetm

Sl.No.	Characteristic	Value
I.	Color	Dark brown
ii.	Odor	Charring wood & chemical
iii.	Temperature	41°C
iv.	pH value	7.5
v.	Total Suspended Solids, mg/l	1340
vi.	Settleable solids, ml/l/hr	120
vii.	Total Dissolved Solids, mg/l	1630
viii.	Dissolved Fixed Solids, mg/l	380
ix.	BOD (5 days at 20°C), mg/l	115
x.	COD mg/l	1490
xi.	Permanganate value (4hrs), mg/l	200
xii.	BOD : Permanganate value	0.575
xiii.	Sulfides (as S), mg/l	NIl
xiv.	Residual chlorine (as Cl_2), mg/l	Nil
xv.	Phenolic substances (as phenol), mg/l	0.96
xvi.	Tannin & Lignin compounds (as tannic acid, mg/l	55
xvii.	Oils & grease (petroleum ether solvent), mg/l	2.8

The characteristics of a sulfite mill effluent are tabulated below (Table 3-2).

Table 3-2.
Characteristics of a Sulfite Mill Effluent

Sl. No.	Characteristics of effluent	Spent liquor effluent	Pulp Mill	Paper Machine
i.	pH value	3.8 to 4.2	4.1 to 6.7	5.8 to 7.2
ii.	Suspended solids, mg/l	4240 to 21000	425 to 2800	950 to 3140
iii.	BOD (5 days at 20°C mg/l)	16000 to 25000	120 to 590	150 to 430

EFFECTS OF PULP & PAPER MILL EFFLUENTS

Color and immediate oxygen demand are the two main objectionable features of these effluents. The oxygen demand results in immediate depletion of oxygen of receiving streams. The lignin compounds which contribute color to the effluent are not so easily biodegradable and hence the dark color imparted to the receiving streams persists for a long stretch and inhibit photosynthesis and other natural purification phenomena. The effluent contains strong wood, chips, bits of bark, cellulose fibers, dissolved lignin and a complex mixture of chemicals including sulfite, phenols and free chlorine (mention has to be made that about 50% of the total wood used for the process is rejected as waste materials). The chemicals may destroy the flora and fauna of receiving waters. Further, the settleable materials sink to the bottom and blankets fish spawning grounds and destroy certain type of aquatic life. The effluent may also contain methyl mercaptan, pentachlorophenol and sodium penta chlorophenate which are considered highly toxic to fish.

TREATMENT

Pulp and paper mill waste treatment includes the following primary and secondary treatments.

1. Chemical recovery
2. Coagulation, sedimentation, and floatation to remove suspended matter.
3. Chemical treatment for the removal of color.
4. Activated sludge treatment to remove dissolved organics.
5. Lagooning, mainly for storage and bio-degradation of organic matter.

Chemical Recovery

Recovery plant becomes an integral part in kraft pulping process. It has already been described in the previous pages.

Coagulation and Sedimentation

Two principal coagulants used are alum and activated silica in combination with polyelectrolytes. About 90% removal of suspended solids is reported. Most of the paper industries have only primary clarifier. The diameter of the clarifiers vary from 5 m to 35 m. Sludge is dewatered, thickened, dried and sent for land-fill or incinerated. The clarified effluent is subjected to further treatment.

Chemical Treatment for Color Removal

Only a few mills follow this treatment. Various patented products have been developed for the removal of color.

Activated sludge Process

This is the most successful method for treating effluents from kraft mill and semi chemical pulp mill. The following modifications over the conventional activated sludge process yield better results.

1. Aeration for longer duration.
2. Increased seed concentration.
3. Aeration using mechanical agitation or sparger or both.

Lagooning

This is a major method employed for treating pulp and paper mill wastes. These lagoons, mostly ponds are used for storage, during which time the effluent stabilizes. The detention period ranges from 10 days to 30 days. The major draw back in lagooning is the requirement of more space.

Although a great deal of research has been done, treatment of pulp and paper mill effluents is still in its early stage. The methods described above are not economical. It is a must to evolve a simple and economical method of treatment.

PARAMETERS TO BE DETERMINED

Important parameters

1. Color
2. Odor
3. Temperature
4. pH value
5. Suspended solids

6. Biochemical Oxygen Demand: Adjust pH and remove chlorine before preparing dilutions. Range of dilutions, 1:49, 1:99 and 1:199. Seeding is necessary.
7. Sulfides: Determine in the field itself.
8. Sulfites.
9. Residual chlorine: Determine in the field itself.
10. Mercury :

    If caustic soda manufactured by mercury cell method is used for cooking / digestion of wood, the residual mercury present in sodium hydroxide during its manufacture may gain access into the waste and it is toxic to fish and other aquatic life and its determination should be made more precisely.
11. Phenolic substances :The phenolic compounds are usual components of pulp effluents. Fishes acquire a phenolic odor when the concentration of phenolic substances in the water course increases above 0.1 mg/l due to the discharge of pulp effluents.
12. Lignin compounds :

    Pulp industries are the only source for lignin and lignosulfonic acids to be let out into water courses. Lignin compounds do not contribute to significant extent to BOD as they are not easily biodegradable. Hence the results of the BOD test should be interpreted carefully correlating the concentration of lignin compounds present in the waste and COD value.
13. Mercaptans

14. Permanganate value
15. BOD : Permanganate value
16. Chemical Oxygen Demand: 25ml sample aliquot and 25 ml 0. 25 N. $K_2Cr_2O7$are generally used.

Additional parameters

17. Sulfate
18. Percent Sodium
19. Boron
20. Residual sodium carbonate
21. Oils & Grease.

REFERENCES

1) LouisKlein, "River pollution - Vol II, Causes and Effects" - 1967, Butterworths, London.
2) U. S. Environmental Protection Agency, "Environmental Pollution Control, Pulp & Paper Industry - Part-I, AIR" - 1976, EPA, Cincinnati, Ohio.
3) R. N. Shreeve & Joseph. A. Brink, "Chemical Process Industries" 4th Edn. - 1977, McGraw Hill International Book Co.,
4) N. L. Nemerow," Industrial Water Pollution - 1978, Addison - Wesley Publishing Co., Massachusetts.
5) N. Manivasakam, "Environmental Pollution", 1984, National Book Trust (India), New Delhi.

Chapter - 4

TEXTILE PROCESSING

In a textile industry, yarn is prepared, woven or knitted and processed. The basic raw materials are cotton, synthetic fibers and wool. The wet processing of each of them is different and hence dealt separately.

COTTON TEXTILES

In a cotton textile mill raw cotton is carded, combed, spun, slashed (sizing) and woven or knitted into cloth. All these processes are dry processes except sizing. The cloth obtained is called as grey cloth and is subjected to various treatments such as desizing, scouring, bleaching, mercerizing, dyeing or printing and finishing. These are wet processes and generate significant volumes of effluents. Mention has to be made that all these processes are carried out in an integrated unit or each wet process may be carried out at separate units.

The processes carried out in a textile mill along with effluent streams are given in Fig 4.1

PROCESSES AND CHARACTERISTICS OF EFFLUENTS

Sizing (Slashing)

It is the process by which warp thread is sized with starch to give necessary tensile strength and smoothness required for weaving. Effluent from this process contain starch and softeners. Though the volume is much less, B.O.D. load of the effluent is high due to starch. In recent times, starch is slowly replaced by corboxy methyl cellulose and polyvinyl-alcohol (low BOD sizes) which reduce the BOD of the effluent by 40 to 90%.

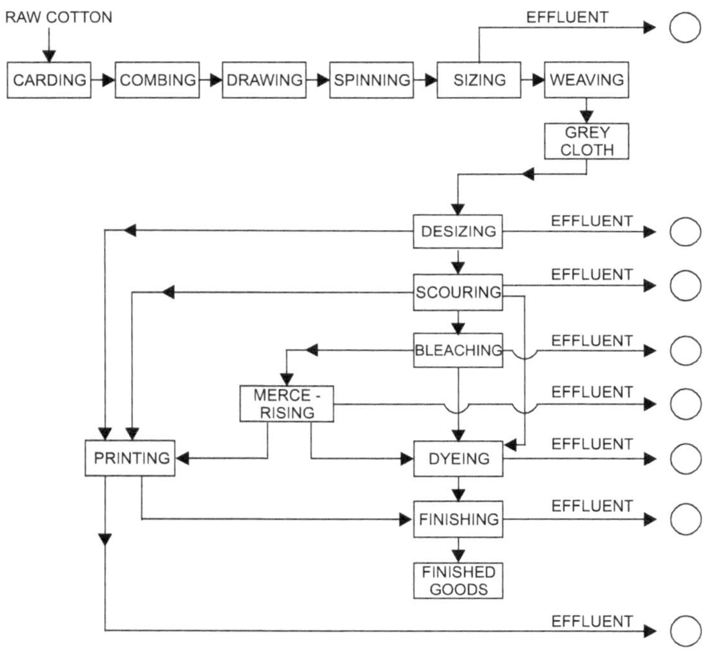

Fig 4-1. Flow Chart Indicating the Processes carried out and Effluent Streams in a Cotton Textile Mill

Desizing

It is a process employed to remove the sizing materials present in the grey cloth to make it suitable for dyeing and further processing. There are two methods of desizing.

a) Desizing with acid: The grey cloth is treated with dilute sulfuric acid (0.5%) and washed well.

b) Desizing with enzymes: The cloth is steeped in baths containing enzymes (1% solution by volume) or malt which hydrolyzes starch and converts to readily soluble substances. Then the clothes are subjected to mild acid treatment.

Desizing effluents arise mainly from the washings and contain the products of hydrolysis of starch and other sizes in dissolved and colloidal form. The pH of the effluent is usually in neutral range and it has a high BOD (about 35 to 50% of the total BOD) and dissolved solids. The suspended solids (mostly of cotton fibres) content is appreciable.

Kier boiling (Scouring)

To remove the natural impurities such as grease, wax, fats and other impurities, the desized cloth is subjected to kier boiling (ie) they are boiled with the aid of steam in an alkaline solution containing caustic soda (1 to 3%) (major component), soda ash, sodium silicate and sodium peroxide, with small amounts of detergents for several hours. After boiling, the spent liquor is discharged as waste. This is a strong waste, dark brown in colour, and highly alkaline. Temperature of the effluent is high. BOD of the liquor is also high, contributing to about 35% of total waste load. The 4 hours permanganate value is also high but BOD: PV ratio is low indicating its resistance towards biochemical oxidation. Suspended solids (mostly fragments of cotton) and dissolved solids content are also appreciable. However the volume of spent liquor is less. Wash waters from kiering contribute significant quantities to the volume of effluents. These wash waters are usually weak solutions of kier liquor containing soap and/or other wetting agents. They contain appreciable quantities of suspended solids much of which is cotton lint.

Bleaching

The clothes are bleached to remove natural coloring materials and to render the clothes white. Alkaline hypochlorite or chlorine is used for bleaching cotton textiles. Bleaching is usually followed by washing with fresh water and then by scouring (treatment with sodium bisulfite) to remove traces of alkali and chlorine.

Some finer fabrics are bleached using peroxide. Peroxide bleaching is carried out during kiering or immediately after kiering. The chemicals used for peroxide bleaching are sodium peroxide, caustic soda, sulfuric acid and certain soluble oils.

Fabrics made from synthetic fibres are bleached using mild bleaching agents such as sodium hypochlorite.

The wastes usually contain unused bleaching agents (free chlorine, hypochlorite, caustic soda and sodium peroxide). Care must be taken in disposing of these wastes, as other acidic effluents react with this, resulting in the evolution of chlorine.

Mercerizing

Mercerization gives a lustre and strength to the fabrics and mercerized fabrics can be easily dyed. Most of the clothes are usually mercerized but rough varieties are not mercerized and after bleaching, they are sent for dyeing without mercerizing.

Mercerization involves treatment of the clothes with cold caustic soda solution (20% w/v) followed by thorough washing with water. Traces of alkali adhering are removed by treatment with a dilute solution of acid followed by rinsing with water (Yarn is also subjected to mercerization in some mills). The casutic soda solution is reused after recovery. Wash waters are the effluents from this section which are strongly alkaline.

Dyeing

The materials may either be dyed or printed. Dyeing is carried out with varied types of dyes and a brief on each dyeing process is given below. The quality and to certain extent quantity of the effluent depends on the type of dyeing.

i) Direct dyes

In most of the small dyeing units, direct dyes (neutral dyes) are used as they are easy to apply and no auxiliary chemicals such as mordants are needed. Though the color is stable for washing, their color fastness is not so appreciable.

ii) Basic dyes

This class of dyes give bright colors. They are applied along with weak organic acids (such as tannic acid)

iii) Sulfur dyes

For dark colors, these dyes are employed. These are sulfur compounds applied usually with sodium sulfide followed by oxidation with chromate. Effluents from this dyeing contains considerable amounts of sulfide.

iv) Vat dyes

These are water insoluble and fast dyes applied along with strong reducing agents (sodium hydro sulfite) and alkali to make the dye soluble. The cloth is then exposed to air for oxidation. (Oxidising agents such as peroxide or chromate are also used instead of air exposure). The excess alkali remaining on the cloth is neutralised by scouring (ie) treatment with dilute solution of acid. This type of dyeing generates more volume of effluents, as the clothes are washed at the end of each step.

v) Naphthol dyes

Betanaphthol is first applied to the fabric, dried and then treated with a developer for coupling and diazotisation after which the color is formed. This is followed by soaping and alkali treatment.

vi) Developing dyes

In this dyeing, the dyes are applied, dried and treated with sodium nitrite and acid and finally with betanaphthol. Effluents from this dyeing contains a number of chemicals.

Yarn dyeing

In some of the mills, yarn itself is dyed after scouring and bleaching.

The volume of effluents generated in dyeing is comparatively more. It contains dyes, mordants, acids (acetic acid) alkalis, nitrites, chromium salts, sodium chloride and soaps. These effluents are usually hot, highly colored and have an alkaline pH. Among the dyeing effluents, those from sulfide and vat dyeing are the strongest. The spent sulfide liquors are dark colored with a high pH and sulfide content. The permanganate value (4 hrs) is also high. Care must be taken while neutralizing these liquors as acid may liberate hydrogen-sulfide gas. Sulfides may be removed by treatment with chlorine or hypochlorites. Spent vat dyes are strongly alkaline and have fairly high permanganate value.

Printing

It is carried out normally using hand blocks or in roller machines. The dye is applied as a thick paste along with starch or printing gums and mordants for fixation of dye. The fabrics are then thoroughly washed to remove the unfixed dyes etc.,

Effluents from printing arises through washing of fabrics and printing machines. The volume of effluent is much less. It contains unfixed dyes, starch, gums, oils, resins and soaps used in preparing dye pastes and for fixation.

Finishing

This is done after dyeing or printing to impart smoothness to fabric. Only a part of the processed fabrics are finished according to the requirements. Starches, dextrins, natural and synthetic waxes and synthetic resins are used for finishing. The quantity of water used and effluent generated from finishing are much less.

Combined Effluent

Depending on the processes carried out, the volume and the nature of the effluents vary. Generally the effluents are highly colored, alkaline with more suspended and dissolved solids. The impurities encountered in all processes are present in the combined effluent. BOD and COD values are also high. Effluent is amenable for biological oxidation.

The characteristics of individual and combined effluents of cotton textile processing are summarized and given. See Tables 4-1 to 4-7.

Table 4-1.
Characteristics of Cotton Textile Processing Wastes*

Process	Wastes in ppm†		Gallon wastes per 1000lb goods
	pH	BOD	
Slashing, Sizing			
Yarn‡	7.0-9.5	620-2,500	60-940
Desizing	......	1,700-5,200	300-1,100
Kiering	10-13	680-2,900	310-1,700
Scouring	....	50-110	2300-5,100
Bleaching	8.5-9.6	90-1,700	300-14,900
Mercerizing	9.5 and above	45-65	27,900-36,950
Dyeing			
Aniline black	....	40-55	15,000-23,000
Basic	6-7.5	100-200	18,000-23,000
Developed Colours	5-10	75-200	8,900-25,000
Direct	6.5-7.6	220-600	1,700-6,400
Indigo	5-10	90-1700	600-6,000
Naphthol	5-10	15-675	2,300-16,800
Sulfur	8-10	11-1,800	2,900-25,600
Vats	5-10	125-1,500	1,000-20,000

* From "An Industrial Waste Guide to the Cotton Textile Industry" U.S.Public Health Service.

† Composite of all wastes connected with each process.

‡ Cloth weaving mill waste.

Table 4-2.
Quantity and Sanitary Characters of Enzyme Desize and Starch Rinse Wastes[§]

Item	Enzyme Desize	Starch Rinse
Volume (gal/100 lb goods)	2.5	720
pH value	7.35	7.1
5-day BOD, ppm	4,375 - 5,063	997
Total Solids, ppm	8,946 - 10,548	1,482 - 1,830
Suspended Solids, ppm	197 - 272	244 - 268
Total Dissolved Solids, ppm	8,682 - 10,276	1,238 - 1,562

Table 4-3.
Raw Kier Waste Characteristics*

Item	Raw Waste
pH value	12
Alakli, ppm as $CaCO_3$	2,174
Color	Brown
Turbidity, ppm	205
Total Solids, ppm	6323
Total Suspended Solids, ppm	55
5-day BOD, ppm	2140
Ratio, Vol/BOD	1.3 : 1

[§] From N.L.Nemerow, "Oxidation of Enzyme Desize and Starch Rinse Textile Wastes, "Sewage and Industrial wastes 26, 1231, (1954)

* From N.L.Nemerow, "Oxidation of Cotton Kier Wastes, "Sewage and Industrial wastes, 25, 1060 (1953)

Table 4-4.
Characteristics of Bleaching Effluent (Grab Sample)

Sl. No.	Characteristics	Result	
		Chlorine Bleach	Peroxide Bleach
1.	Color	Whitish	Whitish
2.	Odor	Higly chlorinous	-
3.	Total Residual chlorine, mg/l	2760	-
4.	pH value	6.0	10.5
5.	Total Solids, mg/l	28500	2820
6.	Total Dissolved Solids, mg/l	22000	2390
7.	Dissolved Fixed Solids, mg/l	9500	1780
8.	Chlorides (as Cl), mg/l	3600	560

Table 4-5.
Characteristics of Mercerizing Effluent (Grab Sample)

Sl. No.	Characteristics	Result		
		Mill "A"	Mill "B"	Mill "C"
1.	Color	Whitish	Whitish	Whitish
2.	Odor	Caustic	Caustic	Caustic
3.	pH value	13.5	12.7	13.5
4.	Total Suspended Solids, mg/l	430	2700	2200
5.	Total Dissolved Solids, mg/l	10940	31700	12000
6.	Dissolved Fixed Solids, mg/l	6660	14900	7380
7.	BOD (5 days at 20°C), mg/l	280	150	-
8.	Chemical Oxygen Demand, mg/l	872	-	-
9.	Chlorides (as Cl), mg/l	640	-	-
10.	Mercury (as Hg), mg/l	0.02	-	0.01
11.	Percent Sodium, (as Na)	94.6	98.0	-

Table 4-6.
Characteristics of Dyeing Effluent
(Grab Sample)

Sl.No.	Characteristics	Result
1.	Temperature °C	50
2.	pH Value	10.5
3.	Phenolphthalein alkalinity (as$CaCO_3$), mg/l	13600
4.	Total alkalinity (as $CaCO_3$), mg/l	16100
5.	Total Suspended Solids, mg/l	10200
6.	Total Dissolved Solids, mg/l	29800
7.	Dissolved Fixed Solids, mg/l	24060
8.	Permanganate value (4 hrs), mg/l	376
9.	Chemical Oxygen Demand, mg/l	1490
10.	Chlorides (as Cl) mg/l	1800

Table 4-7.
Characteristics of a Combined Effluent*
(Composite Sample)

Sl.No.	Characteristics	Result
1.	Temperature °C	35
2.	pH Value	9.5
3.	Total Solids, mg/l	3040
4.	Total Suspended Solids, mg/l	1030
5.	Total dissolved Solids, mg/l	2010
6.	Dissolved Fixed Solids, mg/l	1730
7.	BOD (5 days at 20°C), mg/l	250
8.	Chemical Oxygen Demand, mg/l	390
9.	Chlorides (as Cl), mg/l	420
10.	Total Chromium (as Cr), mg/l	4.5
11.	Zinc (as Zn), mg/l	0.8
12.	Phenolic substances (as phenol), mg/l	0.06
13.	Oils & grease, mg/l (petroleium ether extractables)	3.6

* mercerizing waste not included

EFFECTS OF EFFLUENTS

I. Effect on water courses

1. **Color:** The effluents contain dyes in higher concentration which impart color to the receiving streams and they persist for longer distances. Photosynthesis of phytoplankton is affected by these colors.

2. **pH value:** The extreme alkalinity makes the receiving water unfit for any purpose. Further, it is deleterious to most of the aquatic life.

3. **Suspended Impurities:** The colloidal and suspended impurities produce turbidity in the receiving waters. The turbidity and color along with the oil and scum create an unsightly appearance.

4. **Depletion of Oxygen:** Natural substances such as starch and dextrin and inorganic substances such as sulfide and nitrite present in the effluent exert an immediate oxygen demand. The stream will then be devoid of oxygen and the aquatic life are affected adversely.

5. **Toxic Substances:** Chromium, sulfide, chlorine and aniline dyes present in these wastes are directly toxic to fish and microbial organisms which carryout purification. Thus the self purification of the water body is affected.

6. **Oils:** Various oils (mineral) in the effluent interfere with the oxygenation of stream as they form a blanket on the surface and prevents the entry of oxygen at air/water interface.

7. **Dissolved Minerals:** The mineral materials, mostly sodium salts increase the salinity of the water and consequently it becomes unfit for irrigation.

II. Effects on Agricultural land

See section "Effects on Land" Chapter-1. "Basic Information".

III. Effects on Sewers and Sewage Treatment plants

1. The dissolved solids form incrustation on the surface of the sewers.
2. Presence of sulfide leads to "Crown Corrosion" of sewers.
3. Continuous exposure of the machineries and other structures to textile effluents results in corrosion due to acids and hydrogen sulfide.

TREATMENT

Treatment of textile processing effluents involves many stages and consequently incur much expenditure. To reduce the pollution load, the following steps are suggested.

1. Segregation of the concentrated waste from dilute wash and rinse waters and reuse (e.g) conc. dye waste and mercerizing waste. The dye waste may be reused with suitable adjustment and modifications. Mercerizing effluent (especially second and third wash waters) may be concentrated and reused, or stored and used in kier boiling make up.
2. Substitution of low pollution load chemicals; For example,

starch which is the major material used in sizing may be substituted with polyvinyl alcohol.

3. Economy in water use: The bulk of textile processing effluent is contributed by rinse and wash waters. Use of counter current system of washing, and systematic use of water reduce the waste volume.

A good treatment system includes primary, secondary and tertiary treatments.

Primary Treatment

(I) **Screening:** Coarse suspended matters are removed in preliminary screening.

(ii) **Equalization:** This is an essential step in treatment, as different streams of effluent possess different pH values and characteristics, and further treatment need an effluent of uniform quality. Equalization helps in providing such effluent and prevent the shock loads and sudden increase or decrease in pH values due to mercerizing or kier boiling and other effluents. These effluent streams should be segregated and stored in separate holding tanks and discharged to equalization tanks at a regular and uniform rate.

(iii) **Neutralization:** Most of the secondary biological treatments are effective only if the pH value of the effluent is between 6 to 9, Hence the pH of the equalized effluent should be adjusted to be within this range. Textile processing effluents are usually highly alkaline and addition of dilute sulfuric acid is recommended. Carbon di- oxide or flue gas may also be used.

(iv) **Chemical Coagulation:** To remove color, suspended impurities, colloidal particles (starch and gums), the effluents are treated with coagulants (alum, ferrous sulfate, ferric chloride and chlorinated copperas), and clarified in a clariflocculator. For efficient coagulation and flocculation, coagulant aids such as polyelectrolytes are also employed in conjunction with coagulants. (starch itself acts like a polyelectrolyte). The sludge is separated and dried on sand beds. The treated effluent is subjected to secondary biological treatment.

Secondary Treatment

Textile processing effluents are amenable for biological treatment. They contain the essential nutrients such as carbon, nitrogen and phosphorus. It is reported that combined treatment with domestic sewage is advantageous in terms of dilution, economy and provision of micro organisms. Following are some of the aerobic oxidation methods.

(i) Trickling Filter

(ii) Activated Sludge process

(iii) Aerated Lagoon

(iv) Oxidation Pond

(v) Oxidation Ditch.

For larger textile mills with greater volume of effluents, either trickling filters or activated sludge treatment units may be installed. They may not be economical for medium and small industries. For medium industries, aerated lagoon is more suitable. They are quite efficient and reported to remove about 90% B.O.D. Where space is available, oxidation pond would be a simple and economical method (combined treatment with sewage yield better results). Recent reports indicate that about 80% BOD is removed in this method. (pond depths vary from 0.9 to 1.2 metres with a detention period of 10 to 30 days)

Tertiary treatment

Tertiary treatment is necessary for textile processing effluents, as they contain enormous amounts of both sodium and chloride ions used in various processes. (Their effects are described in detail in the section "Effects of Effluents"). Their removal is a must, before the effluents are discharged into stream or land. Although addition of gypsum to soil may reduce the percent sodium, it cannot remove chlorides and other dissolved solids. Either reverse osmosis or electrodialysis may be employed for the removal of dissolved solids.

PARAMETERS TO BE DETERMINED

Sizing

1) Color 2) Odor 3) Temperature 4) pH value 5) Total suspended solids 6) BOD: Dilutions ranging from 1:199, 1:499, 1:999, 1:1999 may be prepared (seed may be added if necessary)

Desizing

1) Color 2) Odor 3) Temperature 4) pH value 5) Suspended solids 6) BOD: Range of dilutions, 1:499, 1:999, 1:1999 and 1:4999 7) Surfactants.

Kier boiling

1) Color 2) Odor 3) Temperature 4) pH value 5) Alkalinity 6) Suspended solids. 7) BOD: Before preparing dilutions pH should be adjusted. Range of dilutions 1:499, 1:999, 1:1999. Seeding is necessary.

Bleaching

1) Color 2) Odor 3) Temperature 4) pH value 5) Acidity/Alkalinity 6) Residual chlorine. 7) Suspended solids. 8) Sulfide 9) Percent sodium.

Mercerizing

1) Color 2) Odor 3) Temperature 4) pH value 5) Alkalinity 6) Suspended solids. 7) BOD: Adjust pH before preparing dilutions. Following range of dilutions are recommended. 1:49, 1:99 and 1:199. Seeding is necessary 8) COD 9) Percent sodium. 10) Mercury.

Dyeing

1) Color 2) Odor 3) Temperature 4) pH value 5) Suspended solids. 6) Dissolved fixed solids 7) Chlorides 8) Sulfides 9) Sulfites 10) Nitrite 11) BOD: Pretreat the sample and then prepare dilutions of 1:99, 1:199, 1:499, and 1:999. Seeding is necessary. 12) COD 13) Chromium 14) Mercury 15) Percent sodium 16) Surfactants 17) Other specific substances.

Printing

1) Color 2) Odor 3) Temperature 4) pH value 5) Suspended solids 6) Dissolved fixed solids 7) BOD: Observe usual precautions. Range of dilutions, 1:99, 1:199, 1:399 1:499 and 1:999. Must be seeded. 8) COD 9) Surfactants 10) Oils & Grease.

Combined Effluent: Carryout all the tests prescribed for individual effluents.

SYNTHETIC TEXTILES

The most prominent man made synthetic fibres are rayon, acetate, nylon, acrylic and polyester. These fabrics require no processing for the removal of natural impurities as they are man made. However they do require some treatments such as removal of size, antistat and lubricating oils, used in weaving operations. The commonly employed sizes, antistats and lubricants are (i) Polyvinyl alcohol (ii) Styrene - base resins (iii) Poly- alkylene glycols, (iv) Gelatin (v) Polyvinyl acetate.

The wet processes are generally similar for all these fabrics and include 1. Scouring and rinsing 2. Dyeing or bleaching and rinsing and 3. Final finishing (water proofing, shrink proofing etc.,)

The characteristics of effluent from synthetic textile processing differ fibre to fibre and mill to mill. However the effluent volume and pollution load are comparatively lower than that of cotton textiles.

PROCESSES, VOLUME AND CHARACTERISTICS OF EFFLUENTS

Rayon: Rayon is regenerated cellulose. The primary operations scouring and dyeing are performed in the same bath, then rinsed with a salt solution, the process which is termed as salt take-off. The volume of combined effluents from these operations is about 50000 litres/1000 kg of cloth. The pH value ranges from 7 to 9.5 and the BOD is of about 1500 mg/l. Salt concentration varies from 2000 to 6000 mg/l.

Acetate: This fibre is cellulose acetate. Scouring and dyeing yield a waste with a BOD of 2000 ppm. The effluents are similar in character to that of rayon finishing, containing antistats, sulfonated oils, detergents and softener. The rinse wastes will have a low BOD. The volume of combined effluent is estimated to be about 90000 litres/1000 kg of cloth processed. The waste is usually alkaline and its pH value ranges from 7.5 to 10. BOD value is about 700 mg/l.

Nylon: This is a polyamide fibre manufactured by reaction of hexamethylene diamine and adipic acid. Processes include scouring, rinsing, dyeing followed by rinsing and final specialty finishing. The volume of combined effluent is about 1,50,000 litres/1000 kg of cloth processed. The effluent is alkaline. The pH value ranges from 7.5 to 9.0. BOD is moderate (300 mg/l to 400 mg/l)

Orlon: This is an acrylic fibre and normal dyeing on these fabrics can only produce light shades. Hence they are dyed along with carriers (copper, phenols etc.,) at high temperature and pressure, and consequently the effluents contain these toxic substances and thus differ from effluents of processing of other fibres. The effluent is highly acidic (pH 1.5 to 3.0) due to acid dyeing and contains toxic concentrations of copper (25 to 50mg/l) and phenol (100 to 150 mg/l). BOD value is also high ranging from 500 to 600 mg/l.

Dacron: This polyester fibre is manufactured by esterification of ethylene glycol with terephthalic acid. These are also dyed with carriers such as benzoic acid, salicylic acid, phenylphenol and phenyl methyl carbinol. Pressure dyeing without carriers is also carried out. The wet processes include scouring, dyeing and special finishing, each step followed by rinsing.

EFFECTS OF EFFLUENTS

See "Cotton Textiles"

TREATMENT

See "Cotton Textiles"

PARAMETERS TO BE DETERMINED

1) Color 2) Odor 3) Temperature 4) pH value 5) Acidity/Alkalinity 6) Residual chlorine 7) Suspended solids 8) Dissolved Fixed Solids 9) BOD 10) COD 11) Chlorides 12) Sulfides 13) Percent sodium 14) Mercury 15) Copper 16) Phenols

REFERENCES

1) N. L. Nemerow, "Oxidation of Cotton Kier Wastes" Jour. of Sewage and Industrial Wastes, 25, 1060 (1953).

2) N. L. Nemerow "Oxidation of Enzyme Desize and Starch Rinse Textile Wastes" Jour. of Sewage and Industrial wastes, 26, 1231 (1954)

3) Masselli, etal, "A simplification of Textile waste Survey and Treatment" New England Interstate Water Pollution Control Commission - 1959.

4) U. S. Public Health Service, "An Industrial waste Guide to the Cotton Textile Industry"

5) H. F. Lund (Ed), 'Industrial Pollution Control Hand Book" 1971, Mc Graw - Hill Book Company.

Chapter - 5

TANNERY

Tanning is the process of conversion of animal skins and hides to non-putrescible and tough material known as leather. The tanning processes are broadly classified as

A) Beam house processing, and,

B) Tan yard processing.

(A)BEAM HOUSE PROCESSING

The dry matter of the skin is almost protein, reported to be about 85% collagen. The skin, also contains lipids, albumin, carbohydrates and globulin in minor amounts. In beam house processes, all the impurities are removed and collagen of the skin is made ready to absorb the tannin or chromium. All the beam house processes are described in detail as follows.

(i) **Soaking:** The hides and skins are either wet salted or dry salted and are first soaked in water in pits (containing wetting agents) for one day. This removes the dirt, salts, blood, dung and non-fibrous proteins and restore the moisture lost during preservation and storage.

(ii) **Liming:** After soaking, the hides are washed with running water and again soaked in a pit containing lime and sodium sulfide (8 : 1) to soften the hair and thus facilitate its removal from the skin. This soaking swells the hides/skins and loosens the hair. Further the hides and skins become attractive and more amenable for the removal of trace protein impurities. They are also split - up into two layers, one is the valuable "grain" (or leather) layer and the other is flesh layer also called as "split". The hairs and flesh of these limed hides are completely removed by mechanical means and washed well with a fresh supply of water.

(iii) **Deliming:** It is effected by treatment with either ammonium chloride or sulfate and washing again. After deliming, the stock is sent for tanning.

(B)TAN-YARD PROCESSING

There are two types of tanning.

1. Vegetable tanning and
2. Chrome tanning.

Quantities and characteristics of effluents depend on the type of tanning. In some of the tanneries both the tanning processes are carried out.

(1) VEGETABLE TANNING

This type of tanning produces heavy leathers which are plumper, more easily tooled and embossed. They are also less affected by changes in humidity and body perspiration. The following are the sequential operations in vegetable tanning.

(i) **Tanning:** The delimed skins and hides are soaked for 10-15 days in pits containing vegetable tan liquor which comprises vegetable substances (extracts of bark, wood, nut etc., containing natural tannins) and also pyrogallol or catechol base. (It is reported that using vegetable tanning extracts and syntans, the period of tannins is reduced to a few hours only)

(ii) **Myrobing:** The tanned materials are then soaked in myrobalan liquor.

(iii) **Oiling:** After myrobing, oil is applied liberally on the surfaces with suitable vegetable oil (pungam oil) which makes the leather soft, pliable and resistant to tearing. Incorporation of oil is also termed as stuffing.

(iv) **Drying:** The leather is dried and dyed.

(2) CHROME TANNING

It is carried out to prepare light and more resistant leather. It involves more operations than vegetable tanning, but the duration is much less.

(i) **Bating:** Bating is the treatment given to the delimed materials with a bating agent (mixture of prepared proteolytic enzymes-trypsin and chymotrypsin are predominant) which reduces the pH and swelling, peptizes the fibres and removes the protein degradation products. Also, renders the grain silky & slippery, smoother & porous, diminishes wrinkles and increases width.

(ii) **Pickling:** It is done to prevent precipitation of chromium salts on the skin fibres and involves pickling in sulfuric acid and common salt for one day.

(iii) **Tanning:** After pickling, they are tanned by soaking in chrome tan liquor containing chromium sulfate for 6 hours (pickling and tanning are done in the same drum). After tanning, sodium carbonate is added to the drum to fix the chrome. Then they are taken out of the drum and kept for 24 hours for complete absorption and fixation of chromium.

(iv) **Dyeing:** The tanned leather is then washed, neutralised and dyed with different types of dyes (basic dyestuffs are commonly employed).

(v) **Fat-liquoring:** To the leather is then applied an emulsion of sulfonated oil to make the leather soft and pliable.

A flow chart indicative of all the processes described above is given. See Fig. 5-1.

SOURCES & CHARACTERISTICS OF EFFLUENTS

Soaking effluent

The spent soak liquor is discharged intermittently. It contains excess sodium chloride, dirt, dung and soluble proteins like albumin. The most objectionable feature of this effluent is the abnormal sodium chloride content which varies from 7000 to 20,000 mg/l. The BOD value is also high ranging from 800 to 2500 mg/l. It contains much organic matter and suspended solids and hence this effluent is considered as a source for proteolytic and other bacteria. As the pH is in near neutral range and conditions are so favorable for the growth of bacteria, putrefaction occurs rapidly, releasing obnoxious odors.

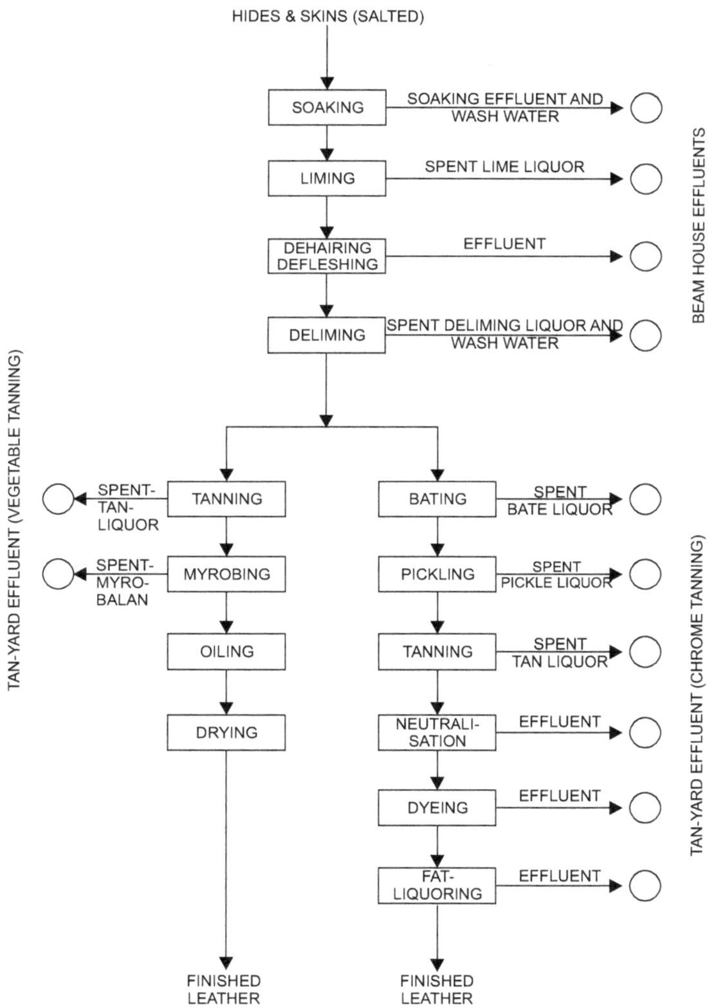

Fig 5-1. Flow Chart of Tanning (Vegetable & Chrome) Processes indicating various Effluent Streams

Liming effluent

It is also discharged intermittently. This effluent is highly alkaline and contains excess amounts of sulfides, suspended and dissolved lime, calcium carbonate, colloidal proteins and their degradation products. It contributes the major fraction of BOD, the value vary from 4000 to 9000. It also contains excess suspended solids. The ammoniacal nirogen and organic nitrogen contents are also high.

Dehairing & Defleshing effluents

As the name implies, it contains mostly hair and flesh and also higher amounts of sulfides and its discharge is almost continuous. Suspended solids content of these effluents is also high.

Deliming effluent

These effluents are slightly acidic and pollution load of them is considerable. BOD varies from 1000 to 2000 mg/l.

Vegetable tanning (and myrobing) effluent

These waste liquors are weakly acidic (pH 4.5 to 5.0) and contain large amounts of organic matter raising the BOD values upto 12,000 mg/l*. Suspended solids content is also high. This waste is the fraction contributing to the high pollution load of tannery effluent. Like pulp mill effluents the

* Note that tannins do not exert such a high BOD. They are not easily biodegradable. They contribute to high COD values. Carbohydrates, sugars and other non-tannin substances are responsible for such a high BOD.

brown color of this liquor is very difficult to bleach and persists in the receiving stream for longer distances. The characteristic foul odor is highly disagreeable.

Bating effluent

Though the BOD of these effluents are not so high, the ammoniacal and organic nitrogen content are high due to the soluble proteins and ammonium salts in it.

Pickling effluent

This waste is discharged intermittently. It is acidic and contains excess sodium chloride. BOD value of this effluent varies from 500 to 1000 mg/l.

Chrome tanning effluent

This is the toxic fraction of tannery effluent and contains excess chromium (in trivalent state) ranging from 100 to 200 mg/l. It has a characteristic chrome green color, acidic in nature with a BOD of about 1000 mg/l.

Dyeing effluent

It contains mostly dyes. The volume is much less.

Fat-liquoring effluent

These waste liquors contain excess quantities of oil and grease.

Combined effluent

Vegetable tanning is carried out in most of the tanneries. Some of the tanneries carryout only chrome tanning.

Both vegetable and chorme tanning are carried out in certain tanneries. Accordingly there are three types of combined effluents.

(i) Beam house effluents and vegetable tanning effluents

(ii) Beam house effluents and chrome tanning effluents.

(iii) Beam house effluents and vegetable and chrome tanning effluents.

Generally the combined effluent will be alkaline, foul smelling, having more BOD and excess quantities of sodium chloride and suspended solids.

The characteristics of vegetable tanning effluent is given in Table 5-1.

EFFECTS OF EFFLUENTS

Most of the constituents present in a tannery effluent are of highly damaging character irrespective of whether they are discharged into river, stream, sewer, or on land or in to ocean.

Effects on streams

1. The dull brown color of the effluent which is very difficult to bleach is imparted to the receiving streams and results in an unsightly appearance. The color persists for a longer stretch.

2. The highly disagreeable foul odor is transmitted to the water.

Table 5-1.
Characteristics of a Vegetable Tanning Effluent (Combined Effluent of Beam House and Tan Yard)

Sl.No.	Characteristics	Results
1.	Color	Brownish
2.	Odor	Highly Disagreeable, Foul odor
3.	pH Value	8.5
4.	Total Solids, mg/l (105°C)	17,640
5.	Total Suspended Solids, mg/l	5,990
6.	Total Dissolved Solids, mg/l	11,650*
7.	Dissolved Fixed Solids, mg/l	10,520
8.	BOD (5 days at 20°C), mg/l	1,200
9.	Permanganate value (4 hrs), mg/l	320
10.	Chlorides (as Cl), mg/l	6,700
11.	Sulfides (as S), mg/l	22
12.	Sulfates (as SO_4), mg/l	360
13.	Ammoniacal Nitrogen (as N), mg/l	21.3
14.	Total Kjeldahl Nitrogen (as N), mg/l	78.9
15.	Calcium + Magnesium (as $CaCO_3$), mg/l	1490
16.	Percent sodium (as Na)	84.7
17.	Oils & Grease, mg/l (Petroleum ether solvent)	16
18.	Tannin and Lignin compounds (as tannic acid), mg/l	50

NOTE-1: Characteristics of the effluent will highly vary from tannery to tannery even for the same capacity.

NOTE-2: When liming effluent mixes with the vegetable tanning effluent, a heavy and bulky precipitate is formed.

* It is to be noted that about 85% of total dissolved solids is only sodium chloride.

3. The individual effluents and combined effluents are either acidic or alkaline and these extremities results in corrosion and deterioration of concrete, metallic pipes etc. The water becomes unsuitable for any use.

4. The most damaging constituent sulfide is usually present in excess concentration. It is toxic to fish and other aquatic life. The concentration of sulfides and oxygen demand are so high that they deplete the dissolved oxygen rapidly resulting in massive fish-kills.

5. Another detrimental factor is the excess concentration of sodium chloride. Sodium in any form makes the water unsuitable for irrigation. Likewise chlorides also. Further, they corrode the treatment structures etc.,

6. The dissolved constituents such as proteins, and albumins putrify slowly evolving highly offensive odors.

7. The suspended solids, mostly hair, fleshings and calcium carbonate, settle to the bottom affecting the fish spawning grounds and also interfere with the aeration and photosynthetic activities of aquatic plants.

8. Chromium is highly toxic to fish and other aquatic life and interferes with the natural purification systems. (Note that the trivalent chromium is not harmful. It is reported that only hexavalent chromium is toxic. The trivalent chromium present in the effluent is usually precipitated as hydroxide when mixed with liming effluent).

9. The effluent may also contain anthrax bacilli which is a pathogen.

Deterioration of ground water quality

As already stated, the abnormal concentration of sodium chloride and chromium affect ground water quality seriously within a short period where the effluent is discharged on land or if stored in unlined tanks. If the ground water is affected once, it cannot be remedied. It is an irreparable damage. There is no other go excepting abandoning the source. It is to be noted that in places where there is a cluster of tanneries, no single source of ground water is available either for human consumption or for irrigation. So many acres of land became infertile due to the seepage caused by stagnation of these effluents.

Effects on sewers

The two major constituents that affect sewers are suspended solids and sulfides. The lime suspension gets precipitated as calcium carbonate by carbon di oxide produced during the decomposition of organic matter. This calcium carbonate, with the hair (which acts as a binding agent) adhere to the surfaces of the sewer gradually reducing the passage and results in complete choking. Further, the concrete structures crumble due to 'Crown Corrosion' by sulfides.

Inhibitory effects on treatment

As chromium is toxic to microbiological and biological organisms, tannery effluents disrupt the operation of trickling filters. Further, sulfides are also toxic to micro-organisms. It is also reported that the lime suspension inhibits biological activities in a sewage treatment plant.

TREATMENT

Treatment of tannery effluents involves primary, secondary and tertiary treatments. The primary treatment includes screening (to remove hair, fleshings, fats etc.) and sedimentation. It is reported that about 90% removal of suspended solids is effected in sedimentation tanks. The sludge is dried over sand drying beds and utilized as manure. Chemical coagulation* (using alum and/or ferric chloride) and biological treatment are the effective, secondary treatments. Activated sludge process with acclimatized organisms yield effluents of better quality. Prior removal of chromium is an essential step when the chrome-tanning waste is subjected to activated sludge process. Chromium is removed by precipitation with lime at a pH of about 6.5. Oxidation pond and anaerobic lagoon are also recommended for small and isolated tanneries.

The effluent from the secondary treatment contains chloride - the most hazardous ion - in excess concentration which makes the disposal of the effluent very difficult. Segregation of spent soak liquor and spent pickling liquor and solar evaporation are the recommended treatments. However disposal of dried salt is also a difficult problem.

PARAMETERS TO BE DETERMINED

Important Parameters

1) Color 2) Odor 3) pH value 4) Total suspended solids 5) Total dissolved solids 6) Dissolved fixed solids. 7) Bio-

* Though ferrous sulphate is a best coagulant for Chrome tan wastes for the removal of chromium, sulfides, BOD and suspended solids, it should not be used for Vegetable tan wastes, as it intensifies the color of the waste.

chemical Oxygen Demand: BOD of a vegetable tanning industry is more than that of a chrome tanning industry. As BOD varies widely from tannery to tannery, a minimum of four dilutions should be made. The range of BOD dilutions are 1:499, 1:999, 1:1999, 1:3999, 1:4999. The 4 hr permanganate value may be taken as a guide in preparing dilutions. The BOD value is approximately 2.5 to 4 times that of permanganate value. The sample should be diluted appropriately before preparing BOD dilutions. The BOD of the effluent of a chrome tannery effluent should be interpreted carefully taking into account the chromium concentration.

8) Chlorides: Sample should be diluted 10 times or 20 times. 10 ml or 20 ml may be taken for estimation.

9) Sulfides: Determine by titrimetric method. If present only in low concentration, colorimetric method is preferable.

10) Percent sodium: Sample may be diluted 200 to 500 times and then estimated.

11) Chromium: Usually present in trivalent form. However determination of both hexavalent and trivalent chromium is necessary to evaluate toxic effects etc., When present in higher concentrations, titrimetric method may be employed. For lower concentrations colorimetric method is suitable. Organic matter must be destroyed whatever the method employed may be.

Additional parameters

12) Ammoniacal nitrogen: Use 25 ml or 50 ml sample for distillation (suppress the sulfide interference before distillation by the addition of lead carbonate to the sample). Distillate volume ranges from 300 to 600 ml. Determine nitrogen by

titrimetric method.

13) Total kjeldahl nitrogen: Use 25 ml or 50 ml sample.

14) Oils and grease.

15) Tannin and Lignin.

REFERENCES

1) R. L. Bolton and L. Klein, "Sewage Treatment - Basic Principles and Trends" 1971 - Butter worths, London.

2) Indian Standards Institution "IS : 2490 (part III) - 1974 - Tolerance limits for Industrial Efflunets discharged into Inland Surface Waters - Part III Tanning Industry (First Revision)" ISI, New Delhi.

3) M. N. Rao and A. K. Datta, "Waste Water Treatment' 1979, Oxford and IBH Publishing Co., Calcutta.

Chapter - 6

CANE SUGAR INDUSTRY

Sugar is produced from sugarcane and sugar beet. Operation of a cane sugar industry is seasonal and starts from November and extends upto May and sometimes June also.

MANUFACTURING PROCESSES

Sugarcane contains 11 to 15% sucrose by weight. The yield of sugar varies from 6 to 15% depending on the quality of cane and efficiency of the mill. Sugarcane is harvested in the field manually by cutting with a knife or machete. The top portions are cut and adhering leaves are stripped. (In some areas mechanical harvesting is done which needs cleaning before shredding. The net result is the generation of more quantity of effluents in sugar industry by washing) The cane is loaded and sent to sugar mill within 24 hours. Sugar loss may occur due to inversion to glucose and fructose if the time exceeds 24 hours.

In the sugar mill, the cane is chopped and shredded (after washing, if necessary). The juice is then extracted by passing it through crushers containing a series of roller mills. The cane is then wetted with water and the remaining juice is also extracted. This operation is termed as maceration. About 93% juice is extracted in all these operations. The fibrous residue left over is called as bagasse (which will be of about 30%

of cane crushed) is used as a boiler fuel and for manufacture of paper and boards.

The extracted juice is screened to remove floating impurities and milk of lime is added to increase the pH to 7.6 to 7.8 thereby preventing the inversion and to aid clarification by coagulating the colloidal impurities along with the addition of bentonite or some other coagulant aid (phosphoric acid is also added some times to aid clarification. In this case excess lime is added). The mixture is heated (to about 102°C) using high pressure steam and allowed to settle. The clarified and bleached juice is sent for evaporation (Bleaching is effected by treatment with sulfurdioxide*). The residue is sent to filter press or rotary drum vacuum filters and the juice in it is reclaimed. The mud from the filter press known as pressmud or limemud (which will be about 1 to 4% of the weight of the cane charged) is used as a manure.

The clarified juice contains about 85% water. It is evaporated to approximately 40% water in multiple effect evaporators to a thick, pale yellow juice. This syrup is then fed into a single effect evaporator where the sugar is crystallised. The later operation is termed as 'pan boiling'. (Sugar-crystal nuclei are added at this stage to imporve crystallisation). Now the syrup contains only 10% water. The mixture of syrup and crystals known as massecuite is passed into a crystallizer (centrifugal baskets) which separates the crystallised sugar from the mother liquor. The spent mother liquor drains away

---

* Carbondioxide is also passed instead of sulfurdioxide which to some extent aids in color removal and controlling the pH. The quantity of limemud would be then 7% of cane crushed.

separately as molasses. The volume of molasses depends on the quantity (and to certain extent quality also) of cane crushed. About 4 to 5% of cane charged is transformed as molasses. The sugar from the centrifugal basket is dried in a drier, graded and bagged. The processes described above are given in the flow chart. See Fig. 6-1.

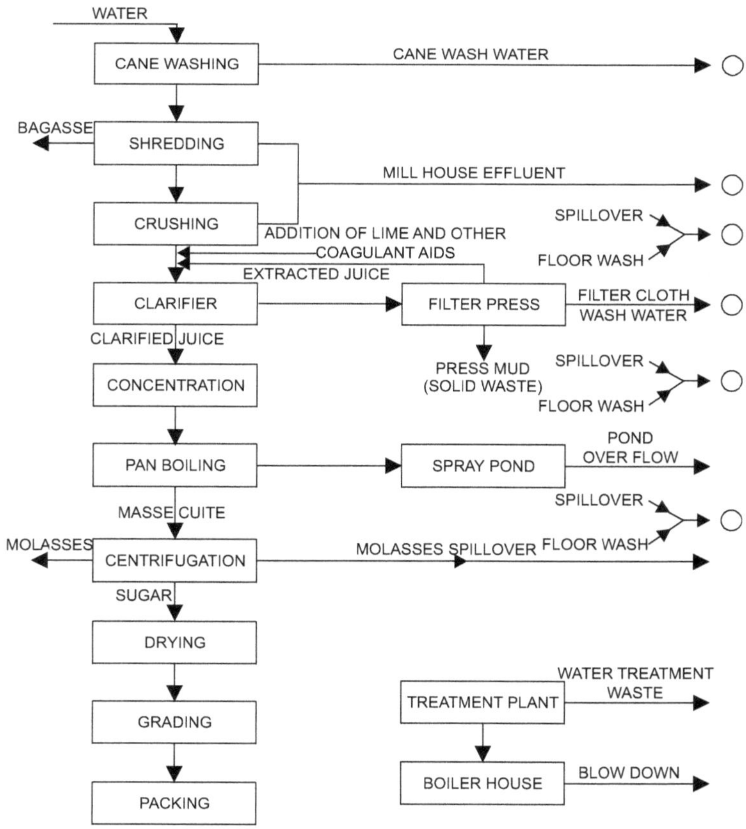

Fig 6-1. Process Flow Chart of Sugar Manufacture with Waste Streams

SOURCES OF EFFLUENTS

1. **Cane wash water:** This is generated only if the cane is washed. (cane washing is done only on mechanically harvested cane). The suspended solid content of cane wash water is high. it may also contain considerable amounts of sugar.

2. **Mill house effluent:** Large amounts of water are used for cooling the bearings of milling machines which pick up lubricants. It also includes spillovers and floor washes.

3. **Filter press cloth wash water:** The clothes of filter press are washed periodically to remove the mud clogging in the pores. This effluent contains appreciable quantities of suspended solids and possesses a high BOD.

4. **Effluent from evaporation:** Cooling and condenser waters are the wastes from this section which contain sugar particles that gain access during concentration of juice at multiple effect evaporators.

5. **Boiling house effluent:** Mostly generated by leakage from centrifuges and periodical floor washings. This waste has an extremely high BOD though small in volume and discharged intermittently.

6. **Spray pond overflow:** The cooling waters are recycled from the pond. However excess water is let out as waste which has low BOD and suspended solids.

7. **Boiler house effluent:** Like other industries boiler blow down also contribute to effluents, but the volume is much less.

8. **Molasses effluents:** Leakage and overflow from molasses storage tanks contribute to the high BOD of sugar factory effluents.

9. During off season, juice heaters and other heating surfaces are cleaned to remove accumulation of scales with hot caustic soda solution, then with dilute hydrochloric acid followed by rinsing with water. These operations produce a large volume of effluent. BOD value of this effluent is high.

VOLUME OF EFFLUENTS

It varies from mill to mill depending on whether cane washing is carried out and also on the availability and management of water, condition of plant and machineries, house keeping etc., The average volume of effluents per tonne of cane crushed is about 1320 litres.

CHARACTERISTICS OF EFFLUENTS

The characteristics of individual effluents and combined effluent vary from mill to mill and from time to time.

All the individual effluents excepting spray pond overflow are acidic and colored, possess disagreeable odor, high BOD and suspended solids. The oil and grease content is also high. The characteristics of effluents of a typical sugar mill is given in the Table 6-1.

Table 6-1.
Characteristics of a Sugar Mill Effluents

Sl.No.	Characteristic	Mill house effluent	Centrifugal and Evaporating section effluent	Filter unit effluent	Spray pond overflow	Combined effluent (excluding Spray pond overflow)
i)	Color	Brownish	Yellowish	Black	Whitish	Yellowish brown
ii)	Odor	Disagreeable (foul odor)	Cane juice	Highly disagreeable	Faint H_2S Odor	Disagreeable
iii)	pH value	4.5	4.5	5.7	7.3	4.4
iv)	Total Suspended Solids, mg/l	1540	52	200	30	630
v)	Total Dissolved Solids, mg/l	1330	846	850	1440 -	2190 -
vi)	Bio-chemical Oxygen Demand, (5 days at 20°C), mg/l	1040	2200	420	125	1000
vii)	Chemical Oxygen Demand, mg/l	4647	3557	1627	1577	3790

Sl.No.	Characteristics	Mill house effluent	Centrifugal and Evaporating section effluent	Filter unit effluent	Spray pond overflow	Combined effluent (excluding Spray pond overflow)
viii)	Permanganate value (4 hrs), mg/l	358	734	162	-	-
ix)	Acidity: (as $CaCO_3$), mg/l					
	a) Methyl orange acidity	-	-	Nil	-	Nil
	b) Phenolphthalein acidity	-	-	64	-	80
x)	Total Kjeldahl Nitrogen (as N), mg/l	5.6	4.48	3.36	2.24	10.08
xi)	Sulfides (as S), mg/l	Nil	Nil	80	0.8	2
xii)	Oils and Grease, mg/l (Petroleum ether solvent)	563	26	32	60	2.4

EFFECTS OF EFFLUENTS

The immediate oxygen demand of sugar factory effluents causes rapid depletion of the dissolved oxygen of receiving streams resulting in anaerobic conditions. This results in the release of foul odors and in the production of hydrogen sulfide which precipitates iron as black sulfide leading to unsightly appearance. All these effects make the water totally unfit for fish and other aquatic life. Also, the dissolved and suspended solids deteriorate slowly resulting in obnoxious odors. Further, suspended impurities block the drainage and ditches. The excess oil and grease content is a nuisance which prevents aeration.

TREATMENT

Conventional biological treatments are not so effective due to the high concentration of carbohydrates and are not economical as the industry is operated seasonally. Volatile organic acids produced during biological oxidation inhibit the biological activity. These acids have to be neutralized by lime.

Treatment of sugar factory effluents by lagooning or in oxidation pond has been found most feasible. Various studies indicate that a two stage oxidation would be effective and economical. After equalization (with a detention period of one day) the waste is sent for anaerobic digestion in open deep pond (minimum depth 2.5 m) for about 5 to 6 days. The second stage is aerobic oxidation in open shallow pond (minimum depth 1 m) and the detention period range is from 10 to 15 days. To improve the efficiency of oxidation pond, the following measures may be implemented.

i) Regular dosing of acclimatized algae to the oxidation pond (for which algal culture tank may be incorporated in this system).

ii) Planting of acclimatized water hyacinths in the inlet bays of the oxidation ponds to clarify the waste water.

PARAMETERS TO BE DETERMINED

Important Parameters

1) Color 2) Odor 3) Temperature 4) pH value 5) Total suspended solids 6) Biochemical Oxygen Demand. Adjust the pH before preparing BOD dilutions. Range of dilutions 1:99, 1:199, 1:499 and 1:999. Before preparing BOD dilutions, sample should be diluted appropriately. Seeding is necessary. 7) Chemical Oxygen Demand : Use 10 ml or 25 ml of sample and 25 ml or 50 ml 0.25 N. $K_2Cr_2O_7$ solution. 8) Sulfides 9) Oil and grease.

Temperature should be measured in the field itself.

Additional Parameters

10) Acidity / alkalinity 11) Total dissolved solids 12) Dissolved fixed solids 13) Percent sodium 14) Boron.

REFERENCES

1) Dr. Parashar, "Treatment of sugar Factory Effluents and their Methods of Analysis", Sugar Ind. Abs. 1972, 34, 155.

Chapter -7

DISTILLERY

Distillery is an industry where ethyl alcohol is manufactured. Industrial alcohol is manufactured from molasses. Beverage alcohol is produced from grains such as barley, wheat and maize. The effluents from the manufacture of both the alcohols are similar in character as their production processes are the same excepting the preparation of fermenting medium.

MANUFACTURE OF INDUSTRIAL ALCOHOL

Molasses* is the exclusively used raw material. About 3 to 10 kg. of molasses is required for producing 1 litre of alcohol. It is diluted first to have a sugar concentration of about 15% to 17% and pumped to large steel fermentors. pH of diluted molasses (termed as mash) is adjusted to lie between 4.0 to 4.5 to suppress bacterial action and multiplication of wild yeasts. To satisfy the nutritional requirements, sometimes magnesium-sulfate, superphosphate and urea are added.

Then selected culture of yeast is inoculated into the mash for fermentation. The following reactions take place during fermentation.

i) Sucrose in molasses is split up into d-glucose and d-fructose by the enzyme invertase.

---

* Molasses is a dark brown liquor obtained on separation of sugar crystals from the mother liquor of processed cane juice at cane sugar industries.

$$\underset{\text{Sucrose}}{C_{12}H_{22}O_{11}} + H_2O \xrightarrow{\text{invertase}} \underset{\text{d-glucose}}{C_6H_{12}O_6} + \underset{\text{d-fructose}}{C_6H_{12}O_6}$$

ii) Conversion of the monosaccharide to alcohol and carbondioxide by the enzyme zymase.

$$C_6H_{12}O_6 \xrightarrow{\text{zymase}} 2C_2H_5OH + 2CO_2$$

The fermentation is an exothermic reaction and the process is kept at controlled temperature (within 38°C) by spraying water at the outer surface of fermentation tank. Fermentation essentially completes at the end of 36 to 50 hrs.

The liquor in the fermentor is pumped to distillation unit for the recovery of alcohol. The residual liquor remaining after the recovery of alcohol is termed as spentwash and discharged as waste. A sludge, settling at the bottom of fermentor termed as "yeast sludge" is also discharged as waste. In some distilleries it is mixed up with the spent wash. In some distilleries it is sold out as manure.

SOURCES OF EFFLUENTS

The spent wash is the major waste in a molasses distillery. Yeast sludge is also a strong waste but now a days it is dried and used as a manure. Some of the distilleries however combine yeast sludge with spent wash stream. In addition, cooling and condenser waters, floor washings, fermentor wash water also contribute to the bulk of effluents.

VOLUME OF EFFLUENTS

It is estimated that about 10 to 15 litres of spent wash is generated per litre of alcohol produced. The quantity of condenser and cooling waters are reported to be about six times that of spent wash and contain only little impurities. Altogether the volume of effluents would be 60 to 100 litres per litre of alcohol produced.

CHARACTERISTICS OF EFFLUENTS

The effluents as such coming from the plant is hot (about 60 to 70°C), highly colored (dark brown), acidic and possess an objectionable odor. The BOD value is extremely high (40000 to 70000 mg/l). Likewise the values for suspended solids, dissolved solids, chloride, sulfate and nitrogen are also too high. The potassium content of the effluent is also high but that makes the effluent usable for irrigation. See Table 7-1.

Table 7-1.
Characteristics of a Distillery Effluent

Sl. No.	Characteristic	Result
i)	Color	Dark brown
ii)	Odor	Characteristic odor of molasses (highly disagreeable)
iii)	pH value	3.9
iv)	Total Suspended Solids, mg/l	63,000
v)	Total Dissolved Solids, mg/l	35,600
vi)	Dissolved Oxygen, mg/l	Nil
vii)	Biochemical Oxygen Demand, mg/l (5 days at 20°C)	40,000
viii)	Permanganate value, mg/l	18,900
ix)	Chlorides (as Cl), mg/l	4,240
x)	Sulfates (as SO_4), mg/l	4,400
xi)	Ammoniacal Nitrogen(as N),mg/l	977
xii)	Potassium (as K), mg/l	10,200
xiii)	Percent potassium	50.5

EFFECTS OF EFFLUENTS

Though distillery effluents do not contain any toxic substances, it creates toxic conditions in the receiving stream by immediate depletion of dissolved oxygen resulting in massive fish kills, destruction of aquatic flora and fauna, production of foul odors and discoloration of streams. Most of the distilleries stagnate their effluents on land as ponds. This type of stagnation is not a recommended practice as there are chances for seepage and subsequent ground water pollution. Further, the obnoxious odor spreads to a few kilometers and it is a serious public health problem.

TREATMENT

Treatment of distillery wastes could be made economical if recovery of potash and vitamin B-12 are made.

As the BOD of the distillery effluent is abnormally high, they are subjected to anaerobic digestion or anaerobic lagooning followed by aerobic biological processes.

Anaerobic digestion

This is reported to reduce the BOD of raw wastes by about 90%. Methane is recovered as a by-product during digestion.

Anaerobic Lagooning

When land is available, anaerobic lagooning is the cheapest and effective method. The major disadvantage is the production of foul odors due to the evolution of volatile gases. The production of odors can be limited by establishing proper anaerobic activity in the lagoon.

Effluents from anaerobic treatment still contain high BOD and cannot be discharged as such into water courses and hence further oxidised either by trickling filtration, or by activated sludge or in aerated lagoon or in oxidation ditches.

Trickling Filters

Either single stage or two stage trickling filtration is adopted. Deep trickling filters with plastic media are reported to be effective.

Activated Sludge Process

Extended aeration yields better results.

Aerated Lagoons

The detention period is longer (i.e.) 20 to 28 days for pretreated distillery effluents. About 90% BOD removal is achieved. Even then, the effluent from the aerated lagoons requires further treatment in a polishing lagoon with a detention period of about one day.

Oxidation Ditches

Better performances are obtained with pretreated diluted effluents. Detention time required is about 50 to 60 hrs.

By-product Recovery

By-products from spent wash include potash, riboflavin and other B vitamins particularly B12. Yeast powder of pharmaceutical grade and animal feed can be obtained from the yeast sludge as it contains dead yeasts, proteins, vitamins, fats and carbohydrates. Details regarding recovery is beyond the scope of this book.

Most of the distilleries dilute the spent wash and discharge it on to cultivable lands but mention has to be made that this is not a good practice as it affects seriously the ground water quality and nature of the soil.

PARAMETERS TO BE DETERMINED

Important parameters

1) Color 2) Odor 3) Temperature 4) pH value 5) Total suspended solids 6) Total dissolved solids 7) Dissolved fixed solids 8) Biochemical oxygen demand: Dilution ranging from 1:4999, 1:9999, 1:19999, 1:39999 should be prepared. Seed is not necessary. BOD dilution should be made using appropriately diluted sample.

Additional parameters

9) Chlorides : Dilute the sample 100 times. Use 25 ml or 50 ml aliquots of the diluted sample 10) Sulfates 11) Ammoniacal Nitrogen: Dilute the sample 10 times. Use 25 ml or 50 ml of diluted sample. 12) Total Nitrogen: Dilute the sample 10 or 20 times. Use 10 ml. or 25 ml aliquots. 13) Potassium: Dilute the sample 1000 times and then determine.

MANUFACTURE OF BEVERAGE ALCOHOL

Grains and cereals such as barley, wheat, maize, rice are used to manufacture beverage alcohols. Wort is prepared from these materials after malting, mixing, adjusting pH and adding nutrients. Then it is allowed to ferment. Wort is then distilled and stored in vats. The effluents consist of waste liquors from wort preparation and spent wash remaining after distillation.

The characteristics of the effluent are similar to the effluent produced from industrial alcohol manufacture.

BREWERY

In a brewery beer and other beverages of low alcoholic content (2 to 7%) are manufactured. A variety of grains and fruits containing carbohydrates are employed for brewing, and among them barley is the most important.

Production of beer consists of two major processes, viz, (i) preparation of malt and (ii) brewing.

The barley is malted to develop the necessary enzymes and desired flavor. For malting, the barley grains are steeped in cold water and spread over on floors or in special compartments for germination. At the proper time (5 to 8 days) after the enzymes are formed, green malt is placed on driers and dried to a predetermined moisture content and stored after screening sprouts. The malt is macerated with hot water such that the resulting mash contains valuable constituents of malt, malt adjuncts and sugars. The mash is then boiled with hops and the resulting wort is cooled and filtered. The wort is then inoculated with a prepared suspension of yeast to ferment sugar to alcohol. After completion of fermentation (7 to 10 days), the yeast and other malt residues are separated by filtration. The remaining liquor (beer) is stored at 32°F for 3 to 6 weeks, carbonated and sent for sale.

SOURCES OF EFFLUENTS

In the malting process, two types of effluents are generated. (i) The liquor remaining in the steep tank after the removal of grains and (ii) the liquor from germinating drum after the removal of green malt.

In brewing plant, the major effluent is fermentation residue. Effluents are also produced from yeast preparation, yeast recovery, washing of brewing vessels, bottles etc., and from floor washings. If distillation is carried out, another highly polluting effluent usually referred as "distillery slopes" or "beer slops" is obtained as residue.

In addition, slightly contaminated cooling waters are also generated.

VOLUME OF EFFLUENTS

The quantity of waste liquor produced from brewing is about 8 to 12 times the volume of beer produced.

CHARACTERISTICS OF EFFLUENTS

Effluents from malting process is slightly alkaline and contain a large amount of organic substances in solution and have a considerable BOD and permanganate value. The effluent from brewing plant is usually acidic and contain appreciable quantities of carbohydrates and yeasts and consequently BOD values are high. Suspended solids are also present in significant quantities. See Table 7-2.

EFFECTS OF EFFLUENTS

The effects of brewery effluents are similar to that of distillery effluents but pollution load is much less. Brewery wastes are amenable for biological treatment. However they cause bulking of sludge in activated suldge process and form growths on percolating filters.

Table 7-2.
Characteristics of Malt House and Combined Brewing Wastes*

	Malt house waste	Brewing waste
pH value	6.9 - 9.5	4.0 - 7.0
Total Suspended Solids, mg/l	22 - 339	16 - 516
Total solids, mg/l	428 - 700	272 - 2724
BOD, mg/l	20 - 204	70 - 3000
Permanganate value, mg/l	31 - 175	30 - 1225
Total Nitrogen (as N) mg/l	14 - 56	7 - 42

PARAMETERS TO BE DETERMINED

Important parameters

1) Color 2) Odor 3) Temperature 4) pH value 5) Total suspended solids 6) Biochemical Oxygen Demand : pH has to be adjusted to the neutral range before preparing BOD dilutions. Seeding is not necessary.

Additional parameters

7) Chemical Oxygen Demand 8) Ammoniacal nitrogen 9) Total nitrogen.

REFERENCES

1) C.S. Boruff, "Fermentation Industries: Industrial wastes, their disposal and treatment" - 1952, Reinhold publishing corporation, NewYork.

2) R.N. Shreeve & Joseph A. Brink, "Chemical Process Industries" - 4th ed, 1977, Mc Graw Hill International Book Co.,

3) M.N. Rao & A.K. Datta, "Waste water Treatment" - 1979, Oxford & IBH publishing Company, Calcutta.

* This table is reproduced from the book "Waste water Treatment" by M.N. Rao & A.K. Datta - 1979, with the kind permission of the publishers, M/s. Oxford & I.B.H. Publishing Co., Calcutta.

Chapter 8

DAIRY

Dairy is an industry where milk is processed and various milk products are manufactured.

DESCRIPTION OF THE PROCESSES

At dairies, raw milk is received from various stations and processed. The processing involves pasteurization, standardization, bottling etc., In addition to processing, cream products (butter, ghee etc.) milk powder and casein, condensed milk and ice-cream are also prepared in some dairies.

Raw milk is usually received in cans and tankers and they are emptied and cleaned immediately in the place itself. The milk is then subjected to pasteurization, cream separation and then bottled.

Pasteurization is carried out by heating the milk at 71°C followed by chilling at 4°C. Cream is separated by centrifugation of the whole milk. The separated cream is churned to butter and the milk separated is used for the production of casein by the addition of hydrochloric acid to start with and then with whey.

At milk powder plant, the whole or non-fat skimmed milk is subjected to preheating, vacuum drying and spray drying.

In the condensery, the whole milk, non fat skimmed

milk, butter milk and whey are condensed by preheating and evaporation.

In a cheese factory, the whole milk or separated milk, is received, preheated, filtered, pasteurized and cooled. Then it is placed in cheese vats to which rennet, acid or other souring agents are added to separate the cheese (casein). To this cheese, cream and other additives are added depending on the end products.

Ice-cream is manufactured by combining milk and milk products with cream in various proportions with the addition of flavoring agents, sugar (or other sweetening agents) and stabilizer. After pasteurization and cooling, fruits, nuts, candies are also added and the whole mixture is frozen.

SOURCES OF EFFLUENTS

The waste liquors generated from dairies consist of:

(I) Waste water generated from milk processing through milk spillage, drippings, washing of cans, tankers, bottles, utensils, equipments and floors. Wash water usually contains alkaline chemicals and syndets used to remove milk and milk products and partially caramelized materials.

(ii) Waste liquors from butter manufacturing (i.e.) process wash water of butter and floor washings which consists of mainly butter milk, cream and a small percent of whole milk.

(iii) Waste liquors obtained by cleaning of powder milk plants and its floor washings.

(iv) Waste waters from the manufacture of cheese including the wash waters and,

(v) Waste liquors from the manufacture of yoghurt.

VOLUME OF EFFLUENTS

The volume of waste liquors generated in an average would be about two to three times the volume of milk processed.

CHARACTERISTICS OF EFFLUENTS

Dairy wastes are white in color and usually slightly alkaline in nature (due to the alkalis and detergents used for floor cleaning etc.,) and they become acidic quite rapidly due to the fermentation of milk sugar to lactic acid. The resulting lower pH leads to the precipitation of casein. All the waste liquors mentioned above have a high BOD as they contain appreciable concentration of carbohydrates. The suspended matter content of milk waste is appreciable mainly due to the fine curd found in cheese wastes. The pollution effects of dairy wastes are attributed to the immediate and high oxygen demand. Decomposition of casein leading to the formation of heavy black sludges and strong butyric-acid odors characterize milk waste pollution.

The characteristics of a dairy effluent depend largely on the quantity of milk processed and type of products manufactured. Table 8-1 shows the characteristics of a typical dairy handling 1,00,000 litres of milk per day and producing butter, ghee, casein, yoghurt and milk powder.

Table 8-1.
Characteristics of a Dairy Effluent
(24 hrs composite sample)

Sl. No.	Characteristic	Result
i.	Color	White
ii.	Odor	Characteristic putrid milk odor
iii.	pH Value	9.3
iv.	Total suspended Solids, mg/l	312
v	Total Dissolved Solids, mg/l	870
vi.	Dissolved Fixed Solids, mg/l	230
vii.	Dissolved Oxygen, mg/l	Nil
viii.	Bio-Chemical Oxygen Demand (5 days at 20°C), mg/l	800
ix.	Chemical Oxygen Demand, mg/l	1232
x.	Permanganate value, mg/l	500
xi.	Ammoniacal Nitrogen (as N), mg/l	17.64
xii.	Total Kjeldahl Nitrogen (as N), mg/l	35.84
xiii.	Sulfide (as S), mg/l	0.1
xiv.	Oils and Grease(petroleum ether solvent),mg/l	105
xv.	Surfactants (anionic) (as ABS), mg/l	2.5

EFFECTS OF EFFLUENTS

Dairy effluents decompose rapidly and deplete dissolved oxygen of the receiving streams immediately, resulting in anaerobic conditions and release of strong foul odors. Due to the nuisance conditions, the receiving waters become the breeding place for flies and mosquitos. It is also reported that higher concentrations of dairy wastes are toxic to certain varieties of fish and algae.

TREATMENT

Dairy wastes are treated efficiently by biological treatments. The low COD:BOD ratio of dairy wastes indicates their amenability for biological oxidation. They also contain required nutrients for micro-organisms in sufficient quantities.

Among the biological treatments, trickling filter and activated sludge process involve more capital expenditure and require maintenance etc. The efficient and low cost treatment methods include, oxidation ditch, aerated lagoon and waste stabilization ponds. Prior to any of these biological treatments, wastes from all sections of the dairy are combined in an equalization tank, the oil and grease are removed either manually or by grease traps. It is reported that 95 to 97% reduction in BOD is achieved in oxidation ditches, 50 to 75% in waste stabilization ponds. Anaerobic lagoons may also be employed which removes about 90% BOD.

PARAMETERS TO BE DETERMINED

Important parameters

1). Color 2). Odor 3). Temperature 4). pH value 5). Total suspended solids (TSS):Determine as early as possible. Storage will lead to precipitation of casein resulting in more TSS. 6). BOD: Prepare dilutions ranging from 1:199, 1:499 and 1:999. Seeding is not necessary. 7). COD: Use 25 ml or 50 ml aliquots of sample and 25 to 50 ml of 0.25 N.$K_2Cr_2O_7$. 8). Sulfides: Determine in the field itself. 9). Oil and grease 10). Surfactants 11). Percent sodium: As soda ash and detergents are used for washing, percent sodium increases than the process water and in certain instances may exceed the tolerance limit (60%). 12). Boron

Note: Percent sodium and boron need to be determined only when the effluent (treated or untreated) is discharged on land for irrigation or in to public sewers.

Additional parameters

13). Dissolved fixed solids 14). Ammonia nitrogen: Use 25 to 50 ml. of aliquot for distillation. 15). Total Kjeldahl nitrogen: Use 25 to 50 ml of aliquot for digestion. 16). Chlorides 17). Sulfates

REFERENCES

1) G.J. Mohan Rao & Subramanyan, "Sources, Flows and Characteristics of Dairy wastes" Ind. Jour. of Env. Health Vol. 14, 207, 1972.

Chapter - 9

STEEL PLANT

In steel mill, iron is produced from its ore, converted to steel and either cold or hot rolled.

An integrated steel mill comprises of the following units:

— Coal washery

— Plants for carbonization of coal and recovery of by-products.

— Iron production (Blast furnace operations)

— Steel making

— Finishing (rolling, pickling etc.,)

The auxiliary and service units include (1) Thermal power plant (2) Oxygen plant (3) Foundry and (4) Slag plant. Each unit generates effluents of varied characters.

PROCESS IN INTEGRATED STEEL PLANT SOURCES AND CHARACTERISTICS OF EFFLUENTS

I. Coal washery

After crushing and screening, coal is washed with water to remove soil, dirt and other unwanted substances. The water used for washing is recycled. However considerable volume of wash water is let out as waste.

The most objectionable impurity is the suspended

solids (coal fine). Other impurities include shale, clay, kaolin, calcite, gypsum, pyrite etc.

II. Carbonization of Coal (Coke manufacture) and By-product Recovery

Coke is prepared from the bituminous coal by heating at a temperature of about 900°C to 1100°C in the absence of air in an oven. The volatile substances released from the coal, leave the oven as gas which is subjected to recovery to yield ammonia, coal tar, benzene, xylene and toluene. In some plants, phenol and naphthalene are also recovered.

The converted coke from the oven is quenched with cold water. About 30% of quench water is evaporated, while the remaining water containing coke dust (called as breeze) is sent for recovery (of breeze), recirculated and then discharged as waste. This waste water from quench tower contains toxic materials and much attention should be paid for its treatment. Large volume of water used in a coke plant is for cooling purpose and the cooling waste water contains tar, ammonia, cyanide, hydrogen sulfide, benzole, phenols, pyridine, oils etc.

The mixture of hot gases leaving coke oven is liquefied, cooled and subjected to various recovery processes. A flow chart indicating the various processes involved in the manufacture of coke and by-products recovery is given in Fig. 9-1.

III. Iron Production (Blast Furnace Operations)

Blast furnace is the basic unit of steel industry where iron is separated from its ore by melting. A mixture of iron ore, coke (fuel as well as reducing agent) and fluxing materials such

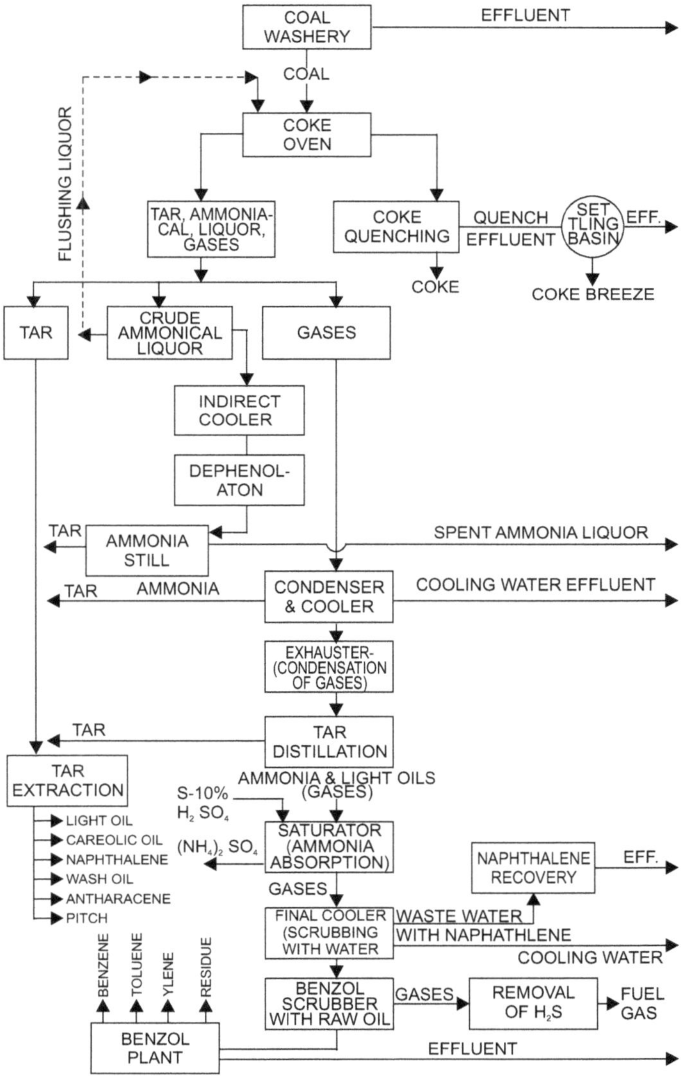

Fig 9-1. Flow Chart for Carbonization of Coal (Coke mfr) and By-product Recovery with Effluent Streams

as limestone and dolomite is charged from the top of the furnace, while a blast of hot air is blown from the bottom. Enormous heat is produced inside the furnace and some portions of the furnace have to be cooled. Iron melts and passes on to bottom. This iron is termed as pig iron.

Gaseous impurities are allowed to pass through dry dust catchers (where flue dust are caught), wet scrubbers (which clean the dust from the upflowing gases) and then finally to electrostatic precipitator where the finest dust particles are absorbed.

Effluents arise from the operation of (i) cooling the furnace (ii) wet scrubbing of flue gases and (iii) flushing of electrostatic precipitators. The furnace cooling water remains uncontaminated and reused.

The wet scrubber effluent is the major and voluminous waste in blast furnace operations. It contains significant quantities of suspended solids comprising iron oxide, alumina, silica, carbon, lime, magnesium, coke and sinter. The concentration of the constituents present in the effluent depends on the quality and quantity of ore, coke and flux materials charged and on the lining materials, condition and number of furnaces. (It is reported that the normal concentration of iron (as Fe_2O_3) is about 70% and silica 12%. Occasionally the iron may present as bicarbonate which gets precipitated in the treatment plant or in the receiving water course and imparts its characteristic rusty brown color).

Some of the characteristics of the wet scrubber waste are tabulated in Table 9-1.

Flushing of electrostatic precipitator contributes only little quantity of effluents, but the concentration of finest dust particles in it is appreciable.

Table 9-1.
Characteristics of Wet Scrubber Effluents

Sl. No.	Characteristic	Value
i)	Volume of effluent	about 500 litres per tonne of iron produced
ii)	pH Value	6.0 - 8.2
iii)	Total Suspended Solids, mg/l	500 to 4500 (sometimes extends upto 10000 mg/l)
	a) Percent by wt. passing 100 mesh sieve	86 - 99
	b) Percent by wt. passing 200 mesh sieve	74 - 97

IV. Steel making

It is a process by which most of the impurities of iron are removed to form steel.

It is carried out in any one of the three types of furnaces, namely (i) open hearth furnace, (ii) basic oxygen furnace and (iii) electric furnace, depending on the final product required and availability of raw materials.

The major effluent from steel making is the scrubbing water or wash water arising due to the separation of dust from hot gases evolved during loading and heating of the charge in the furnace. This effluent contains excess suspended solids and it is usually acidic. The particle size of suspended solids is much smaller than that from blast furnace.

Tremendous amount of water is also used for cooling furnaces, which usually is not contaminated and reused.

To improve the physical properties, molten material is degassed under vacuum obtained through barometric condenser. The effluent from vacuum degassing contains iron, manganese and fluoride.

The excess heat produced in the furnace is used by waste heat boilers. Consequently wastes arise from boiler blow down and regeneration of ion exchange resins. The volume of these wastes is also high.

V. Finishing

1. **Rolling:** In rolling mill, steel ingots are converted to blooms, billets and slabs, during while the scale or metal chips are dropped in to the scale flume situated below the roll tables. The scales are then flushed to a collecting pit, termed as 'scale pit'. During rolling, the rolls are heated and cooled by water. Waste water from this process containing excess oil and grease (used for bearings) also joins 'scale pit'.

   Cold rolling effluents contain more oils as emulsified rolling oils are added directly to the water sprayed on the metal during rolling.

2. **Acid Pickling:** This is an important treatment given before applying final finish to steel products. It removes dirt, grease and especially iron oxide scale which accumulate on the metal during fabrication. Pickling is done by dipping steel in dilute sulfuric acid or hydrochloric acid

solution. (about 20% by wt). As pickling progresses, the acid will become weak and there will be a gradual build up of ferrous sulfate or chloride and after a certain stage, the pickle liquor will be unsuitable for use and rejected as waste. This spent pickle liquor usually contains 5 to 15% free acid and 2 to 10% ferrous iron. This pickling liquor waste with high pollution potential is one among the most strongest wastes. The volume of pickle liquor waste ranges from 100 to 1000 liters per ton of steel.

Rinsing which follows pickling contributes significant volumes of waste water. It is estimated that about 500 to 20000 litres of water may be required for rinsing 1 ton of pickled steel. (It is also reported that in some mills the rinse water requirement is 250 to 500 litres per ton of steel). These rinse waters contain about 0.02 to 0.5% free acid.

The final finishing operations include tinning, plating, galvanizing etc., Effluents are generated from all these operations and they contain acids, alkalis, metals such as chromium, zinc, tin, copper etc., along with cyanides and fluorides.

EFFLUENTS FROM AUXILIARY AND SERVICE UNITS

1. **Thermal power plant:** Boiler blow down (concentrated hot saline) is the major effluent from thermal power plants. It is usually alkaline (pH about 11) and contains antifoam materials and/or phosphates and other organic agents. Regenerant wastes containing brine, acids (HCl, H_2SO_4) alkalis (NaOH), phosphates, hydrazine (arise from regeneration of ion exchange resins, cleaning and flushing

of boilers) are also important. Burning of pulverized coal in boilers releases significant quantities of fly ash which is scrubbed and the effluent contains a higher concentration of suspended solids.

2. **Oxygen plant:** Large quantities of oxygen are needed for blast furnace and other steel making furnaces. The source of oxygen is air. Carbon dioxide and moisture of air are removed by passing through 0.5 N sodium hydroxide solution which ultimately is disposed of as waste.

3. **Foundry:** To control dust in the air (in foundry area) it is scrubbed with water spray type unit or filtered through bag filters. If bag filters are used, the collected dust will be wetted and sent for disposal. Effluent from these operations contains excess suspended solids.

4. **Slag plant:** In this plant, blast furnace slag is converted to useful material by quenching. Washing of the air to control dust, and washing conveyor belt generate effluents rich in suspended solids.

EFFECTS OF EFFLUENTS

The steel industry effluents are the most hazardous as they contain various pollutants in excess concentration. The effects of the individual contaminants are described below.

Acidity: The extreme acidity of pickling liquor is more dangerous. The acidic effluents destroy micro organisms and thus inhibit self purification of stream. Fish and other aquatic animals are adversely affected by a sudden change in pH conditions of the river. The pickling liquors also cause damage to metals, concrete structures, pumps etc., by their corrosive action.

Suspended and colloidal impurities: These are not toxic but interfere with the self purification of streams by reducing the photosynthetic activity of water plants and by smothering benthic organisms. Moreover, abrasive materials present in the suspended matter can choke the gills of fish and thereby cause damage to fisheries. The coal washery effluents which are black and opaque, discolor a stream and render it unfit for most uses. The suspended impurities along with organic materials will undergo putrefaction during warm weather. As a result, solids buoyed by gas may rise to the surface leading to the formation of floating masses which emanate evil odors.

Phenolic substances: Coke oven effluents containing phenolic substances impart characteristic unpleasant taste and odor to receiving waters. Chlorination enhances this effect and even a concentration of 0.002 mg/l of phenol causes chlorophenol taste and odor. They also taint fish by imparting characteristic phenolic taste to the flesh of fish. Higher concentration of phenol is toxic to fish.

Ammonia: Free ammonia present in spent ammoniacal liquor is also toxic to fish. It is to be noted that the toxicity of ammonia and ammonium salts to aquatic animals is directly related to the amount of undissociated ammonium hydroxide in the solution, which in turn is a function of pH. Thus even a high concentration of ammonium ions at a low pH may not be toxic, but the toxicity would be severe at high pH values.

Cyanide: The toxicity of cyanide need not be emphasized. The coke even effluents and spent ammoniacal liquor contains cyanide in excess concentration.

Sulfide: Generally, sulfides deplete the oxygen immediately resulting in anaerobic conditions and foul odors. Usually the hydrogen sulfide is removed by scrubbing. Otherwise, it is a serious toxicant.

Organic matter: In addition to the numerous inorganic substances the coke oven effluents usually contain biodegradable organic matter. They deplete the dissolved oxygen at a faster rate and consequently the water course becomes devoid of oxygen resulting in the release of obnoxious odors and affecting fish and other aquatic life.

Coal tar: This is an important component of coke oven effluents. Heavy tar from the effluent usually settles down and choke sewers and waste mains and its removal is essential in any treatment process. Light oil will float on the surface and gives an ugly appearance to the receiving water.

Oils & grease: They interfere with aeration and diffusion of oxygen and affect seriously the natural purification of water courses. They produce an unsightly appearance by the formation of ugly oil slicks and irridiscent colors.

Iron: Soluble iron (and manganese) in the effluents give rise to unpleasant taste to water if present in concentrations more than 0.1 mg/l.

Heat: The effluents from steel plant are usually hot as the major portion of water is being used for cooling. The effect of heated effluents are numerous. To mention a few are (i) decreasing of dissolved oxygen content (ii) decreasing of viscosity of water thus speeding up the settlement of suspended particles (iii) causing excess algal growth and (iv) increasing the virulence of pathogens.

Summary of the effects of steel plant effluents

1. They render the receiving water unfit for human consumption and for any other purpose.
2. Toxic to fish and other aquatic life.
3. Cause foul odors and tastes.
4. Choke pipes mains and other waste treatment structures.
5. Raise the temperature of receiving waters.
6. Leads to heavy siltation.

TREATMENT

I. Effluent from Coal Washery

Clarifloccualation is the best method for the reduction of high suspended solids content of these wastes. Alum, ferric sulfate and ferric chloride are the coagulants used in combination with coagulant aids such as starch and polyelectrolytes.

II. Effluent from Carbonization of Coal (Coke mfr)

Ammonia and phenols are usually removed from the coke oven liquor and then subjected to biological treatment–either trickling filtration or activated sludge process. The latter is reported to be more effective than the former. Extended aeration, satisfying nutritional requirements (especially addition of phosphorous) are essential conditions for effective treatment by activated sludge process.

III. Effluents from Iron production (Blast furnace operations)

The initial treatment involves the removal of flue dust. Either plain sedimentation or sedimentation preceded by chemical coagulation is employed. Aluminium sulfate, ferric-chloride and ferric sulfate are the coagulants used. Sludge is dewatered in vacuum filter and disposed of for land fill. The clarified effluent is recirculated for cooling etc., Since these cooling waters acquire heat, they must be cooled in a cooling tower included in the treatment system. Before sending in to a cooling tower, they are acidified to a pH of about 6.5 to 7.5 to prevent deposition. Instead of acidification, phosphates may also be added.

IV. Effluents from Steel making

The effluents are settled in primary settling tanks, and subjected to coagulation and flocculation in clariflocculators and reused.

V. Effluents from Finishing Operations

i) The cooling water from reheating furnaces is relatively clear but hot and requires no treatment except cooling.

ii) Waste waters from roughing stand contains mainly course scale particles which settle out rapidly. The treatment system involves primary settling tank with no compartment. A detention period of few minutes is allowed so that most of the particles settle, after which the effluent is passed to secondary settling tank with two or three compartments and oil skimming facilities. The settled mass is sent to furnace. The settled skimmed oils are usually burnt.

iii) Effluents from finishing stand and scarfer mechines contian fine scales. This effluent is allowed to settle as described above and then subjected to coagulation and flocculation in clariflocculator. Complete purification is achieved by passing through sand filters.

iv) Waste waters from cleaning lines and secondary cold reduction are passed through primary gravity separators for the removal of oils and settleable matter after which they are treated with ferric sulfate or ferric chloride.

Pickling waste

The most polluting effluent from steel mill is the spent pickling liquors. They are neutralized and treated with lime and aerated to form ferric hydroxide, which is removed as sludge. The pickling rinse water is usually treated with other cold mill wastes containing oils so that the oils are best removed by the floc [$(Fe(OH)_3$] formed.

In some mills, sulfuric acid is recovered from spent pickle liquor by the recent Blaw-Knox-Ruthmer process in which the pickle liquor is concentrated by evaporation and placed in a reactor in to which hydrogen chloride gas is bubbled. Ferrous sulfate is converted to ferrous chloride and sulfuric acid. The $FeCl_2$ is roasted to obtain iron oxide and the released hydrogen chloride is scrubbed and reused.

Effluents from Auxiliary and Service units

Power plant wastes containing brines, acids and alkalis

are collected in equalization tanks and neutralized if necessary. Neutralized effluents are combined with the other effluents for disposal. Other wastes are subjected to plain sedimentation and then recirculated or disposed of.

PARAMETERS TO BE DETERMINED

Important parameters

1) Color 2) Odor 3) pH value 4) Total suspended solids 5) Biochmical Oxygen Demand: Dilutions ranging from 1:99 to 1:1999 have to be made. Seeding is necessary. Acclaimated seed is preferable. The value obtained is only approximate as there are so many interferences. 6) Cyanide 7) Thiocyanate 8) Phenols 9) Sulfide: If the concentration is high, determine by titrimetric method. If it is much less such as in treated wastes, colorimetric method is applicable. 10) Free ammonia (present as undissociated ammonium hydroxide) 11. Saline ammonia (other than undissociated ammonium hydroxide) 12) Oils and grease.

Additional parameters

13) Acidity/Alkalinity 14) Total dissolved solids 15) Chemical Oxygen Demand 16) BOD: COD value 17) Percent sodium 18) Boron.

REFERENCES

1) W. Rudolf Ross, "Industrial Wastes, their disposal and treatment" 1953, Reinhold publishing corporation, New York.
2) A.D. Henderson & J.J. Baffa "Waste Disposal at a steel plant treatment of flue dust waste" Proceedings of American Society of Civil Engineers, 80,494, (1954)
3) Indian Standards Institution, "IS:8073-1976, Guide for Treatment

and Disposal of Steel Plant Effluents" ISI, New Delhi.

4) M.N. Rao & A.K. Datta, "Waste water treatment"-1979, Oxford & IBH Publishing Company, Calcutta.

5) N.Manivasakam, "Environmental Pollution"-1984, National Book Trust (India), New Delhi.

Chapter - 10

ELECTROPLATING INDUSTRY

In electroplating industry, as the name implies base metals are plated for protection from corrosion, for brightness, and for fair appearance. The platings usually given are chromium, nickel, zinc, cadmium, copper, silver and gold. The plating operations are preceded by cleaning and stripping which contribute strong effluents in significant quantities.

PROCESSES & EFFLUENTS

Cleaning

The materials to be plated when received in plating shop usually contain oils or greases on its surface given as a protective coating. They are removed by saponification with warm alkali. (This process is also called as 'alkaline degreasing') Greases of petroleum and mineral oil origin are either removed using solvents such as trichloro ethylene (non-flammable), gasoline and benzene, or by emulsification with alkaline mixtures containing caustic soda, soda ash, trisodium-phosphate, sodium silicate, borax etc.,

The spent alkaline liquors have high pH, suspended solids and contain soaps, grease and oil. Usually they are stored in steel tanks for controlled discharge or for blending with acid wastes.

After degreasing, the materials are given washing in stationary baths or in running water which are discharged continuously, The rinse and washing effluents are usually alkaline.

Stripping (Acid Pickling)

It is a process usually carried out on iron materials by which the rust and scales adhering to the metallic surface are removed by treatment with sulfuric or hydrochloric acid. The spent liquors are not discharged continuously. Though the volume of these spent liquors is much less, they are strong wastes and cause serious effects. They contain mostly ferrous sulfate and excess free acids with a pH below 2.5

The pickled or acid dipped materials are washed in vats with continuous flow of water. These wash and rinse waste waters are also acidic and their pH value lie in the range of 4.5 to 5.5. These effluents are also stored in either bitumen or acid proof brick lined tanks for mixing with alkaline wastes.

Plating

The pickled materials are placed as cathode in suitable electrolytic cells. The plating baths are usually acidic in nature and contain sulfuric, hydrochloric or nitric acids. Alkaline baths containing carbonate, hydroxide and cyanide are also used. Cyanides are used in plating baths as they are good oxide solvents and they yield a brighter and less porous plates.

After plating, the objects are dipped in stationary water baths, than rinsed with running water. The solution in the stationary water baths are not discharged as waste but utilized to make up solution for plating baths. The rinse waste waters are discharged as waste.

Among the plating wastes, cyanide concentrates and chromate wastes are noteworthy due to their toxicity.

Cyanide Concentrates

This includes cyanide plating solution and cyanide dips with an excess concentration of cyanide. The toxic effects of cyanide need no explanation. These wastes are stored carefully for treatment. Rinse waters contain a very small concentration of cyanide.

Chromate wastes

The spent liquor of chromium plating and dipping contain fairly large quantities of chromium. These liquors are not discharged but segregated and stored for separate treatment. However drag out and rinse operations contribute chromium to the effluent in significant amounts, (ranging from 3 to 30 mg/l). The pH of this effluent is usually in the near neutral range. It is discharged after dilution or after conversion of the hexavalent chromium to trivalent state.

Other plating wastes

The other metallic plating spent liquors (i.e) from cadmium, copper, lead nickel, and zinc and rinse waters contain significant quantities of these metals. All these metals are usually present as ions and their toxicity need not be emphasized.

Spill overs & Floor washes

These contain the metals, cyanides and other chemicals employed in plating operations. The volume of these wastes is much less. The spill overs and splashing may be reduced considerably if proper attention is bestowed.

Other wastes

These include mainly cooling water, the quality of which is unaffected and hence they are either recirculated or discharged without treatment.

CHARACTERISTICS OF EFFLUENTS

Cleaning operations contribute alkali wastes. As indicated already, they contain sodium hydroxide, phosphates, silicates, carbonates, organic emulsifiers and wetting agents. If cleaning is carried out by organic solvents, the effluent from the operation contains petroleum or coal-tar solvents coupled with an emulsifier.

Stripping effluents contain sulfuric, hydrochloric and nitric acids and are highly acidic in nature.

The strength, nature and constituents of plating wastes vary considerably depending on the plating done and type or rinsings, volume of wash water used etc., The combined effluent may be either acidic or alkaline, depending on the type and quantity of baths used and on cleaning and stripping effluents. Mention has to be made that the concentration of chromium in chromium plating effluents will be several times more than that of other metals present in other plating bath effluents.

The volume of effluents generated and characteristics vary from one plating shop to another, and even within the same shop it may vary day to day.

EFFECTS OF EFFLUENTS

Effects on Sewers

The important characteristic of plating effluents is their toxicity. Hence they are not disposed of into rivers or other water courses as other effluents. Due to their toxicity and low volume, they are discharged into sewers. The most dangerous component of plating waste is cyanide. It is not fully removed by sewage treatment and hydrogen cyanide gas affects the workmen in the sewer system and sewage treatment plant. Solvents from cleaning operations exert toxicity and may cause explosion in the sewer system. Oils and greases coat the sewer lines and are not amenable for biological sewage treatment.

Acidic plating effluents are highly corrosive and attack concrete structures. Alkaline plating wastes are also corrosive but not aggressive as acid wastes.

Suspended impurities such as precipitated metallic hydroxides, metallic chips etc., may settle and clog the municipal sewer system.

Effects on Streams

Even very small concentrations of cyanide, chromic acid and chromates, heavy metal compounds of cadmium, copper lead, nickel and zinc are toxic to aquatic life. In higher concentrations, they are toxic to human-beings also. As they are toxic to microbiological organisms self purification is inhibited. Further, iron and tin impart color to the receiving stream. Iron would also tend to deplete the dissolved oxygen. Nitrates and phosphates present in the effluent results in excess algal growth. The colloidal and other suspended impurities render an unsightly appearance to the stream.

TREATMENT

The treatment pattern is as follows:

1. Segregation of cyanide wastes, chromium effluents and other metal bearing wastes.
2. Treatment of cyanide waste by alkaline chlorination.
3. Treatment of chromium effluents with ferrous sulfate or other suitable reducing agents to reduce chromium (VI) to chromium (III) and precipitation with lime. (For details see page 133)
4. Mixing of treated effluents of cyanide and chromium with other metal bearing wastes followed by pH adjustment, precipitation at pH 8.5 to 9.0 and settling. (sludge may be dried on sand beds and used for land filling)
5. Floating oils and grease are scraped from holding tanks. Emulsified oils are removed by pre- treatment with alum and caustic soda prior to the entry into holding tanks.

Alkaline Chlorination for Cyanide Removal

pH of the waste is raised to about 11 by the addition of lime or caustic soda. Then chlorine gas is passed into the waste accompanied by violent agitation to prevent precipitation of cyanide complexes. The probable reaction (with excess chlorine) under alkaline condition is represented below.

$$2NaCN + 12NaOH + 5Cl_2 \rightarrow 2Na_2CO_3 + 10NaCl + 6H_2O + N_2\uparrow$$

(Approximately, 6 kg each of caustic soda and chlorine are required to decompose 1 kg of cyanide). Duration of chlorination depends on the type of waste and it may even extend for a 24 hr period for complete oxidation.

Reduction & Precipitation of Chromium

The first step is the reduction of Cr (VI) to Cr (III) with reducing agents such as ferrous sulfate, sulfur dioxide or sodium bisulfite. The pH of the medium should be maintained at about or below 3. If necessary, sufficient amount of mineral acid may be added. After completion of reduction, Cr (III) is precipitated as its hydroxide with lime slurry. The sequence of reactions is represented below.

$$H_2Cr_2O_7 + 6FeSO_4 + 6H_2SO_4 \rightarrow Cr_2(SO_4)_3 + 3Fe_2(SO_4)_3 + 7H_2O$$

$$Cr_2(SO_4)_3 + 3Ca(OH)_2 \rightarrow 2Cr(OH)_3 + 3CaSO_4$$

$$3Fe_2(SO_4)_3 + 9Ca(OH)_2 \rightarrow 6Fe(OH)_3 + 9CaSO_4$$

It is calculated that to precipitate 1 kg of Cr (VI), 16Kg of copperas, $FeSO_4.7H_2O$, 6Kg of H_2SO_4,10Kg of lime are required. The sludge contains 2 Kg of $Cr(OH)_3$ and about 0.5 Kg of $Fe(OH)_3$ and 2 Kg of $CaSO_4$.

The rinse waters may be reused after passing through cation and anion exchange resin columns. Oil is removed by activated carbon filters before passing into ion exchangers. The regenerant waste waters from cation exchanger is treated with ferrous sulfate for reduction of Cr(VI) to Cr(III) if chromium is present. pH adjusted and treated with lime. Anion exchanger regenerant waste waters containing cyanide is subjected to alkaline chlorination, pH adjusted and mixed with cationic effluent and treated with lime.

PARAMETERS TO BE DETERMINED

Plating effluents contain only a little organic matter. On the other hand heavy metals are present in toxic concentrations. Hence all the metals that would probably be present are determined. As already stated, type of plating may vary from shop to shop and the metals and constituents to be determined will also vary. However, the metals and other inorganic and organic constituents to be determined are as follows:

1) Hexavalent chromium and trivalent chromium 2) Cadmium 3) Copper 4) Lead 5) Iron 6) Zinc 7) Tin 8) Silver 9) Ammonia 10) Cyanide 11) Sulfide 12) Chloride 13) Nitrate 14) Fluoride 15) Sulfate 16) pH value 17) Acidity/Alkalinity 18) Suspended Solids 19) Dissolved Fixed Solids 20) Oils and Grease 21) Chemical Oxygen Demand 22) Mercury (if alkaline degreasing is carried out with the caustic soda manufactured by electrolytic cell process).

Note – 1

If metals are present in small quantities, colorimetric methods may be employed. If their concentration is in excess, titrimetric/ gravimetric methods are more suitable. Care must be taken to eliminate interferences.

Note – 2

Determine cyanide and sulfide immediately, or preserve them with suitable preservatives as described in part IV of this book. Employ colorimetric or titrimetric method according to the concentration.

Note – 3

BOD is not recommended for these effluents due to the toxicity and inhibition of microbial activity by the metals and toxic substances.

REFERENCES

1) Gurnham "Disposal of Metal Finishing Wastes: Metal finishing Guide book" -1950, McGraw Hill Book Co., Newyork.

2) Rudolf Ross," Industrial Waste Treatment" -1953. Reinhold Publishing Corporation, New York.

3) H.F.Lund (Ed) "Industrial Pollution Control Hand book" -1971 Mc-Graw Hill Book Co, New York.

4) Indian Standards Institution, "IS:7453-1974, Guide for Treatment of Effluents of Electroplating Industry".

5) N.L.Nemerow, "Industrial water pollution - Origins, Characteristics & Treatment -1978, Addison-Wesley Publishing Co.

Chapter - 11

OIL FIELD & OIL REFINERY

Mineral oil wastes are classified as originating from (1) Oil Field and (2) Oil Refineries.

At oil fields, wastes result from pumping and desalting operations. They contain drilling muds, saline water and oil (free and emulsified). Among these, salt water poses difficult problems in treatment. The characteristics of an oil field brine sample is summarized in Table 11-1.

Table 11-1.
Mineral Analysis of Oil Field Brine Samples*

Constituents	Concentration range, ppm
Sodium	8260 - 63,275
Calcium	1507 - 12,888
Magnesium	346 - 4290
Chloride	12750 - 1,27,220
Bromide	32 - 633
Sulfate	0 - 1578
Bicarbonate	43 - 644
Carbonate	0
Total Solids	25,210 - 2,48,600

* From 78th US Congress, First Session, House document 266. "Industrial waste guide to the oil Industry". Ohio River Pollution Control Survey, Supplement D. Appendix - IX (1943) p 1175.

Crude oil is a complex mixture of hydrocarbons of varying molecular weights and structures. In an oil refinery, it is subjected to distillation, fractionation, alkylation and polymerization to obtain various products such as gasoline, kerosene, lubricants, gas, oil, and fuel oil, wax, asphalt and petroleum coke.

Large volumes of water are used in refineries for cooling, steam generation and for refining and subsequently discharged as waste. It is estimated that refineries with once through cooling system use water between 18000 and 27500 m^3[†] per 1000 tonnes of crude processed. Almost this entire volume of water is discharged as waste.

In recirculated systems, the quantity of water used is much less ranging from 1350 to 5650 m^3 per 1000 tonnes of crude processed, and the volume of waste discharged is about 320 to 1800 m^3 per 1000 tonnes.

The various processes in a refinery include topping, catalytic cracking or thermal cracking, acid treatment, alkaline treatment, sweetening treatment, hydrode sulfurization and coking. Detailed discussion on the processes is beyond the scope of this book. The wastes contain a number of materials, the important among them are given below.

1. Oil (free or emulsified) from leaks, spills, tank draw off and other sources. The quantity of oil let-out as waste is estimated to be about 3% of the weight of crude processed.

2. Alkalis from alkaline treatment of oil for the removal of

† 1 m^3 = 1000 litres.

acidic components and sweetening process for the removal of mercaptans. These alkaline wastes are highly odoriferous and difficult to treat.

3. Acids and acidic sludges obtained on acid treatment of oil.
4. Condensate from steam stripping operations.
5. Emulsions from barometric condenser water.
6. Chromates, phosphates and zinc in cooling water.
7. Coke from equipment tubes and towers and other locations.
8. Waste catalyst and filtering clays.
9. Chemicals from by-product manufacture.

The effluents also contain significant quantities of suspended matter, cyanides, sulfides, mercaptans and phenolic compounds, and ammonia.

Table 11.2 gives the general characteristics of a composite waste from a typical oil refinery.

EFFECTS OF EFFLUENTS

1. Oils form non-destructible layers on the surface of water resulting in an unsightly appearance. It also interferes with the aeration/oxygenation of the stream as the oily layer prevents the entry of oxygen. Thus the natural purification is inhibited. Further, light penetration is also inhibited and consequently photosynthesis is affected.

   There are chances for fire hazards if the concentration of oil is in excess. Oils, also coat the gills of fish and affect the physiological phenomena of biological organisms.

2. Hydrogen sulfide, phenols and mercaptans present in the effluent imparts disagreeable odor to the receiving waters. They are also toxic to aquatic life.
3. Further, cyanide and chromates present in significant concentration are extremely toxic.
4. A great demand for oxygen is exerted by the above substances.
5. Acids/alkalis are toxic to aquatic life and affects self purification processes.
6. Fluoride is well known for its effect on human beings and animals (i.e.) dental and skeletal fluorosis.

Table 11-2.
Characteristics of a Composite Oil Refinery Effluent*

Sl. No.	Characteristic	Result
1.	pH Value	6.5 - 9.0
2.	Suspended solids, mg/l	200 - 400
3.	Biochemical Oygen Demand (5 days at 20°C), mg/l	100 - 300
4.	Oil (Free), mg/l	2000 - 3000
5.	Oil (Emulsified), mg/l	80 - 120
6.	Sulfides (as H_2S) and marcaptans, mg/l	10 - 220
7.	Phenolic compounds (as C_6H_5OH), mg/l	12 - 30
8.	Chromates (as CrO_4) and Zinc (as Zn) or phosphate (as PO_4) from Cooling water, mg/l	5 - 20

* R.N. Chakravarthy and T. R. Bhaskaran, "Treatment and Disposal of Oil Refinery Wastes". IAWPC 10th Anniversary commemoration volume - 1973.

TREATMENT OF OIL FIELD BRINE WASTES

Brine wastes are discharged into sea. Controlled discharge into surface waters is also practised. Solar evaporation is another method. Injection of the brine wastes into subterranean formation after pretreatment is also followed but the pre-treatment itself is expensive.

TREATMENT OF OIL REFINERY WASTES

Oil free wastes are usually segregated and discharged without any treatment. If necessary, separates treatment is given to traces of oil present.

Oil bearing effluents are given elaborate treatment. They include,

i) Primary treatment - oil removal.

ii) Secondary treatment- involving chemical and biological methods, and

iii) Tertiary treatment.

(i) Primary Treatment - Oil Removal

Oil is removed inside the production unit itself by stripping* and extraction†. The free oil remaining is removed by passing through gravity separators. API‡ separators are commonly employed for this purpose. This pretreatment removes 50 to 99% of free oil and 10 to 85 % of suspended solids.

* Stripping is done with steam to remove H_2S, mercaptans and ammonia. Cyanides and phenols are also removed to some extent.

† Extraction is carried out with solvents like tricresyl phosphates and mixed organic esters to remove phenol from wastes.

‡ American Petroleum Institute.

(ii) Secondary Treatment

a) Chemical methods

These are employed to remove emulsified oil, suspended solids and toxic substances thereby conditioning the effluents for further biological treatment. Air floatation with or without chemical coagulants remove most of the emulsified oil.

Sedimentation is normally employed to remove suspended solids after chemical treatment.

Neutralization of acidic and alkaline wastes are carried out by judicious mixing of various effluent streams. The soluble substances are removed by precipitation with suitable chemicals.

b)Biological Treatment

The soluble organics and phenols are removed by biological treatment. The common biological methods include trickling filtration, activated sludge process, aerated lagoons and oxidation ponds. Among these, the activated sludge process is reported to be more effective (i.e.) removal of 70 to 95% BOD, 30 to 70% COD, and 65 to 99% phenols and cyanides.

iii) Tertiary Treatment

It includes activated carbon filtration and ozonation, which are highly effective in the removal of taste and odor and refractory organics from biological treatment effluents.

The treated effluents which satisfy the relevant tolerance limits are finally disposed of into stream, river or sea.

PARAMETERS TO BE DETERMINED

1) Color 2) Odor 3) Temperature 4) pH Value 5) Acidity/Alkalinity 6) Suspended solids 7) Dissolved fixed solids 8) BOD 9) COD 10) Undissociated ammonium hydroxide and other saline ammonia 11) Oils (both free and emulsified) 12) Sulfides 13) Chlorides 14) Cyanides 15) Phenolic substances 16) Chromium (hexavalent and trivalent) 17) Iron 18) Percent sodium 19) Zinc 20) Mercury (This may occur in the effluent due to the use of the caustic soda in alkali treatment) 21) Phosphates 22) Fluorides 23) Mercaptans.

REFERENCES

1) American Petroleum Institute, "Manual on Disposal of Refinery Wastes- Vol III - Chemical Wastes" 2nd ed-1951.
2) American Petroleum Institute, "Manual on Disposal of Refinery Wastes" Vol-1 Waste water containing Oil" 5th ed- 1953.
3) Indian Standards Institution - IS 10044 - 1981. Guide for Treatment and Disposal of Effluents of Petroleum Refining Industry".
4) N. Manivasakam, "Environmental Pollution"- 1984, National Book Trust, New Delhi.

Chapter - 12

PETRO CHEMICAL INDUSTRY

In a petrochemical industry intermediate chemicals used in the manufacture of solvents, resins, fibres, fertilizers, synthetic rubbers, plastics etc., are produced. After the crude oil is fractioned to different products, they are sent to petrochemical industry. The range of petrochemical products is very wide including both organic and inorganic substances. A broad spectrum of chemicals are prepared from petrochemicals. It is estimated that more than 80% of the entire chemical industry is based on petrochemicals.

The basic raw materials supplied by petroleum refineries are natural gas, gas from cracking processes, LPG, liquid distillate (C_4 to C_9) distillates from special cracking processes, and cyclic fractions of aromatics. They include alkenes, ethylene, propylene, butylene and higher olefins and aromatics. These mixtures are separated into their constituents and subjected to chemical conversions in a petrochemical industry, where basic intermediates such as ethylene dioxide, dichloro ethane, ethyl benzene, acetaldelyde, acetic acid, acetic-anhydride, isoprene, butadiene, adipic acid, styrene, cumene, cyclohexane, phenol, benzoic acid and phthalic anhydride are produced. These basic intermediates are used as raw materials for the preparation of solvents, synthetic resins, synthetic fibres, plastics, fuel additives, and synthetic detergents.

Some of the processes carried out in a petrochemical

complex are alkylation, dealkylation, hydro-dealkylation, amination, cracking or pyrolysis (thermal and catalytic), esterification, halogenation, hydro-halogenetion, hydration and hydrolysis, hydrogenation, nitration, oxidation and hydroformylation.

CHARACTERISTICS OF EFFLUENTS

In most of the processes stated above, water is formed as a byproduct in large quantities. In addition, water is being used as a medium of transport for the chemical reaction. In both the cases, water is contaminated heavily. (This waste streams are uncommon to other industries).

Another important waste is the bottom streams and sediments from the columns in which distillation is carried out. These include a definite proportion of the unreacted raw materials, impurities present in the raw materials, residual chemicals used for the purification of them and catalysts and other byproducts formed during reaction. Hence in a petrochemical waste, the presence of hydrocarbons, alcohols, aldehydes, ketones, acids, esters, salts, phenols and oils is quite common.

In characterization and treatment view point each petrochemical unit is a different industry and each class of chemicals such as hydrocarbons, aldehydes etc., has to be studied in detail to evaluate the pollution potential of the waste. The specific toxic constituents should be given due importance.

No definite characteristics can be assayed for petro chemical industry, due to the wide range of chemicals employed and produced. Some of the general characters are given below.

1. High B.O.D. and C.O.D. values.
2. High C.O.D./B.O.D. ratio indicating its resistance towards biological oxidation.
3. Higher dissolved solids content.
4. Heavy metals and other toxic substances due to the use of catalysts.
5. Other compounds both toxic and inhibitory to biological treatment.

The Table 12-1 gives the characteristics of the waste from a petro-chemical complex which produces caprolactum using benzene, ammonia, hydrogen, carbondioxide, oleum (fuming sulfuric acid) and sulfuric acid as raw materials and process chemicals.

Table 12-1.
Characteristics of a Typical Petrochemical Waste*

Sl. No.	Characteristics	Result
1.	pH Value	5.3 to 10.0
2.	Total Solids, mg/l	1956 to 14478
3.	Biochemical Oygen Demand (5 days at 20°C), mg/l	650 to 5800
4.	Chemical Oxygen Demand, mg/l	947 to 9000
5.	C.O.D : B.O.D.	1.4 to 2.0
6.	Total ammoniacal nitrogen, mg/l	150 to 600
7.	Phosphates (as PO_4) mg/l	Traces to 15
8.	Oils and Grease, mg/l	10 to 85

* M.D.Patel and D.R.Patel, "Biodegradability and Treatability of Caprolactum waste" Indian Journal of Environmental Health Vol.19 No.4, 1977.

EFFECTS OF EFFLUENTS

The effects are similar to that of oil refinery effluents. In addition, it causes foam on surface waters due to the presence of surface active agents. When discharged into sewers it upsets the biological activity and hinders purification. Further, the low pH damages the sewers and allied concrete structures in the treatment plant. Suspended solids may also clog the sewers.

TREATMENT OF EFFLUENTS

Chemical recovery is the first-step in treatment carried out inside the plant itself as the effluent of one unit becomes the influent of other unit. Most of the valuable chemicals are recovered in the process, simultaneously reducing the pollution load. After recovery, the waste water is reused for cooling etc.,

The remaining waste streams are now sent to treatment plant. Preliminary treatments include equalization, pH adjustment, coagulation and flocculation, primary settling and oil separation. Secondary biological treatment involves trickling filtration, activated sludge process, oxidation ponds, aerated lagoons or anaerobic digestion.

Some of the modern treatment plants include hydrocylone process in which solid-liquid separation is achieved. In another process, molecular sieves are used for absorption. Ultra filtration and reverse osmosis are also developed to significant extent to treat petrochemical wastes.

PARAMETERS TO BE DETERMINED

1) Color 2) Odor 3) Temperature 4) pH value 5) Total suspended solids 6) Dissolved fixed solids 7) Biochemical-

-Oxygen demand. There are so many toxic substances present in the waste which inhibit biological activity and BOD results must be interpreted carefully. Seeding is necessary. Acclaimated seed yields better results. 8) Chemical Oxygen Demand 9) COD/BOD Ratio 10) Undissociated ammonium hydroxide and other saline ammonia 11) Oils (both free and emulsified) 12) Specific substances such as cyanide, phenols and heavy metals etc.

REFERENCES

1) U.Guthrie, "Petroleum Products Hand Book" 1960, Mc Graw Hill Book Co., New Delhi.

2) G.T.Waggy etal "Identification and Control of Petrochemical Pollutants inhibitory to Anaerobic Process" - EPA-R2-73-194 (1973) US EPA Office of Monitoring and Research.

3) M.N. Rao & A.K.Datta, "Waste water Treatment " -1979, Oxford & IBH Publishing Co., Calcutta.

Chapter - 13

FERTILIZER INDUSTRY

The fertilizer industry can be conveniently classified in to three groups and studied.

I. FERTILIZER RAW MATERIALS

The raw materials, phosphorus, potash, and sulfur are obtained by mining and no effluent is generated.

II. FERTILIZER INTERMEDIATES

1) Sulfuric acid 2) Phosphoric acid, 3) Nitric acid and 4) Hydrochloric acid. These are manufactured in the fertilizer industry itself. Details regarding effluents of these products are dealt separately in Chapter-14 "Acid Wastes".

III. FERTILIZER PRODUCTS

These are further classified into solid and liquid fertilizers.

SOLID FERTILIZERS

a) **Nitrogen Fertilizers**

1) Urea [NH_2CONH_2]

2) Ammonium sulfate [$(NH_4)_2SO_4$]

3) Ammonium nitrate [NH_4NO_3]

b) **Phosphorus Fertilizers**

1) Normal superphosphates [$Ca_2H_2(PO_4)_2 + CaSO_4$]

2) Triple superphosphate [$Ca_2H_2(PO_4)_2$]

c) **Nitrogen -Phosphate Fertilizers**

1) Mono ammonium phosphate (MAP) [$NH_4H_2PO_4$]

2) Diammonium phosphate (DAP) [$(NH_4)_2HPO_4$]

d) **NPK Fertilizer**

e) **Blended Fertilizer**

LIQUID FERTILIZERS

1. Ammonia
2. Other liquid formulations
3. Other slurry formulations

NITROGEN FERTILIZERS

Ammonia Production

Needless to say that the important raw material required for the manufacture of nitrogen fertilizers is ammonia. It is synthesized using nitrogen and hydrogen. The required nitrogen is obtained from the atmospheric air. No effluent is generated in obtaining nitrogen. The important sources of hydrogen are naphtha, fuel oil, natural gas, coke-oven gas, coal and electrolysis of water.

There are four methods of ammonia production.

1. Steam reforming of natural gas.
2. Partial oxidation process.
3. Coal gasification process.
4. Combustion of hydrogen - nitrogen mixtures.

Among these, steam reforming of natural gas is the most widely employed method. In this process, naphtha or natural gas is used as raw material. After desulphurization these are passed through a suitable catalyst along with steam to form hydrogen, carbon monoxide and carbon dioxide. Air is added as a source of nitrogen to this partially reformed gas. Oxygen in the supplied air burns to form water and provides additional heat to complete the reforming reaction. The reformed gas is a mixture of hydrogen, CO, CO_2 and water vapour. Carbon monoxide is removed by passing with steam over a suitable catalyst, and converted to CO_2. Additional hydrogen is also produced in this reaction. CO_2 produced is removed by absorption with hot potash or other alkaline solutions such as mono, di, or triethanolamine. The absorbed carbon dioxide is recovered and reused. The final traces of CO and CO_2 are removed by catalytic conversion to methane which is a valuable fuel. The final gas mixture containing pure hydrogen and nitrogen are made to react over activated iron oxide catalyst under elevated temperature and pressure to form ammonia.

Boiler blow down is the major waste produced in this process. Other effluent streams contain spent absorbent liquor and wash solutions employed in the removal of final traces of CO and CO_2.

In partial oxidation process, natural gas and steam are burned and quenched to produce hydrogen, CO, and CO_2. Methane and argon are also produced simultaneously. CO along with steam is converted to carbon di oxide and hydrogen, by passing through an iron oxide catalyst. CO_2 is scrubbed with monoethanolamine. Effluents from scrubbing of atmospheric CO_2 (with caustic soda), and excess quench water (containing dissolved CO, CO_2 methane and argon), are the two effluents generated in this process.

In another process employing coke oven gas as the raw material, gas is compressed and high molecular weight carbons are removed by condensation. CO_2 and H_2S are removed by scrubbing. The processed gas is finally washed with sodium hydroxide solution to remove the traces of CO_2. This wash solution is the only effluent. Liquid nitrogen is used for synthesis of ammonia. In the combustion process, compressed air and hydrogen are burned to produce ammonia. The CO_2 in the air prior to combustion is removed by sodium hydroxide / carbonate scrubbing which is discharged as waste.

As already mentioned, boiler blow down and cooling water are major effluent streams contributing to the bulk. A small stream of effluent from raw material handling (mainly due to spill) is also discharged which may contain fuel oil or naphtha or coal dust depending upon the feed stock used. The scrubber effluents though not contribute to the bulk, they contain hazardous substances such as cyanide, chromium, arsenic, vanadium and absorbents such as organic amines, methanol etc., Urea, ammonium sulfate, ammonium chloride, ammonium nitrate and calcium ammonium nitrate are produced using ammonia.

Urea

Liquid ammonia and carbon dioxide (obtained during ammonia production) are made to react at about 160° to 200°C at high pressure (100-250 atm) to produce a mixture of urea, ammonium carbamate and water. Carbamate is recycled. The urea solution is concentrated and prilled to produce solid urea. Practically no effluent is generated from urea production (excepting spills) as it is a complete recycle system.

Ammonium sulfate

1. Ammonium sulfate is produced by the reaction of synthetic ammonia and sulfuric acid. The crystals of ammonium sulfate are separated from the mother liquor by centrifugation. No waste water is produced in this process.

2. Ammonia generated as a waste during the production of coke from coal is made to react with sulfuric acid to obtain ammonium sulfate.

Ammonium chloride

Though the reaction between ammonia and hydrochloric acid gas can yield ammonium chloride, it is obtained as a by-product in the production of soda ash. In the direct process an effluent of considerable volume (containing ammonia) is produced from the washing of the gases (NH_3 and HCl).

Ammonium nitrate

Three processes are employed at present for the manufacture of ammonium nitrate.

Stengle process:

Ammonia vapor and nitric acid (60%) are allowed to react in a packed tower. The resulting product is condensed and dried. Water from the final condenser is discharged as waste which contains ammonium nitrate. The dried ammonium nitrate is broken to suitable size and bagged.

Prill process:

Ammonia and nitric acid are made to react in the agitated vessel and the resultant solution is concentrated and agitated to produce ammonium nitrate.

Crystallization process:

Same as that of Prill process except drying. In both the processes no effluent is discharged, as all products are collected and recycled.

Calcium ammonium nitrate

The ammonium nitrate liquor obtained in the above process is sprayed into a granulator filled with calculated quantities of lime stone powder. The resulting granules are dried, screened, cooled and coated with soapstone powder. In the production of all fertilizers described above, boiler blow down and cooling water are the major effluents discharged continuously. They contain conditioning agents, biocides and a little amount of fertilizer. Spillage and leakages also form another effluent stream which usually contain the respective fertilizers. For details of effluents from sulfuric acid, nitric acid and hydrochloric acid plants refer Chapter-14 "Acid Wastes".

PHOSPHATIC FERTILIZERS

Phosphoric acid is the base material used for the production of phosphatic fertilizers. Most of the phosphatic fertilizer production units have their own phosphoric acid plants.

Phosphoric acid production

Phosphoric acid is obtained by treating ground phosphate rock with either H_2SO_4 or HNO_3 or HCl. Each acid generates a specific type of effluent.

When sulfuric acid is used, calcium sulfate is produced in large quantities. It is estimated that five tons of calcium sulfate (gypsum) is produced per every tonne of phosphoric acid produced. Gypsum is filtered and the effluent from filtration contains suspended matter, phosphorus pentoxide and fluorine.

With nitric acid, soluble calcium nitrate is formed along with phosphoric acid. To the mixture, ammonia and carbondioxide are passed such that calcium nitrate is converted to insoluble calcium carbonate and soluble ammonium nitrate. The entire mixture is then converted directly into solid fertilizer and no effluent is generated in this process.

When hydrochloric acid is used, highly soluble calcium chloride is formed which is separated from the phosphoric acid through liquid- liquid extraction.

Whatever may be the acid used, an appreciable quantity of fluoride is evolved as silicon tetrafluoride vapor and hydrofluoric acid gas. They are absorbed in water and discharged as waste. In the concentration and purification of raw phosphoric acid, small quantities of acid and still more fluoride are wasted along with the water condensate and solid impurities.

Normal Super Phosphate [$Ca_2H_2(PO_4)_2$ +$CaSO_4$]

Well ground phosphate rock and sulfuric acid (60 to 70%) are mixed rapidly (within seconds). The reaction is highly exothermic, the mass solidifies rapidly with the evolution of fluorine gas which is scrubbed and discharged as waste. In addition to fluorine, the scrubber waste contains phosphate and dust. After about 2 hours, the mass is kept in a curing room (for about 2 to 3 weeks), then sized and bagged.

Triple Super Phosphate [$Ca_2H_2(PO_4)_2$]

It is manufactured by treating ground phosphate rock with phosphoric acid (64-75% H_3PO_4 in one process and 56% H_3PO_4, in another process). The reaction is exothermic but solidification is comparatively slow. As in normal super phosphate, fluorine gas evolves which is absorbed and discharged.

NITROGEN - PHOSPHATE FERTILIZERS

Mono Ammonium Phosphate (MAP) [$NH_4H_2PO_4$]

Ammonia and phosphoric acid (30 to 40% P_2O_5) are allowed to react in a rection vessel with or without agitation. The slurry formed is distributed over dried recycled material, passed on to rotary driers and then sent for storage. Practically this process does not generate any effluent.

Diammonium Phosphate (DAP) [(NH_4)2HPO_4]

Monoammonium phosphate and ammonia react to form diammonium phosphate. Dust particles produced during the reaction are scrubbed out and reused in the process. Secondary scrubbing is discharged as waste which contains ammonia, fluoride, phosphate and dust.

NPK FERTILIZERS

NPK fertilizers are produced in three ways. Dried fertilizers are blended in suitable proportions. No waste is produced in this process. In another process, potash particles are coated with DAP. Effluent from this process is similar to that of from DAP manufacture, The third process is identical with phosphoric acid acidulation, nitric acid is used in place of phosphoric acid and at later stages potassium chloride is added. The effluent from this process is similar to that obtained from phosphoric acid acidulation.

BLENDED FERTILIZERS

These fertilizers are prepared by just blending various fertilizers and no effluent is generated.

VOLUME OF EFFLUENTS

Quantity of effluents vary from one industry to another depending on the raw materials used, process adopted, and finished product. Some units will have their own acid (sulfuric, nitric and phosphoric) manufacturing plants, while others do not possess those plants. Some industries produce both nitrogenous and phosphatic fertilizers while some other produce either one of the above two. Some other units just blend the fertilizer products. The volume and characteristics will vary accordingly. As already mentioned, the boiler blow down and cooling water are the major effluent streams. It is estimated that about 8 million to 12 million litres of effluent is discharged in a urea plant producing 1000 tonnes of urea per day. A phosphatic fertilizer unit manufacturing 100 tonnes of P_2O_5 per day discharges about 3 million to 6 million litres of effluent.

CHARACTERISTICS OF EFFLUENTS

The two major effluent streams are from (i) ammonia production and (ii) phosphoric acid production. Effluent from ammonia production is highly alkaline (due to sodium hydroxide or carbonate from gas cleaning) and contains excess ammonia from gas scrubbing and gas cleaning operations. Table 13-1 and 13-2 give the characteristics of effluents generated from two different nitrogenous fertilizer industries.

Table 13-1.
Characteristics of Effluent from Urea Production

Sl. No.	Characteristic	mg/l
i.	Urea	560
ii.	Ammonia	315
iii.	Ammoniacal Nitrogen	260
iv.	Total Dissolved Solids	1435

Table 13-2.
Characteristics of Effluent from Calcium - Ammonium Nitrate Production

Sl. No.	Characteristic	mg/l
i.	pH Value	8.5
ii.	Suspended Solids	325
iii.	Dissolved Solids	2090
iv.	Biochemical Oxygen Demand	280
v.	Chemical Oygen Demand	535
vi.	Ammoniacal Nitrogen (as NH_3)	720
vii.	Organic Nitrogen (as N)	102
viii.	Nitrate Nitrogen (as NO_3)	65
ix.	Sulfate (as SO_4)	980
x.	Phosphate (as PO_4)	4.2

The phosphoric acid effluent is acidic and contains high amounts of phosphates and suspended solids. Fluoride concentration is extremely high and ranges from 1900 to 2200 mg/l.

EFFECTS OF EFFLUENTS

When discharged into inland surface waters, fertilizer effluents (especially phosphatic) causes eutrophication. Due to this algal bloom, whole stretch of water becomes choked, plants rot, all oxygen used up and fish die. In addition, the movement of the fish is interrupted, the extent of fish spawning grounds reduced considerably affecting reproduction. The decayed algae imparts disagreeable odors and tastes to water.

As the phosphatic fertilizer effluents contain exorbitant amounts of fluoride (above 1000 mg/l) receiving waters become rich in fluoride causing dental and skeletal fluorosis to the consumers. The fluoride enriched waters also affect cattle and other livestock by way of causing abnormal calcification of bones. Collapse of leaf tissues of plants is also attributed to fluorides which act in them as a cumulative poison.

Ammonia when present in free state is an irritant and highly toxic to fish.

The toxic effects of cyanide and chromium need not be emphasized.

TREATMENT OF EFFLUENTS

Treatment of fertilizer effluents require much attention. Careful planning is necessary to reduce pollution load and to reduce the capital expenditure and recurring expenditures.. Unlike other industrial effluents, fertilizer effluent streams should be segregated and treated separately. One effluent stream may be used to neutralize or dilute or to treat other stream. As far as possible recovery and recycling should be practised. The following scheme of segregation may be highly useful.

i) Highly acidic and alkaline effluents.

ii) Effluents containing high concentration of ammonia.

iii) Effluents containing high concentration of urea.

iv) Effluents containing excess fluorides and phosphates.

v) Effluents with high suspended solids.

vi) Effluents containing chromium (cooling water)

vii) Cyanide bearing effluents.

viii) Arsenic bearing effluents.

ix) Oil bearing effluents.

x) Effluents containing carbon slurry.

xi) Ash slurry.

As already stated, this individual effluents may be treated separately, and combined together for final disposal or some of the streams may be combined judiciously after assessing their volume, characteristics, frequency of discharge etc.,

i) Acidic/alkaline effluents

When both acidic and alkaline effluents are available they may be mixed taking necessary precautions. If only one is generated it may be neutralized using acid or alkali accordingly.

ii) Effluents containing high concentration of ammonia

These may be treated either by physico-chemical means or by biological methods. Air stripping or steam stripping of ammonia and ion exchange are some of the physico-chemical methods. The ammonia separated is either reused in the plant or converted into other useful products during treatment itself. When ion exchange is employed, the treated effluent may be used for boiler feed make up. Biological treatment involves nitrification of ammonia to nitrite and nitrate (aerobic process) and denitrification to nitrogen, oxides of nitrogen and carbondioxide (anaerobic process) using nitrifying and denitrifying organisms. Algal uptake is also an effective biological process. Algae is cultured and grown in waste water containing ammonia. These algae are capable of extracting ammonia (as their nutrient) from the effluent. The grown up algae is used as a manure.

iii) Effluents containing high concentration of urea (and nitrate)

Inbuilt plants for thermal hydrolysis of urea with recovery of ammonia are set up in modern fertilizer units. Hydrolyzing urea with the enzyme urease (secreted by a group of soil bacteria) is also practised. The enzyme converts urea into ammonia and carbondioxide. Nitrates are removed by biological denitrification as mentioned above.

iv) Effluents containing excess Fluorides and Phosphates

Most of the fluorides are present as fluosilicic acid, and phosphate as phosphoric acid with some soluble calcium phosphate. A two stage treatment with chalk and lime or double lime treatment, is adopted. In the former treatment, chalk precipitates fluosilicic acid to calcium fluoride and silica and phosphates as $Ca(H_2PO_4)_2$. Traces of fluoride are removed and calciumhydroxy apatite $[Ca_5OH(PO_4)]_3$ is formed on treatment with lime. The double lime treatment is essentially the same as that of two stage treatment but lime is used in the first stage itself instead of chalk.

v) Effluents with high suspended solids

The effluents are allowed to settle in suitably designed settling basins. When the particle size is much smaller to settle, they may be coagulated and clarified.

vi) Effluents containing chromium

Chromium in these effluents is usually present as chromates (hexavalent). Reduction of (hexavalent) chromates to trivalent chromium is the first step of treatment. Reduction is carried out after acidifying the effluent (usually with sulfuric acid to pH 2 to 4) and treating with either ferrous sulfate or sodium sulfite or sulfur dioxide. It is estimated that for the reduction of 1 g. chromate (as CrO_4) 10g. ferrous sulfate or 2.5g. sodium-sulfite or 1.5 g. SO_2 is required. Treatment of the reduced effluent with lime, precipitates chromium (as chromium hydroxide) and partially precipitates the phosphates present in cooling water. Lime may be replaced by effluent containing excess ammonia thereby reducing the cost of treatment and pollution load considerably.

vii) Cyanide bearing effluents

The satisfactory method for treating cyanide effluents is alkaline chlorination whereby cyanides are converted to non-toxic nitrogen and carbon dioxide. After raising the pH of the effluent to about 9 using caustic soda, chlorine is passed For details, refer Chapter - 10, "Electroplating Industry".

viii) Arsenic bearing effluents

Arsenic is a major constituent of spent carbon dioxide scrubber liquid. This arsenical solution is segregated carefully from other effluent streams, filtered and concentrated. Sometimes a secondary filtration through activated carbon is also practiced. The concentrated liquor is reused in the process. If not used, is evaporated to dryness and solids are placed in concrete drums, sealed and buried.

ix) Oil bearing effluents

As oils and grease are present in free state they float on the surface and are removed by gravity type or mechanical oil separators after the effluents are allowed to settle in settling tanks. The treated effluent still contains about 50 mg/l oil which may further be reduced to about 5 mg/l by passing through a bed of activated carbon or porous coke.

x) Effluents containing carbon slurry

If partial oxidation process is employed for the manufacture of ammonia, carbon is formed in the process and let out as carbon slurry. Recovery of this carbon is the best means of treatment, which also brings revenue as the carbon could be utilized for the manufacture of activated carbon, rubber and carbon black.

PARAMETERS TO BE DETERMINED

I. NITROGENOUS FERTILIZERS

A. Important Parameters

1) Color 2) Odor 3) Temperature 4) pH value 5) Acidity/Alkalinity 6) Total suspended solids 7) Dissolved volatile solids 8) Dissolved fixed solids 9) B.O.D: BOD is usually not high. Dilutions ranging from 1:49, 1:99,1:999,1:499, may be made. Permanganate value (4 hrs) will be highly useful in deciding the dilution. Seeding is necessary. Acclimated seed yields better results. 10) C.O.D. 11) Ammoniacal Nitrogen 12) Oils and Grease

B. Additional parameters

13) Phosphates (from cooling water). When the phosphate concentration in the effluent is high, gravimetric method may be employed. Phosphates may first be precipitated as ammonium molybdophosphate in dilute ammonia solution and reprecipitated as magnesium pyrophosphate and weighed. (For details, refer Quantitative Inorganic Analysis by A.I. Vogel). 14) Chromium (from cooling water 15) Chlorides 16) Sulfates 17) Sulfites 18) Nitrates 19) Arsenic 20) Cyanides 21) Vanadium (used as a catalyst) 22) Calcium 23) Phenolic Substances. A separate determination of urea would be useful.

II. PHOSPHATIC FERTILIZERS

All the parameters prescribed for Nitrogenous fertilizers, except nitrate, cyanide, arsenic and vanadium. In addition, fluoride has to be determined. Since fluoride occurs in abnormally high concentrations (above 1000 mg/l titrimetric or gravimetric methods may be employed. (Refer A.I. Vogel. "Quantitative Inorganic Analysis including Elementary Instrumental methods of Analysis")

III. NITROGEN - PHOSPHATE FERTILIZERS

All the parameters prescribed for Nitrogenous fertilizers. Include fluoride also.

REFERENCES

1) G.L.Culp and R.L.Culp; "Advanced Waste water Treatment'-1971-Van Nostrand Reinhold Publishing corporation, New York.

2) S.R.Alagarsamy, B.B.Bhalerao and S.Rajagopalan "Treatment of waste from fertilizer Plants" Ind. Jour. of Env.Health, Vol.15 No.1,1973.

3) Bela G.Liptak., "Environmental Engineers Handbook" Vol.-I Water Pollution "-1973, Chilton Book Co. USA.

4) G.J.Mohan Rao and P.V.R.Subrahmanyam "Aspects of Water Pollution in Fertilizer Industry" Ind. Jour. of Env. Health, Vol.15 No.4-1973.

5) The Water Prevention and Control of Pollution Act, 1974 Govt. of India.

6) H.C.Arora and E.N.Chattopadhya. "A Study on the Effluent Disposal of Superphosphate Fertilizer Factory" Ind. Jour. of Env. Health, vol. 16 No.2,1974.

7) S.R. Varma etal "Studies on the characteristics and Disposal Problems of Industrial Effluents"-Ind. Jour. of Env. Health, Vol.19 No.3,1977.

8) S.K.Gupta etal, “Treatment of Nitrogenous Fertilizer waste by Flocculating Algal-Bacterial System ' Ind.Jour. of Env.Health Vol.21, No.2, 1979.

9) IS:1981, “ Guide for Treatment and Disposal of Effluents of Fertilizer Industry”- Indian Standards Institution, New Delhi.

PART -III

OTHER INDUSTRIAL EFFLUENTS

Chapter - 14

ACID WASTES

(Also refer chapters on Coal Mining, Metal Pickling etc.,)

Acid wastes are usually not discharged from acid manufacturing units as we think. Most of the acid production processes do not generate effluents and the products if any generated are recycled. The possible waste waters arise from spill overs, cooling water and wash waters. Most of the acid wastes are generated during the production of other chemicals and materials (eg.rayon) manufacturing processes, desulfurization of oils at refineries, drying of chlorine gas in chlor-alkali manufacture, production of pesticides, pickling etc.,

The three important mineral acids are sulfuric acid, hydrochloric acid and nitric acid. The manufacturing processes are described below.

Sulfuric Acid Production

There are two methods used for the manufacture of sulfuric acid, (i) contact process and (ii) chamber process. The former is the widely employed process. The principle of the method is burning elemental sulfur to sulfurdioxide which is converted to sulfur trioxide in the presence of a catalyst (vanadium pentoxide). The sulfur trioxide is absorbed in 99% sulfuric acid and hydrolyzed to proper concentration.

Essentially no acid is let out as effluent. Production of sulfurdioxide and conversion to sulfur trioxide generates heat and must be cooled. The cooling water is the only effluent produced in sulfuric acid production.

In oil refineries, sulfuric acid is employed as a catalyst, in alkylation used for refining lubricant oils and for desulfurization. In DDT manufacture the spent acid cannot be discharged as such and either it must be neutralized with alkalis or burnt. (See the section, "Treatment").

Hydrochloric Acid Production

The major use of hydrochloric acid is in steel pickling. It is also used in the manufacture of chemicals (especially chloro derivatives) and pharmaceuticals and food processing. A large percentage of hydrochloric acid is obtained as a byproduct in the manufacture of other chemicals (e.g.) in the manufacture of chloromethanes and polyvinyl chloride. Direct production of hydrochloric acid is also carried out. In one process sodium chloride and sulfuric acid are made to react together under elevated temperature to form HCl.

$$NaCl + H_2SO_4 \rightarrow HCl + NaHSO_4$$

$$NaCl + NaHSO_4 \rightarrow HCl + Na_2SO_4$$

In other process both hydrogen and chlorine are allowed to combine which is an exothermic reaction (The component gases hydrogen and chlorine are obtained as by products in the manufacture of caustic soda).

In both the process, the hydrogen chloride gas is absorbed in water in a tantalum and Karbate absorber.

In the processes described above, no effluent is

generated. Cooling water is deemed to be the only effluent.

Nitric Acid Production

Nitric acid is used to manufacture inorganic nitrites and nitrates and nitro derivatives of organic compounds and in pickling (of brass), in parting of gold and silver, and photo engraving. The modern and cheapest method of manufacture of nitric acid is the catalytic oxidation of ammonia with air. Gaseous ammonia is first oxidised to nitric oxide which is further oxidised to nitrogen peroxide and is absorbed in water to form nitric acid. All theses reactions are carried out in a closed system, the intermediate products if any discharged are recycled to the system itself and practically no waste arises from production, excepting cooling water and lubricating oils used for machineries.

Phosphoric Acid Production

See Chapter 13, "Fertilizer Industry"

CHARACTERISTICS OF EFFLUENTS

Needless to say that acid wastes are low in pH and highly corrosive. They must be discharged after proper treatment and adjusting the pH to neutral range. There is no direct limit on acidity but it is indirectly controlled by pH values. Acid wastes are generated from various industries as acids are utilized in different processes. They are let out either as separate streams or in combination with other effluents. The nature and characteristics will vary accordingly. Refer chapters on Coal industry, Electroplating industry, Explosives manufacture, Fertilizer industry, Pesticide manufacture and Pickling wastes.

EFFECTS OF EFFLUENTS

See Chapter 1. 'Basic Information'

TREATMENT

(Also refer "Metal Pickling wastes and Electroplating Industry Wastes")

In some of the plants, acids are recovered; in others they are reused. Some industries use the spent acid as the raw material for the production of certain chemicals (e.g. sulfuric acid used in the drying of chlorine gas in chlor-alkali industries is utilized for the production of alum.) Rest of the industries neutralize their acid wastes.

In oil refineries, sulfuric acid waste is subjected to spray burning. This method involves spraying of waste acid in to a hot combustion chamber (950 to 1200°C) where it is converted to sulfurdioxide and later absorbed to prepare new sulfuric acid.

The most common method of treating acid wastes is neutralization with alkalis. For detailed discussion refer Chapter 45.4.1 "Neutralization of Acid wastes" in Part V.

PARAMETERS TO BE DETERMINED

Acid wastes are dealt in various chapters (Acid mine drainage in Coal industry, Electroplating Industry, Explosives manufacture, Fertilizer industry, Pesticide manufacture and Pickling wastes) Refer these chapters for necessary parameters to be determined.

REFERENCES

1) Cooper, J.E. "How to Dispose of Acid Wastes" Chem Ind. 66, 684 (1950)

2) Dickerson, B.W. and R.M.Brooks "Neutralization of acid wastes" Ind.Eng. Chem. 42, 599 (1950).

3) R.N. Shreeve and J.A.Brink Jr "Chemical Process Industries" 4th ed, 1977, Mc Graw Hill International Book Co.

Chapter - 15

ASBESTOS INDUSTRY

Asbestos is a fibrous non-combustible mineral. It is a magnesium silicate mixed with varying quantities of calcium and iron silicates. It is used in the manufacture of fire proof roofing sheets and insulating materials.

Asbestos after mining is processed to remove the impurities. In dry process it is cleaned by forced air. No liquid effluent is produced in this process, but the dust problem is severe. In wet process, the crushed ore is cleaned with water. This wash water is the effluent which is filtered and recycled or treated and discharged. Another possibility of effluent is the rain water that passes through the wasted asbestos tailings at asbestos mines.

CHARACTERISTICS OF EFFLUENTS

The waste water is highly alkaline, turbid with excess suspended impurities and possesses high BOD and COD.

EFFECTS OF EFFLUENTS

If asbestos enters the human body via drinking water, it produces cancer. The asbestos dust can cause 'asbestosis' the dreaded disease.

TREATMENT

At asbestos manufacturing plants, the effluent is

treated with sulfuric acid and allowed to settle in settling ponds. The sludge containing asbestos is dredged periodically and used for land fill. Chemical coagulation with iron salts in combination with polyelectrolytes is reported to be an efficient method but comparatively costlier. At asbestos mines, drainage ditches and dams are constructed to prevent rain water from entry into waste asbestos tailings.

PARAMETERS TO BE DETERMINED

1) Color 2) Odor 3) pH 4) Total suspended solids 5) BOD 6) COD 7) Alkalinity 8) Calcium and Magnesium hardness 9) Silicates (For analysis refer "Quantitative Inorganic Analysis" by A I. Vogel).

Chapter - 16

BLEACH-LIQUOR MANUFACTURE

Bleach liquor manufacture once thought to be a large scale industry has become a cottage industry at present and a number of units have been installed. Manufacturing process is much simpler but safety is highly essential.

Bleach Liquor Manufacture

The basic principle in the manufacture of bleach liquor is the rection between hydrated lime and gaseous chlorine. Chlorine is received in cylinders from chlor-alkali industries.

Hydrated lime is made to a solution and placed in a reaction tank into which regulated amounts of chlorine is passed and mixed thoroughly using electrically operated stirrers. After completion of the reaction, the bleach liquor is pumped to tankers. After a batch operation is over, the reaction tank is washed. These wash waters and spill overs constitute a characteristic stream of effluent. The free chlorine content is abnormally high and it is a highly toxic effluent. The characteristics are given in Table 16-1.

The effluent from bleach liquor manufacture has to be handled carefully as any means of acidification will liberate chlorine. The chlorine concentration of this effluent is exceedingly high (varies from 500 to 8000 mg/l). The effluent should not be discharged without treatment into water courses or on land.

Table 16-1.
Characteristics of Bleach Liquor Manufacturing Effluent

Sl. No.	Characteristics	Unit A	Unit B
i.	pH Value	10.5	11.5
ii.	Total Suspended Solids, mg/l	23100	22270
iii.	Settleable Solids, ml/l/h	150	-
iv.	Total Dissolved Solids, mg/l	36500	10900
v.	Free Chlorine (as Cl_2), mg/l	7660	576
vi.	Chlorides (as Cl), mg/l	9200	2300
vii.	Alkalinity (as $CaCO_3$)		
	a) Phenolphthalein alkalinity, mg/l	7300	5500
	b) Total alkalinity, mg/l	10800	12400

TREATMENT

This effluent is allowed to settle stage by stage and the supernatant after proper dilution has to be discharged. The sludge may be reused or dried and disposed of at remote places. This effluent may also be used to bleach dyeing effluents arising from textile processing.

PARAMETERS TO BE DETERMINED

1) Color 2) pH value 3) Alkalinity: Use smaller aliquots 4) Total suspended solids. 5) Settleable solids. 6) Total dissolved solids. 7) Free chlorine: use smaller aliquots. Titrate with N/10 or N/4 Thio.

REFERENCES

1) Scones, J.S. (ed) "Chlorine: Its manufacture, properties and uses" ACS Monograph 154, 1962, Reinhold Publishing Corporation.

2) Szilard, J.A. "Bleaching Agents and Techniques" - 1973, Noyes Data Corporation.

Chapter - 17

CANNERY

In cannery, vegetables and fruits are processed and canned. It is one of the major industries at United States. Most of the processes in a cannery utilize large volumes of water and consequently volume of liquid effluents discharged is also high. The effluent is organic in nature amenable for biological oxidation and hence can be discharged into sewer.

The vegetables canned are beans, carrots, beets, potatoes, mushrooms, peas, corn and pumpkin. Fruits include apple, apricot, grapes, tomato and citrus family fruits such as lemon, orange etc.,

PROCESSES, SOURCES AND CHARACTERISTICS OF EFFLUENTS

Cannery wastes are classified according to the products being processed and three major items are (i) vegetables (ii) fruits, and (iii) citrus fruits. The volume and characteristics of waste waters vary considerably from one plant to another and within the same plant from day to day. Large volumes of waste water arise from the preliminary processes of washings of the raw materials to remove toxic chemical residues and natural contaminants and due to hydraulic transportation of the produce from one equipment to another. The processes of trimming, juicing, blanching (or lye treatment) and pasteurizing of the materials, cleaning of equipments, floor washings, and cooling of finished products also contribute significant quantities of effluents.

Waste from peeling operation contains excess suspended matter (primarily organic in nature), the quantity of which depends on whether the vegetables have been blanched or lye-treated prior to peeling. Blanching is a process applied to vegetables to expel air and gases, to remove the color, to soften for easy precooking and to inactivate enzymes which cause undesirable odors and color changes. The effluent from blanching is so strong as it contains the leached sugars, starches and other soluble organics, but the volume is much less. Lye treatment also contributes a strong waste which is highly alkaline.

Effluent from citrus fruit processing differs from the effluents of other canneries. It contains pulp press liquors, pectin wastes, processing plant wastes, cooling waters, floor washings etc.,

Table 17-1 indicates the characteristics of cannery wastes.

TREATMENT

Cannery wastes are screened, neutralized, chemically precipitated and subjected to lagooning or spray irrigation. Screening removes large solids, which are dried by spreading on grounds and burned or used in animal feed formulations. Chemical precipitation is the most effective treatment for apple, tomato and cherry wastes. Ferric salts or aluminate are the commonly used chemicals in conjunction with lime. (app. dose :

0.1 kg to 0.5 kg of alum and 0.2 to 0.5 kg of lime/1000 litres) About 50% BOD is removed by precipitation. The volume of sludge produced on chemical precipitation is high ranging about 10 to 15 percent the volume of waste. Lagooning is the economical and practically simple treatment for cannery wastes. It involves both aerobic and anaerobic biological action, sedimentation, soil absorption, evaporation and dilution. Two major disadvantages encountered in lagooning are : i) evolution of bad odors, ii) lagoons serve as a breeding ground for flies, mosquitoes and other insects, if proper aerobic conditions are not maintained. Spray irrigation is considered to be an economical and unobjectionable method that can be applied to cannery waste that do not contain pathogens and toxic materials.

Table 17-1.
Cannery Waste Characteristics*

Product	Flow gal/case	BOD ppm	Suspended solids, ppm
Tomatoes	4.5 -78.0	616-1870	550-925
Corn	30-116	885-2936	530-2395
Green beans	104.5	93	291
Green beans & corn	99.5	270	264
Mixed vegetables	12.2	750	593
Pears	32.4-42.5	238 - 468	340-637
Peaches	37.5	1075	250
Apples	26.8	1600	300
Cherries	16.0	800	185

* Eckenfelder W.W.etal, "Study of fruit and vegetable processing waste disposal methods in the eastern region, "Final Report, New York State Canners Association, September, 1958.

PARAMETERS TO BE DETERMINED

1) Color 2) Odor 3) pH value 4) Temperature 5) Suspended solids 6) Dissolved fixed solids 7) Bio-chemical Oxygen Demand: Dilutions depend on the cannery and its products 8) Permanganate value 9) BOD : Permanganate value 10) Chemical Oxygen Demand 11) Ammoniacal Nitrogen 12) Organic Nitrogen 13) Percent Sodium, 14) Boron 15) Other specific substances (including preservatives).

REFERENCES

1) Rudolf W. "Industrial Waste Treatment"-1953, Reinhold publishing Corporation, New york.

2) H.F.Lund (Ed), "Industrial Pollution Control Hand book"-1971, Mc Graw Hill Book Co.,Newyork.

3) N.L. Nemerow, "Industrial Water Pollution - Origins, Characteristics, and Treatment"-1978, Addison wesley Publishing Co., Masschusetts.

Chapter - 18

CEMENT MANUFACTURE

Cement industry contributes much to air pollution and liquid effluents are not a problem. Cement is a compound made up . of predominately CaO and SiO_2 along with minor constituents such as Al_2O_3, Fe_2O_3 and MgO. Raw materials required for the manufacture of cement are lime, sand clay, shale, iron-ore and blast furnace slag. These raw materials are crushed, pulverized and blended in appropriate proportions. Then the mixture is subjected to either wet or dry process of fine grinding. In wet process, materials are ground under moist condition. In dry process, they are ground in dry state. The ground mixture is then introduced into the kiln where it is subjected to a temperature of about 1650 to 1670°C and fused to a hard mass called clinker, which is cooled rapidly by air and then mixed with a small amount of gypsum (4 to 5% to regulate the setting time) and again ground to a fine powder and cooled by water jacketed heat exchangers.

In most of the cement industries, this cooling water is not harmful and can be recycled after cooling. Apart from this, wet scrubbing of kiln dust yields an effluent that has a high pH value, alkalinity, suspended and dissolved solids (potassium and sulfate are predominant ions).

The residue or sludge obtained on clarification of this effluent is recycled to the kiln.

The treatment systems include (i) cooling towers or ponds to reduce the temperature of cooling water and (ii) clarifiers for separation of solids from wet scrubber effluents or from dry dust leaching.

REFERENCES

1) Taylor, H.F.F. (ed) "Chemistry of Cements" (2 vols), 1964, Academic Press, New york.

2) Witt J.C. "Portland Cement Technology" 2nd ed, 1965, Chemical Publishing Co.

Chapter - 19

CHLOR - ALKALI INDUSTRY

Caustic soda is an important chemical invariably used for many purposes and manufactured by electrolysis of brine. Hydrogen and chlorine are obtained as by-products. The three major steps involved in the manufacture of caustic soda are brine purification, electrolysis and caustic fusion (concentration of dilute caustic soda)

Brine purification

Purification of brine is an essential step to obtain pure caustic soda. Sodium chloride is made to an aqueous solution which contains calcium, iron and magnesium compounds as impurities. They are removed by the addition of soda ash and caustic soda. Sulfate is removed by adding barium chloride. All these reagents are added to brine in a clarifier and the precipitates are allowed to settle. The clear brine is neutralized with hydrochloric acid, stored and fed continuously to electrolytic cells.

*Electrolysis**

Different types of cells were used in the past and at present mercury cells are employed. In the mercury cell method, liquid mercury acts as a cathode and graphite as

---

* The electrolysis technic also becomes obsolete which is replaced by the most efficient and wasteless membrane cell technology.

anode. The average electrolysis temperature is about 65°C. The voltage applied is 3.0 to 4.5 V and current is about 10,000 amps. Concentration of sodiumchloride solution is 190 g/l. During electrolysis, the following reactions take place.

$$2NaCl \rightarrow 2Na^+ + 2Cl^-$$

At anode: $2Cl^- \rightarrow Cl_2\,(g) + 2e^-$

At cathode: $2Na^+ + 2e^- \rightarrow 2Na + 2Hg \rightarrow \underset{\text{Sodium amalgam}}{2NaHg}$

The sodium amalgam continuously flows to the second compartment of the cell where it acts as anode. Graphite is the cathode. Electrolyte is sodium hydroxide solution. Purified water is fed counter current to the sodium amalgam.

Anodic reaction : $2Na(Hg) \rightarrow 2Na^+ + 2Hg + 2e^-$

Cathodic reaction: $2H_2O + 2e^- \rightarrow 2OH^- + H_2(g)\uparrow$

Overall reaction: $2Na(Hg) + 2H_2O \rightarrow 2NaOH + 2Hg + H_2$

Caustic fusion

The dilute caustic (5%) obtained is concentrated stage by stage, cooled and packed in steel drums or sent for flaking.

Chlorine drying and liquefaction

The important by product chlorine, evolved at anode contains much water vapor which is removed by passing through conc. H_2SO_4. The dried chlorine is compressed to liquid and stored in cylinders.

Hydrogen

Another by-product hydrogen is utilized for the production of acid and / or ammonia. It is also used for the manufacture of hydrogenated oils intended for cooking.

SOURCES & CHARACTERISTICS OF EFFLUENTS

Brine plant Effluent

Brine purification yields only a sludge. But the cooling water of pumps, motors and leakage in brine section constitutes a separate effluent stream which is usually alkaline, and contains high amounts of sodium chloride. Suspended solid content is also high. It also contains barium. (Mercury is also present in the effluents of some units. (See next paragraph) (See Table 19-1).

Electrolysis section effluent

Proper maintenance of electrolytic cells is essential otherwise leaks may occur which contribute a concentrated effluent containing significant quantities of mercury. A little quantity of waste waters join this stream from chlorine and hydrogen condensation. This combined effluent is usually stored, mercury recovered and sent to brine clarifier (See Table 19-1).

Chlor-alkali effluent

The combined effluent from brine plant and electrolysis section (described above) is termed as the chlor-alkali effluent which is highly alkaline and contains excess suspended and dissolved solids. Mercury content is also appreciable (See Table 19-1)

Caustic fusion

Cooling water and spill overs constitute the caustic fusion effluent. Their characteristics are given in Table 19-1.

All the chlorine produced is not cylindered. Usually the chlor-alkali industries have their own bleaching powder and bleach liquor manufacturing units. Some of the industries also have chloro-methane plants so that most of the chlorine produced can be utilized. No effluent is generated in bleaching powder manufacture but chloro-methane production yields characteristic effluents. Bleach liquor manufacture and effluents are dealt separately in Chapter-16.

Table 19-1.
Characteristics of Various Effluents of Caustic Soda Manufacturing Industry

Sl. No.	Characteristic	Brine plant	Electrolysis section effluent	Chlor-alkali effluent (Brine plant effluent and Electrolysis section recycled effluent)	Caustic fusion effluent
i.	Color	Whitish	Brown	Brownish	Whitish
ii.	pH Value	9.2	13	12.2	9.5
iii.	Total Suspended Solids, mg/l	6900	2000	580	140
iv.	Total Dissolved Solids, mg/l	38600	27300	4620	456
v.	Dissolved Fixed Solids, mg/l	36200	26600	4270	400
vi.	Chlorides (as Cl) mg/l	22400	11400	2300	120
vii.	Mercury (as Hg), mg/l	1.25	120	3.2	0.5
viii.	Barium (as Ba), mg/l	32.96	-	-	-
ix.	Permanganate value (4 hrs), mg/l	11	24	22	-

Chloromethanes Production:

Chloro methanes are derivates of methane. They are chloro-methane or methyl chloride (CH_3Cl), dichloro- methane or methylene chloride (CH_2Cl_2), trichloro-methane or chloroform ($CHCl_3$) and carbon tetrachloride (CCl_4). Methanol is used as the raw material which is chlorinated to obtain the above four products. During the production of each chloro derivative, hydrochloric acid is produced as a byproduct which is discharged as waste. This is a highly acidic waste with excess concentration of chloride. Typical characteristics of this waste are given in Table 19-2.

Table 19-2.
Characteristics of Effluent from Chloromethanes Manufacture

Sl. No.	Characteristic	mg/l
i.	pH Value	1.4
ii.	Total Suspended Solids, mg/l	12
iii.	Total Dissolved Solids, mg/l	680
iv.	Acidity (as $CaCO_3$)	
	a) Methyl orange acidity, mg/l	82500
	b) Total acidity, mg/l	82500
v.	Free Chlorine (as Cl_2), mg/l	7.1
vi.	Chlorides (as Cl_3), mg/l	62000
vii.	Mercury (as Hg), mg/l	0.04

EFFECTS OF EFFLUENTS

1) Caustic soda manufacture Effluents

Untreated effluent contains toxic levels of mercury. The effluents when discharged into surface waters or into sea, mercury enters and accumulates in the body of fish as methyl mercury and passed on to human beings, resulting in affliction and death. Removal of mercury is an essential step in the treatment of this effluents. Further, the sodiumchloride present is highly hazardous to plants if the effluent is discharged on land. The effects of sodium and chloride ions are already described in various contexts and need not be emphasized again. It has to be mentioned that the newly developed membrane cell technology is far more superior as it does not produce any effluent at all.

2) Chloromethanes manufacture effluent

This is a highly acidic effluent containing abnormal concentration of chloride. Effects of acid wastes are described in detail in Chapter - 1 "Basic Information".

TREATMENT

1) Caustic soda manufacture effluent

Suspended solids are removed by settling, pH adjusted and sent for mercury removal. Specific ion exchange resins for mercury are developed at reasonable cost. They remove mercury from the effluent effectively. As with other resins, they also have to be regenerated after exhaustion. Though the mercury removed effluent is suitable for disposal, higher concentration of chloride is a limiting factor. To overcome this difficulty, the effluent may be diluted if sufficient water is available.

2) Chloromethane effluent

This acidic effluent is neutralized with dolomite or

lime, allowed to settle and discharged. (For details see Chapter 14 "Acid waste" and Chapter 45.4.1 Neutralization). It is to be noted that the dissolved solids and suspended solids increase considerably due to the production of soluble salts on neutralization. This is a problem of prime concern which needs tertiary treatment (reverse osmosis or electro dialysis).

PARAMETERS TO BE DETERMINED

Caustic soda manufacture effluents

1) Color 2) Odor 3) Temperature 4) pH value 5) Alkalinity 6) Total suspended solids 7) Total dissolved solids 8) Chlorides : Sample has to be diluted atleast 10 times. Use the diluted sample for estimation after pH adjustment. 9) Mercury 10) Barium 11) Chemical Oxygen Demand.

Note: BOD is not an important parameter for these wastes. The value obtained will not be the true BOD as the estimation suffers so many interferences, especially due to metals which suppress bacterial activity. Further, chemical wastes do not contain organic matter and the BOD values will not be useful.

Chloromethane manufacture effluent

1) Color 2) Odor 3) Temperature 4) pH value 5) Acidity: Use smaller aliquots. Titrate with IN. alkali 6) Total suspended solids 7) Total dissolved solids 8) Free chlorine 9) Chlorides; Dilute the sample 10 times. Use smaller aliquots of diluted sample 10) Mercury.

REFERENCES

1) Buchn, W. "Caustic soda production Technique"-1962, Noyes Data Corporation.

2) Gould, R. F. (ed) "Literature of Chemical Technology" - 1967, ACS Monograph.

3) Jones, H. R. "Mercury pollution control"-1971, Noyes Data Corporation.

Chapter - 20

COAL MINING

(ACID-MINE DRAINAGE WASTE)

During coal mining, water is pumped out of mines. These waters are usually referred as coal mining waste water or acid mine drainage waste. Sometimes rain water also enter into the mine which is also regarded as waste. In short, water that comes out of coal mines are referred as acid mine drainage waste. The stratum of coal mine is usually composed of iron and sulfur in the form of various compounds. During mining, ferrous sulfide (the most predominant form) is exposed to air and water and is converted into ferrous sulfate and sulfuric acid. This process is hastened by sulfur bacteria.

$$2FeS_2 + 7O_2 + 2H_2O \rightarrow 2FeSO_4 + 2H_2SO_4$$

The ferrous sulfate is further oxidised to ferric sulfate in the presence of air and water with the additional production of sulfuric acid. As indicated above, bacterial action plays an important role in the production of acid.

CHARACTERISTICS OF EFFLUENTS

This is a highly acidic waste containing sulfuric acid and ferrous sulfate. The acidity varies from mine to mine depending on the stratum. It may range from 100 to 50,000 mg/l. Ferrous sulfate is present in amounts 10 to 1500 mg/l. Manganous sulfate and aluminium sulfate are also present in concentration ranging from 0 to 250 mg/l and 0 to 350 mg/l respectively. Characteristics of two different mine wastes are given in Table 20-1.

It is to be noted that the mine drainage A is a typical waste while that of B exhibits only the characteristics of ground water.

Table 20-1.
Characteristics of Two Different Mine Waste Waters*

Sl. No.	Mine drainage characteristics	Mine A	Mine B
i.	Drainage flow (gallons per minute)	2500	900
ii.	pH value	3.7	6.2
iii.	Free acid, mg/l	124	4
iv.	Total acid, mg/l	466	13
v.	SiO_2, mg/l	14	9.6
vi.	Fe, mg/l	22[†]	0.2
vii.	Mn, mg/l	10	4.6
viii.	Al, mg/l	17	1.9
ix.	Ca, mg/l	95	34
x.	Mg, mg/l	55	12
xi.	SO_4, mg/l	746	172
xii.	Cl, mg/l	9	2.4

EFFECTS OF EFFLUENTS

The high acidity results in sudden changes of pH of

* Reproduced from N.L.Nemerow, "Industrial Water Pollution-Origin Characteristics and Treatment"-1978, with the kind permission of publishers M/s. Addison-Wesley Publishing Co. Massachusetts.

† Apparently a considerable amount of iron originally present in solution has been precipitated out in suspended form.

receiving water and makes the water unsuitable for survival of fish and other aquatic organisms. The water is also made hard and sulfated. High concentration of iron, manganese and aluminium makes the water unfit for domestic, industrial or other uses without treatment (which is expensive). For other effects see the section "Acids" in Chapter 1. "Basic information".

TREATMENT

Due to the effects described above, the acid mine drainage must be treated before discharge in to stream. Some of the suggested treatments are:

i) Neutralization

Two stage neutralization with lime stone and lime followed by aeration and sedimentation. The process is expensive.

ii) Lagooning

Prolonged lagooning results in a compact ferric-hydroxide residue as well as reduction in ferrous sulfate content due to oxidation and precipitation to ferric hydroxide.

iii) Proportioning

Use of equalization basins and proportionate discharge with stream flow minimizes the down stream effects.

iv) Reuse for Coal Washing

The raw coal contains about 30% $CaCO_3$ and $MgCO_3$ and washing of coal with acid mine drainage results in neutralization of the waste and in the removal of this alkaline materials. It is reported that the pH of an acid mine drainage waste rises from 3 to about 7, if used for coal washing. Simultaneously the iron content is reduced considerably from about 500 to 1 mg/l.

Other treatments include aeration, ion-exchange, activated sludge and deep-well disposal.

PARAMETERS TO BE DETERMINED

1) Color 2) Odor 3) Temperature 4) pH value 5) Acidity 6) Total suspended solids 7) Total dissolved solids 8) Dissolved fixed solids 9) BOD 10) COD 11) Iron: Digest and make up to a suitable volume. Use aliquots of the digested sample for estimation. Titrimetric method is preferable to colorimetric methods. 12) Manganese 13) Aluminium 14) Calcium 15) Magnesium 16) Sulfate 17) Chloride.

COAL PROCESSING WASTES

(Also refer Coal Washery Effluent in Chapter-9. "Steel Plant")

Coal processing involves the removal of impurities and cleaning of raw coal. The processing includes crushing, screening, classification and washing. Wash water is the only liquid effluent containing large quantities of suspended coal fines. The clays and shales present in coal are broken down to fine particles and they appear as semi colloidal particles in the effluent. The waste water also contains calcium and magnesium sulfates and iron. Treatment of the waste aims in reduction of suspended solids. Methods now practiced are i) settlement and impounding to collect the fine-size coal (silt) ii) clariflocculation and iii) froth floatation.

REFERENCES

1) U.S.Bureau of Mines, "Acid Mine water in the anthracite region of Pennsylvania", Technical Paper No. 710.
2) "Acid Mine drainage, summary report", Sewage Ind. wastes 26, 655 (1954).
3) Hall E.P. "Industrial Waste Water Control"-1965, Academic press, New York.
4) Deul, M. and E.A. Mihok, "Mine Water Research-Neutralization", U.S.Bureau of Mines. Chem. Abstr. 66, 1186262 (1967).
5) Dillon, K.E, "Waste Disposal made Profitable". Chem. Eng. 74, 146 (1967).

Chapter - 21

COFFEE PULPING

The coffee seed or coffee bean is obtained from coffee fruit (coffee cherry) after a process known as pulping. It is a process by which the skin, fleshy parts of the fruit, and the parchment* are removed.

In a coffee pulping house, coffee cherry is received from various places and placed in vats and discharged hydraulically to the pulping machine. Pulper separates the seeds from the fruit. The outer skin and fleshy parts of the fruit are discharged along with water as waste. The seed known as coffee bean with remaining little fleshy part is sent for washing. This wash water constitutes a major effluent.

The next stage is fermentation of bean (i.e.) the washed beans are heaped or placed in a tank as such under moist condition for about 12 to 48 hours. Fermentation is necessary to remove the flesh adhering to the parchment which otherwise cannot be removed. During fermentation, the insoluble protopectin is converted to pectin which in turn is converted to pectic acid (with some methanol). These conversions are carried out by the enzymes protopectinase and pectase-pectinase. The pectic acid is finally converted to calcium pectate which is soluble and can be easily removed by washing. These wash waters are discharged as waste.

---

* Parchment is the layer that immediately surrounds the silver skin of the coffee bean.

After washing, the beans are taken to drying yard either hydraulically or manually and dried out for several days in sunlight. Mechanical drying in addition to sun drying is also employed in some units. When the beans are dried sufficiently, they are milled to remove the parchment, then graded and packed.

EFFLUENTS AND THEIR CHARACTERISTICS

Three types of effluents are generated in coffee pulping.

1. Pulping waste.
2. Coffee bean wash water after pulping.
3. Fermentation waste (waste water from fermentation tank and wash water after fermentation)

Pulping Waste

This is an acidic effluent (pH varies from 5.0 to 5.5) with high suspended and settleable solids mostly the outer skin and pulp which are recovered and used as fuel or fertilizer. BOD of the waste is also high as it contains sugars. It possesses an offensive odor due to the fermentation of pulp. See Table 21-1.

Coffee bean Wash Water after Pulping

It is an acidic waste with high BOD and permanganate values. As the effluent contains completely organic matter, that too of plant origin, it is amenable for biological treatment (BOD: PV ratio = 3 : 1). See Table 21-1.

Coffee bean Wash Water after Fermentation

This is also an acidic waste with high BOD and

suspended solids. Like other wastes, it is also amenable for biological oxidation. See Table 21-1 for characteristics.

Table 21-1.
Characteristics of Coffee Pulping Effluents

Sl. No.	Characteristic	Pulping waste water	Coffee bean wash water after pulping	Fermentation waste (waste water from fermentation tank + wash water after fermentation)
i.	pH Value	5.2	5.8	5
ii.	Total Suspended Solids, mg/l	154	800	500
iii.	Total Dissolved Solids, mg/l	800	2600	2000
iv.	Dissolved Fixed Solids, mg/l	100	280	220
v.	Biochemical Oxygen Demand, mg/l	1100	4700	1900
vi.	Chemical Oxygen Demand, mg/l	2880	6800	2800
vii.	Permanganate value (4 hrs), mg/l	460	1520	650
viii.	Ammoniacal Nitrogen (as N), mg/l	2.24	7.84	-
ix.	Organic Nitrogen (as N), mg/l	5.6	16.8	-
x.	Soluble Tannin & Lignin (as Tannic acid), mg/l	10	15	-
xi.	Petroleum ether extractable matter, mg/l	8.4	2.4	-

TREATMENT

Preliminary screening followed by sedimentation is practised at present. For effective treatment, chemical coagulation after sedimentation may be employed. For still better results, secondary treatment with biological filters with high rates of application and recirculation may be applied.

PARAMETERS TO BE DETERMINED

1) Color 2) Odor 3) pH value 4) Total suspended solids: Effluent is very difficult to filter. Use glassmicrofibre filter paper and smaller aliquots of the sample 5) Settleable solids ml/l/hr 6) Total dissolved solids 7) Dissolved fixed solids 8) BOD : Dilutions ranging from 1 : 199, 1 : 499, 1 : 999 and 1 : 9999 have to be made. Seeding is not necessary. Adjust the pH of the sample. Remove suspended solids prior to BOD determination. Use appropriately diluted (about 10 times) sample for BOD dilutions 9) COD : Use 10 ml or 25 ml aliquots of the sample and 50 ml or 75 ml of 0.25 N.$K_2Cr_2O_7$ 10) Permanganate value (4 hrs) : Dilute the sample 10 times. Use 50 ml sample and 10 or 20 ml of $KMnO_4$ 11) Oils and Grease 12) Percent sodium: Need to be done only if the effluent is discharged on land.

Chapter - 22

DYE MANUFACTURE

Growth of dye industry is remarkable especially in the past 25 years and a variety of new dyes are introduced. These dyes are mainly used to dye fabrics. A small percentage is used to color foods, drugs and cosmetics. Dyes are classified in two ways, (i) according to the chemical composition and (ii) according to the application. Manufacturers view the dyes from the chemical aspect while the user of dyes group them according to methods of application. A classification of dyes according to the application is as follows.

(1) Acid dyes (2) Basic dyes (3) Azoic dyes (4) Direct dyes (5) Disperse dyes (6) Mordant dyes (7) Sulfur dyes (8) Vat dyes and (9) Solvent dyes.

Acid dyes are used to color wool, silk and other synthetic polyamide fibres from acidified solution. The fundamental chemical structure is that of the azo, triarylmethane groups such as- NO_2,-SO_3H, and-COOH.

Basic dyes are mostly amino or substituted amino derivatives frequently from the triaryl methane or xanthene class, and are applied mostly to paper. Azoic dyes are essentially colorless azo dye intermediates and they require further chemical treatment for coupling and diazotisation during dyeing of textiles. Direct dyes are used to dye cotton and other vegetable fibres and they are frequently from the azo class. Disperse dyes used for dyeing cellulose acetate and some of the synthetic fibres are simple azo dyes or insoluble amino anthraquinone dyes. Mordant dyes are applied principally to wool along with a mordant such as chromium, aluminium or iron. These dyes contain-OH or -COOH group attached to azo or anthraquinone complexes. Sulfur (or sulfide) dyes are employed to dye cotton. This class of dyes contain a chromophore with sulfur and dyed from sodium sulfide bath. Vat dyes possess a complicated chemical structure as indanthrenes and they are most widely used dyes for cotton as they exhibit exceptional fastness to light and alkaline washing. Solvent dyes are simple azos, triaryl methane bases or anthraquinones used to color cosmetics, shoes, oils and waxes.

The organic raw materials required for dye manufacture are numerous which are derived from petroleum and coal. The raw material sequence is represented below:

Petroleum and coal → *Hydrocarbons* → *Intermediates* →*dyes*

In addition to the petroleum and coal derivatives, the following inorganic and organic chemicals are used extensively.

Acids

Sulfuric, nitric, hydrochloric, hydrocyanic, acetic, formic etc.,

Alkalis

Caustic soda, soda ash,ammonia, hydarted lime, quick lime, caustic potash and alkyl amines etc.,

Salts

Sodium nitrite, sodium chloride, sodium sulfate, sodium hydrosulfite, sodium cyanide, potassiumchloride, aluminium chloride, copper sulfate etc.,

Gases

Chlorine, bromine, iodine, hydrogen, acetylene etc.,

Others

Ethyl alcohol, methanol, formaldehyde, alkyl halides iron, sulfur etc.,

The reactions involved in the manufacturing processes are amination, nitration, halogenation, sulfonation, oxidation, hydrolysis (or alkali fusion), alkylation, condensation and addition. Details regarding the manufacturing process of dyes are beyond the scope of this book. The characteristics of dye manufacturing effluent vary from one unit to another. Effluent from one unit may be entirely different from the other. The general characteristics are the intense color, residual dyes, presence of acids/alkalis and excess salts. Dye house effluents contain both organic and inorganic matter. Usually they posses high BOD, COD, and suspended solids. Some effluents have only low BOD, but 4 hrs permanganate value and COD will be exceedingly high which indicates the resistance of the effluent towards biological treatment. Table 22-1 indicates the characteristics of an azo dye manufacturing effluent.

EFFECTS OF EFFLUENTS

The most objectionable features of the effluent are color, BOD and dissolved solids which mineralize the receiving stream. The chlorides and sulfates are highly corrosive. For other details refer section, "Effects of Effluents" in Chapter-1. "Basic Information."

Table 22-1.
Characteristics of Effluent of an Azo Dye Manufacturing Unit

Sl. No.	Characteristic	mg/l
i.	pH Value	9.1
ii.	Total Suspended Solids	4430
iii.	Total Dissolved Solids	150000
iv.	Dissolved Fixed Solids	135000
v.	Permanganate value	3000
vi.	Ammoniacal Nitrogen (as N)	6.72
vii.	Chlorides (as Cl)	8465
viii.	Sulfates (as SO_4)	915
ix.	Percent sodium (as Na)	99.1
x.	Boron (as B)	2
xi.	Phenolic substances (as phenol)	0.12

TREATMENT

Treatment of dye manufacture effluent involves many stages and consequently expensive. Primary treatment includes screening, equalization, neutralization, chemical coagulation and clarification. It is to be mentioned that primary treatment removes only little color as the coloring materials are present in dissolved condition. Trickling filtration and activated sludge process are the commonly employed secondary biological treatments, but these also reduce the color by about 40 to 50% only. The remaining color has to be removed by chlorination (which needs enormous quantities of chlorine) or by activated carbon adsorption. Both methods are expensive. Inspite of all the treatments, the effluent still contains dissolved solids in excess concentrations which can only be removed by reverse osmosis or by electro dialysis.

PARAMETERS TO BE DETERMINED

1) Color 2) Odor 3) Temperature 4) pH value, 5) Acidity / Alkalinity (Potentiometric titration) 6) Total suspended solids 7) Total dissolved solids 8) Dissolved fixed solids 9) Permanganate value : Use 0.1250 N $KMnO4$ 4 hrs & 3 mts) 10) Chemical Oxygen Demand: only if the effluent contains chlorides less than 2000 mg/l) 11) Ammoniacal nitrogen 12) Organic nitrogen 13) Nitrate 14) Sulfite 15) Chloride 16) Sulfate 17) Mercury 18) Chromium 19) Zinc 20) Other heavy metals as per use in manufacture 21) Percent sodium 22) Boron 23) Phenolic substances 24) Residual chlorine (if necessary) 25) Oils and grease

REFERENCES

1) Gore T.S. etal (ed), "Chemistry of Natural and Synthetic Colouring Matters" - 1962. Academic press, New York.

2) Venkatraman, K (ed) - "The Chemistry of Synthetic Dyes" - Vol - 7 - 1974, Academic Press, New York.

3) N. Manivasakam, "Environmental pollution", 1984, National Book Trust, New Delhi.

Chapter - 23

EDIBLE OIL REFINERY

In an edible oil refinery, vegetable oils are subjected to alkali refining, bleaching and deodorization for refinement.

Alkali Refining

Oil is received in tankers and the free fatty acid content is determined. Then they are treated with calculated amount of caustic soda solution, in large mixing tanks. Excess free fatty acid reacts with caustic soda to form soap (commonly called as foots) and it settles down to the bottom. After settling, it is collected and allowed for further settlement as the foot contains considerable amount of unsaponified oil which is sent back to reaction tank. The soap is either sold out (to soap manufacturers) or discharged as effluent. This is the only effluent arise from edible oil refining. The oil (now free from free fatty acid) in the reaction tank is sent to bleach tank.

Bleaching

Bleaching is effected by passing the oil through a column containing a special type of adsorbent clay. No liquid effluent arises from bleaching. The exhausted clay is disposed of as solid waste.

Deodorization

It may be a batch or continuous process accomplished by blowing superheated steam (200 to 260°C) through the oil under a high vacuum. The steam is condensed and cooled in cooling spray pond for reuse. The over flow of cooling spray pond is another effluent. It contains oil in emulsified form.

(For salad oils, prior to deodorization, the bleached oil is subjected to winterizing, a treatment that consists of cooling the oil to a low temperature at about 5°C and filtering the crystallized stearin. Shortening is another process in which bleached oil is hydrogenated and bleached again before deodorization. These processes do not produce any waste).

Boiler blow-down and softener regenerant waste constitute separate effluent streams but their volume is much less.

CHARACTERISTICS OF EFFLUENTS

The liquid waste of edible oil refinery is highly alkaline, containing free and saponified oils. BOD and COD values of the effluent are abnormally high. Likewise suspended and dissolved solids contents are also high. Characteristics of a typical edible oil refinery are given in Table 23-1.

Table 23-1.
Characteristics of an Edible Oil Refinery Effluent

Sl. No.	Characteristic	mg/l
i.	pH Value	10.2
ii.	Total Suspended Solids	10500
iii.	Total Dissolved Solids	2500
iv.	Dissolved Fixed Solids	2000
v.	BOD (5 days at 20°C)	11500
vi.	Permanganate value (4 Hours)	1850
vii.	Chemical Oxygen Demand	22600
viii.	Percent sodium (as Na)	80
ix.	Oils and grease	6900

EFFECTS OF EFFLUENTS

See the section, "Effects of effluents" in Chapter-1 "Basic Information" under the heads alkalis, suspended solids, and dissolved solids.

If the effluents are disposed of into surface waters, free oils float on the surface of the water giving an unsightly appearance. The surface films also affect aeration. Further, putrefaction of these oils, results in anaerobic condition, releasing offensive odors.

If the untreated effluent is allowed to stagnate for a long period, it seeps slowly and affects the ground water in the vicinity. The sodium content of the effluent is exceedingly high and is deleterious to crops if discharged on to land. If the refining includes salting out, then the effluent contains enormous amounts of chlorides leading to soil deterioration and consequently affect crops (Refer also the section, "Effects of Effluents" in Chapter-5 "Tannery")

TREATMENT

The following treatment measures would be effective.

Primary Treatment

Plain sedimentation, coagulation with alum (if necessary pH may be adjusted prior to the addition of alum), settlement or clarification in a flocculator.

Secondary Treatment

Anaerobic digestion is preferable to aerobic methods. Needless to say that anaerobic treatment should be followed by aeration.

Tertiary Treatment

As the primary and secondary treatments cannot remove the dissolved solids, tertiary treatment is essential. Reverse osmosis and Electro dialysis are the effective and practicable methods.

PARAMETERS TO BE DETERMINED

1) Color 2) Odor 3) pH value 4) Alkalinity 5) Total suspended solids 6) Total dissolved solids 7) Dissolved fixed solids 8) Bio-chemical Oxygen Demand : Adjust the pH to about 7 before preparing dilutions. The following dilutions would be useful. 1:499: 1:999, 1:1999. 1:4999. Use appropriately diluted sample (10 times) for preparing BOD dilutions. Seeding is necessary. 9) Chemical Oxygen Demand. Dilute the sample 10 times or 20 times. Use 20 or 50 ml of the diluted sample and 10 ml, 25 ml or 50 ml of 0.25 N.$K_2Cr_2O_7$. If the chloride content of the sample exceeds 2000 mg/l do not conduct COD test. 10) Permanganate value: Use 10 ml or 20 ml of the sample and 25 to 50 ml 0.1250 N. $KMnO_4$. 11) Chlorides: If salting out is carried out, dilute the sample appropriately 10 to 50 times) and use the diluted sample for estimation. Adjust the pH to about 7 if necessary. 12) Sulfides 13) Mercury 14) Percent sodium 15) Oils and grease.

REFERENCES

1) Devine, J and P.N. Williams (ed), "Chemistry and Technology of Edible oils and Fats" - 1961, Pergamon press.
2) Andersen, A.J.C. and P.N. Williams (ed) "Refining of Oils and Fats for edible Purposes, "2nd edn. -1962, Pergamon press.

Chapter - 24

EXPLOSIVES MANUFACTURE

There are three fundamental types of explosives, namely, chemical, mechanical and atomic. This chapter describes only chemical explosives. Classification of explosives is given in the following Figure 24 - 1.

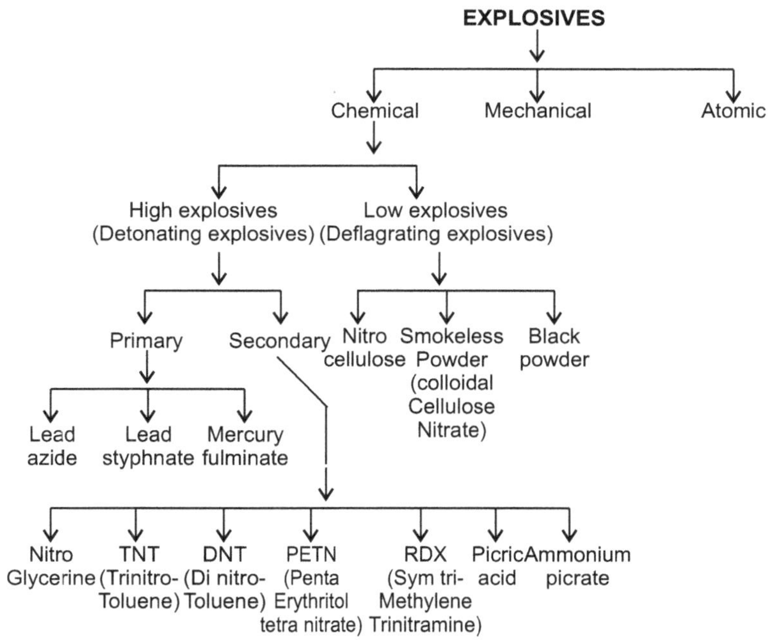

Fig 24-1. Classification of Explosives

MANUFACTURING PROCESSES, SOURCES & CHARACTERISTICS OF EFFLUENTS

Cellulose (tri) nitrate

It is produced by nitration of cotton linters (which are already purified by boiling with dilute caustic solution in vats and bleached) with nitric acid in presence of sulfuric acid as per the reaction given below.

$$C_6H_7O_2(OH)_3 + 3HONO_2 + H_2SO_4 \rightarrow C_6H_7O_2(ONO_2)_3 + 3H_2SO_4 + 3H_2O$$

Cellulose *Nitric acid*

The spent acid is separated by centrifugation and either fortified for reuse or sold out or discharged as waste.

The nitrated cellulose is drowned in water, boiled for the removal of acids and other impurities and again washed. This boiling and washing produce an acidic effluent.

In order to prepare smokeless powder which is more stable on storage, the cellulose nitrate is boiled for 40 hours with atleast four changes of water and pulped by means of a beater followed by poaching (i.e.) boiling with a dilute solution of soda ash (2 kgs per ton of cellulose nitrate) to remove the sulfate esters and washing many times with water. Residual water from the poached nitrocellulose is removed by centrifugation and subsequent alcohol percolation under pressure dehydration. Then it is disintegrated and colloided by mixing with alcohol and ether. A stabilizer such as diphenyl amine and plasticizer (dibutyl phthalate) are added to improve storage life.

Cordite, another, important explosive is cellulose

nitrate treated with a mixture of acetone, nitroglycerine and petroleum jelly.

The effluent streams in the production of cellulose nitrate and smokeless powders are:

1. Alkaline effluent from purification of cotton linters
2. Acidic effluent from boiling and washing of nitrated cellulose
3. Alkaline effluent from boiling of cellulose nitrate during the production of smokeless powder.
4. Effluent from solvent recovery system (i.e.) alcohol and ether used for colloiding cellulose nitrate.

Other effluents are from the following auxiliary units :

a) Nitric acid production b) Sulfuric acid production c) Diphenyl amine production (aniline is one of the constituents of the waste from this unit).

The characteristics of smokeless-powder plant wastes are given in Table 24-1.

Nitroglycerine

Pure glycerine is nitrated using a mixture of nitric acid and sulfuric acids, spent acid separated, impure nitro glycerine is washed first with water and then with 2% soda ash solution and again with water. These wash waters with some of the spent acid constitute the effluent.

Table 24-1.*
Characteristics of Effluents from Smokeless Powder Plants

Sl. No.	Characteristic	Plant A	Plant B
i.	pH Value	< 1.6	< 1.6
ii.	Acidity, mg/l		
	Methyl orange	1860	1540
	Phenolphthalein	1990	1610
iii.	Total suspended Solids, mg/l	53	190
iv.	Total Dissolved Solids, mg/l	-	851
v.	Dissolved Fixed Solids, mg/l	-	218
vi.	Bio-chemical Oygen Demand, mg/l	49.1	47.6
vii.	Oxygen consumed, mg/l	76.2	99.8
viii.	Nitrite Nitrogen, mg/l	2.7	1.9
ix.	Nitrate Nitrogen, mg/l	530	470
x.	Sulfate (as SO_4) mg/l	1280	1100

TNT (Trinitro toluene)

This important military explosive is produced by multiple stage nitration of toluene with a mixture of nitric and sulfuric acids. The nitrated product is then washed for removing the free acid, crystallized and purified with sodium sulfate. The impure beta and gamma isomers are washed out (as soluble sulfonates) from alfa trinitro toluene. The purified product is remelted and flaked. Wash water after nitration and sulfite purification are the two major effluents. The nitration water is highly acidic and slightly yellow colored. Sulfite

* Table 24-1 is reproduced from N.L.Nemerow "Industrial Water Pollutio-Origins, Characteristics and Treatment" with the kind permission of the publishers M/s. Addison - Wesley Publishing Company, Massachusetts.

purification waste water, also termed as red water is alkaline. Combined effluent is acidic and highly colored. Characteristics are given in Table 24-2.

In the production of arms and ammunition, metal pickling wastes, cutting oils and soaps are discharged. Acid pickling is done to clean the brass or steel cartridge. The effluents are colored, turbid and possess oily or soapy odor. The characteristics of these effluents are given in Table 24-3.

Table 24-2.
Characteristics of Effluents from TNT Plants*

Sl. No.	Characteristic	Plant A	Plant B
i.	pH Value	2.4	2.7
ii.	Acidity, mg/l		
	Methyl orange	291	134
	Phenolphthalein	485	178
iii.	Total suspended Solids, mg/l	166	29
iv.	Total Dissolved Solids, mg/l	2111	1129
v.	Dissolved Fixed Solids, mg/l	1780	1108
vi.	Oxygen consumed, mg/l	795	551
vii.	Nitrite Nitrogen, mg/l	15	20
viii.	Nitrate Nitrogen, mg/l	107	-
ix.	NH_3 Nitrogen, mg/l	5.3	-
x.	Sulfate (as SO_4) mg/l	672	604

* Table 24-2 is reproduced from N.L.Nemerow "Industrial Water Pollution-Origins, Characteristics and Treatment" with the kind permission of the publishers M/s. Addison - Wesley Publishing Company, Massachusetts.

Table 24-3.
Characteristics of Effluent from Small Arms Ammunition Plants*

Sl. No.	Characteristic	Plant A	Plant B	Plant C
i.	Acidity, (Methyl orange)	-	6	60
ii.	Alkalinity (Total)	213	-	-
iii.	Total Suspended Solids, mg/l	276	408	408
iv.	Total Dissolved Solids, mg/l	1301	684	1345
v.	Dissolved Fixed Solids, mg/l	1061	551	1044
vi.	Biochemical Oxygen Demand, mg/l	141	138	295
vii.	Oxygen consumed, mg/l	67	103	163
viii.	Copper (as Cu), mg/	31	22	86
ix.	Sulfate (as SO_4) mg/l	244	397	647
x.	Grease, mg/l	184	419	502

EFFECTS OF EFFLUENTS

1. These are highly acidic and exhibit the characters of acidic wastes (see the section, Effects of Effluents" in Chapter-1 "Basic Information". Also refer Chapter-14 "Acid Wastes").
2. They may contain organic solvents which cause fire hazards.
3. They contain organic substances which are extremely toxic.

* Table 24-3 is reproduced from N.L.Nemerow "Industrial Water Pollution-Origins, Characteristics and Treatment" with the kind permission of the publishers M/s. Addison - Wesley Publishing Company, Massachusetts.

TREATMENT

Smokeless Powders

Equalization and neutralization are the required pretreatment. Aeration and biological oxidation are reported to be the effective secondary treatments. The concentration of nitrate is an important factor if the waste is disposed of into surface waters as it may cause algal bloom.

TNT

The primary treatment involves neutralization. The color of TNT manufacturing waste is so strong and can be bleached at reasonable cost by chlorine only. Activated carbon treatment though effective is much costlier but it could simultaneously remove the residual TNT.

Ammunition waste

The essential pretreatments are grease removal and neutralization. Chemical coagulation is reported to be more effective in the removal of BOD and suspended solids and to a certain extent useful in increasing the pH.

PARAMETERS TO BE DETERMINED

For Smokeless powders & TNT

1) Color 2) Odor 3) pH value 4) Acidity: Use smaller aliquots of sample. Titrate with 1N. NaOH if acidity is more 5) Total suspended solids 6) Total dissolved solids 7) Dissolved fixed solids 8) Bio-chemical Oxygen Demand 9) Chemical Oxygen Demand 10) Permanganate value 11) Nitrite 12) Nitrate 13) Ammonia and organic nitrogen 14) Sulfates 15) Oils & grease.

For small arms and ammunitions

All parameters described above and in addition, the following: 16) Alkalinity 17) Copper 18) Zinc 19) Chromium, and 20) Soaps and detergents

REFERENCES

1) Urbanski, T., "Chemistry and Technology of Explosives" - 3 Vols-1964-1967, Pergamon Press.

2) Toolay, P., "Fuels, Explosives and Dyestuffs"-1971-J. Murray London.

3) Smit R. S. and W. W. Walter "Surveys of Liquid wastes from Ammunitions Manufacturing" U. S. Public Health Service, Reprint No.2506, Public Health Reports Nos. 58, 194, 1365 and 1393

4) N. L. Nemerow "Industrial Water Pollution-Origins, Characteristics, and Treatment-1978, Addison Wesley Publishing company, Massachusetts.

Chapter - 25

GELATIN MANUFACTURE

Gelatin is an organic colloidal protein obtained by the hydrolysis of collagen-white fibres of the connective tissues of animal body especially of bones and skin. There are four different types of gelatin, viz, edible, photographic, pharmaceutical and technical. The edible variety is used as a food, particularly for desserts. For preparing emulsions, photographic gelatin is used. Pharmaceutical gelatin is the basic material for making capsules. The technical grade is used for sizing paper, textiles etc.,

MANUFACTURING PROCESS

From Bones

The bones are degreased by heating with steam under pressure and then running off the grease layer. (Solvent extraction is also practiced in some industries). The degreased bones are treated with lime and water to make the gelatin preparation easier. The bones are then crushed and passed on to wooden tanks and treated with cold dil. hydrochloric acid (7-15%). The constituents of the bone, namely, calcium phosphate, carbonate and mineral matter are dissolved by the acid leaving collagen alone. The spent HCl is used for calcium phosphate recovery. The soft collagen termed as ossein is taken and soaked in milk of lime in concrete vats. Then they are washed thoroughly with fresh water and then with slightly acidulated water such that the pH of ossein is adjusted and made suitable for hydration.

The next important step is the series of extractions with water. The reaction takes place as indicated below.

$$C_{102}H_{149}N_{31}O_{38} + H_2O \rightarrow C_{102}H_{151}N_{31}O_{39}$$

Collagen Water Gelatin

The pH of the hydrolyzing liquid is adjusted to be 3 to 4 and the first extraction made with water kept at 60 to 65°C. After 8 hrs of contact (during which 10% gelatin extracted) the liquid is run into trays to gell the solution. The ossein is extracted atleast five times, every time increasing the temperature. From the third extraction itself the product will be colored and is bleached either with hydrogen peroxide or with sulfuric acid. The first extraction is the best material and used for photographic purposes and last extract yields material suitable only for glue. The cooled solid gelatin is cut and dried in dry air at about 40°C. It is estimated that to produce 1 ton of gelatin, about 3 tons of bones, 1.2 tonnes of HCl and 0.75 tons of lime are required. About 1.7 tons of calcium phosphate is obtained as by-product.

From skin fleshings

The fleshings obtained on defleshings of hides and skins during tanning or skins unsuitable for making leather are used for gelatin manufacture. The skins are soaked in water for the removal of sodium chloride. All other processes are the same as described for bones except the initial acid treatment.

SOURCES OF EFFLUENTS

1. Waste water from soaking of skins and hides for the removal of salt
2. Effluent from calcium phosphate recovery from the spent acid.
3. Effluent from lime treatment.
4. Ossein wash water (both fresh and acidulated water)
5. Bleach effluents
6. Spill overs and floor-wash waters and equipment cleaning waters etc.,

CHARACTERISTICS

The combined effluent is usually cream colored and possesses a disagreeable odor. pH value usually lies between 6 and 7, BOD and COD values are moderate, chloride content is high. The BOD: PV ratio is high indicating its amenability for biological treatment. See Table 25-1 for characteristics of a typical gelatin manufacturing industry.

EFFECTS OF EFFLUENTS

1. As the raw materials are bones, skins and fleshings they may harbour pathogenic organisms such as Anthrax bacilli which may eventually be discharged into water courses.
2. It imparts a disagreeable odor to the receiving waters.
3. As phosphates are the best nutrients, algal blooms occur when disposed of into surface waters.

4. Though the effluent is in near neutral range it exhibits corrosive properties due to excess chloride concentration.

Table 25-1.
Characteristics of a
Gelatin Manufacturing Industry Effluent

Sl. No.	Characteristic	mg/l
i.	pH Value	6.6
ii.	Total Suspended Solids	190
iii.	Total Dissolved Solids	4480
iv.	Dissolved Fixed Solids	3600
v.	Biochemical Oxygen Demand	340
vi.	Permanganate value (4 Hours)	120
vii.	Chemical Oxygen Demand	470
viii.	Ammoniacal Nitrogen	53.8
ix.	Organic Nitrogen	6.2
x.	Chlorides (as Cl)	2030
xi.	Phosphate (as PO_4)	40
xii.	Oils and grease	13.2

TREATMENT

These effluents as indicated already are easily biodegradable. Hence after primary coagulation and clarification, secondary biological oxidation treatment is applied. Either trickling filtration or activated sludge process is employed. The latter is reported to be more effective and economical. The effluent from secondary treatment, before discharge should be chlorinated to ensure hygienic safety. Mention has to be made that the chloride concentration of the effluent is not altered by the primary and secondary treatments and tertiary treatment is required.

PARAMETERS TO BE DETERMINED

1) Color 2) Odor 3) Temperature 4) pH value 5) Acidity/Alkalinity 6) Total suspended solids 7) Total dissolved solids 8) Dissolved fixed solids 9) BOD: Adjust the pH of the sample to neutral range if necessary. Prepare BOD dilutions ranging from 1:19 to 1:499. 10) Permanganate value 11) Ammoniacal Nitrogen: 50 ml of the aliquot is enough for estimation 12) Organic Nitrogen: use 50 ml aliquot for digestion 13) Chlorides: Dilute sample 10 times or more and use diluted sample for determination 14) Phosphates: Digest and dilute the sample (10 times) and use diluted sample for estimation.

Chapter - 26

METAL CONTAINER MANUFACTURE

Metal container is made from metallic sheets (usually of steel). The sheets received are normally coated with grease and oil on the surface which are removed by alkaline degreasing (i.e) treatment with alkali. The degreased metallic sheet is then subjected to acid pickling and subsequent washing. Then they are galvanized or electroplated or painted. For details of surface pretreatment and characteristics of the effluents, refer Chapter-9 "Steel Plant" and Chapter-10 "Electroplating Industry".

The finished metallic sheets are then cut into proper size, welded or joined. All these processes are purely dry processes. Final paint coating and container decoration do not generate any effluent except spillages. But the cleaning of equipments and scrubbing of stack gases from baking ovens generate wastes.

CHARACTERISTICS OF EFFLUENTS

These effluents are either extremely acidic or alkaline. They contain metal fines, oils and greases (both in free and emulsified form). The final paint coating waste contains organic or inorganic pigments, resins and solvents, latex and driers.

EFFECTS OF EFFLUENTS

Described in detail in Chapter-9 "Steel Plant".

TREATMENT

Good house keeping and process modifications help to reduce the pollution load to significant extent. For example, the sheets may be coated with some other agents that could be easily removed by water spray such that alkaline degreasing or solvent cleaning may be eliminated. Non-phosphatic detergents may be used for cleaning. In acid pickling, reuse of bath after replenishing with required chemicals, avoiding of dragout, improved rinsing with agitation and vibration, would be highly useful in reducing the volume of effluents and pollution load. Likewise final coating may be changed from oven-baked finish to ultraviolet cured finish to reduce substantially the pollution load due to paints and other surface coatings.

The metallic particles, pigment and resins are present in finely divided form, and chemical coagulation followed by clarification is the suitable method. It is reported that ferric chloride and lime are the most effective coagulants for these wastes. Refer Chapter-9 "Steel Plant" and Chapter-10. "Electroplating Industry" for details.

PARAMETERS TO BE DETERMINED

Alkaline degreasing

1) Color 2) Odor 3) Temperature 4) pH value 5) Alkalinity 6) Total suspended solids 7) Total dissolved solids 8) Dissolved fixed solids 9) BOD : Adjust the pH before preparing BOD dilutions. Seeding is necessary. 10) Permanganate value 11) COD 12) Phosphates 13) Percent sodium 14) Boron 15) Oils and Grease 16) Mercury.

Acid Pickling

1) Color 2) Odor 3) pH value 4) Acidity 5) Total suspended solids 6) Total dissolved solids 7) Dissolved fixed solids 8) Iron 9) Zinc 10) Sulfate 11) Chloride 12) Permanganate value 13) COD : If the chlorides exceeds 2000 mg/l, do not perform this test 14) Oils and grease.

Electroplating

See Chapter-10 "Electroplating Industry".

REFERENCES

1) Environmental Protection Agency, "Upgrading Metal Finishing Facilities to reduce Pollution, EP 710 M56/No. 1/974 - 1974.

Chapter - 27

METAL PICKLING WASTE

For iron pickling see Chapter-9 "Steel Plant" and Chapter-10, "Electroplating Industry" in Part II.

Copper pickling

In brass* and copper industry, during the metal fabrication, they are annealed and then pickled to remove the oxide scale and impurities to prevent damage to the dies and entry of the scale into the finished product.

Pickling is done in a bath containing 5 to 10% H_2SO_4 (by volume) so that oxide scales are removed. The metal still contains stains which are removed by dipping in "bright dip" solutions which contain about 60 gms sodium dichromate per litre apart from 5 to 10% H_2SO_4. The metal after pickling and bright dip are rinsed well with fresh water. The drag out and rinse waters are the major effluents. The pickle liquors after exhaustion become waste. They are not discharged but dumped. Characteristics of pickle-bath waste and bright dip waste are given in Table 27-1. The metals are recovered by ion-exchangers or precipitated under alkaline conditions. The dilute rinse wastes are neutralized and disposed of.

---

* Brass is an alloy of copper (about 65%) and zinc (35%) with small amounts of tin and lead.

Table 27-1.*
Characteristics of Pickle-Bath Waste and Bright Dip Waste

Sl. No.	Characteristic	Amount in gm/litre	
		Pickle bath waste	Bright-dip waste
i.	H_2SO_4	59.7 - 163.5	5.6 - 85.8
ii.	Copper	4.0 - 22.6	6.9 - 44.0
iii.	Zinc	4.3 - 41.4	0.2 - 37.0
iv.	Chromium (hexavalent)	0 to 0.56	4.3 - 19.1
v.	Chromium (Total)	-	13.5 - 47.7
vi.	Iron	0.1 - 0.21	0.03 - 0.36

REFERENCES

1) Wise, W., "Character, Treatment and Disposal of Metal Processing Waste" - Sewage, Industrial wastes 20,96 (1948).

*Reproduced from N.L. Nemerow, "Industrial Water Pollution- Origins, Characteristics and Treatment-1978, with the kind permission of the publisher M/s. Addison-Wesley Publishing Company, Massachusetts.

Chapter - 28

PESTICIDE MANUFACTURE

Pesticides are broadly classified into three groups (1) Chlorinated hydrocarbons (2) Organo phosphorus compounds and (3) Carbamate pesticides. In addition, some of the inorganic fluorine compounds and mercuric compounds are also classified as pesticides.

DDT, BHC, dieldrin, aldrin, endrin, heptachlor, 2, 4-D are some of the chlorinated hydrocarbons.

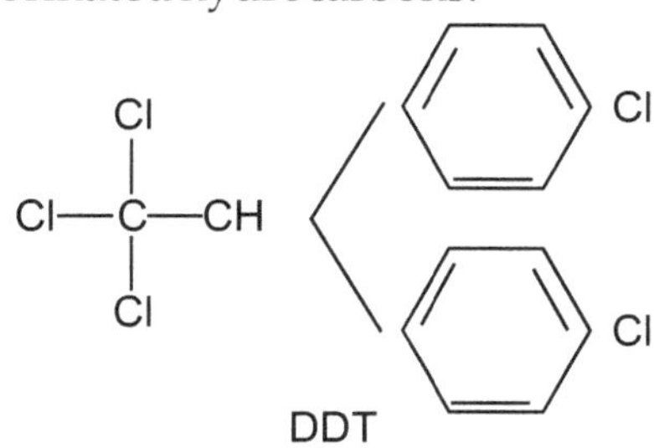

DDT
Dichloro Diphenyl trichloro ethane

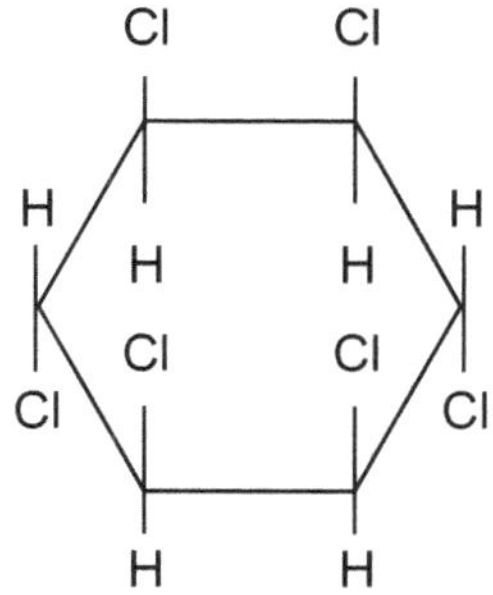

BHC
Benzene hexachloride (lindane)

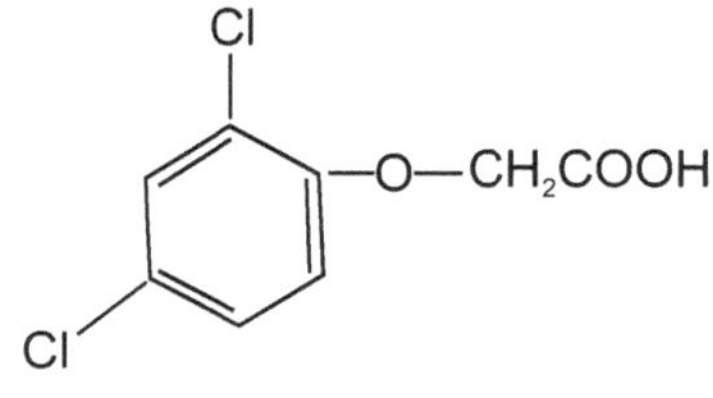

2, 4-Dichlorophenoxyacetic acid
(2,4-D)

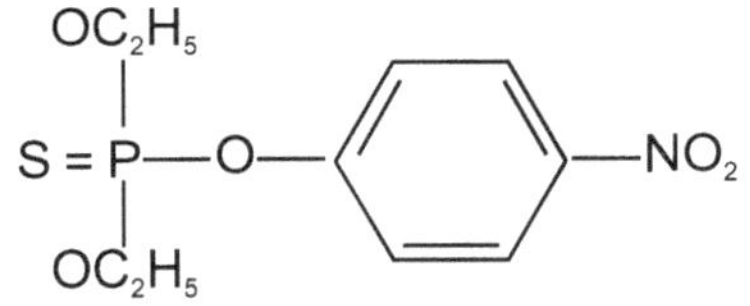

O, O - Diethyl - O - p-Nitrophenyl thiophosfate (Parathion)

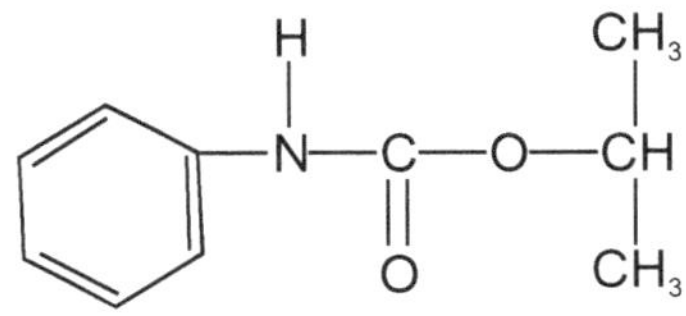

IPC
Isopropyl N-phenyl carbamate

Parathion is a typical example for organophosphorus compounds. The widely used carbamate pesticide is IPC.

Description of the manufacture of each pesticide is beyond the scope of this book. Most of the reactions are carried out in closed systems. Effluents arise only from purification of crude pesticides for the removal of other unwanted contaminants and from cooling. Spill overs and wash waters are also important effluents. They are combined with the production effluents. Generally, the pesticide manufacturing wastes are acidic and contain excess concentration of the pesticide and solvent used for its manufacture. It is estimated that about 2500 litres of acid is produced per ton of DDT manufactured which is discharged as waste. This spent acid contains 50 to 60% sulfuric acid, 20% ethyl hydrogen sulfate and the remaining is chloro benzene sulfuric acid. Wash waters are also reported to contain acid.

Stutz[2] reports that wastes from parathion manufacture are strong, highly acidic (pH about 2 to 2.5) and mineralized.

Table 28-1 illustrates the characteristics of a BHC manufacturing waste.

**Table 28-1.
Characteristics of Effluent of a
BHC Manufacturing Unit**

Sl. No.	Characteristic	mg/l
i.	pH Value	3.5
ii.	Acidity (as $CaCO_3$)	
	a) Methyl orange acidity	360
	b) Total acidity	440
iii.	Total Suspended Solids	10
iv.	Total Dissolved Solids	880
v.	Dissolved Fixed Solids	250
vi.	Chemical Oxygen Demand	70
vii.	Chlorides (as Cl)	530
viii.	Benzene Hexa Chloride (BHC)	56.3

EFFECTS OF EFFLUENTS

(Also refer Chapter 1. "Basic Information")

The toxicity of the pesticides is well known and need no explanation. However the following aspects should be borne in mind. When the effluents are discharged into surface waters, the pesticides tend to concentrate in aquatic plants and animals and are passed on to human beings via food chain. Most of them are proven carcinogens (cancer producing agents). As the effluents are acidic in nature, they exhibit the characteristics of acid in wastes. See Chapter-14, "Acid wastes" and Chapter-1, "Basic Information"

TREATMENT

The efficient methods like activated carbon adsorption, and alkaline chlorination are expensive. Dilution with water and disposal into sea is the cheapest method. But places at where this type of disposal is not possible, concentrated wastes are dumped. In some units neutralization and plain sedimentation are employed, but this will not remove the residual pesticides effectively. Disposal into municipal sewer is also not advisable as the toxic pesticides and other constituents may hamper the biological treatment.

PARAMETERS TO BE DETERMINED

1) Color 2) Odor 3) pH value 4) Acidity/Alkalinity 5) Suspended solids 6) Total dissolved solids 7) Dissolved fixed solids 8) Chemical Oxygen Demand 9) Chlorides 10) Sulfates 11) Oils and grease 12) Residual Pesticides : i) Chlorinated Hydrocarbons ii) Organophosphorus compounds iii) Carbamates iv) Other compounds.

The most perfect method for the determination of pesticides is gas chromatography. It is also easy to carryout. In the absence of gas chromatography, colorimetric methods may be employed which are tedious, time consuming and less accurate. For colorimetric determinations refer "Official methods of Analysis"-11th ed, (1970) by Association of Official Agricultural Chemists.

REFERENCES

1) Grindlay, J."Effluent Disposal in DDT manufacture" Ind Chem., 26, (1950)

2) Stutz, C.N., "Treating Parathion Wastes" Chem.Engg. Prog 62, 82 (1966).

Chapter - 29

PHARMACEUTICAL INDUSTRY

Pharmaceutical industry produces varied type of products. They range from vitamins, synthetic drugs to antibiotics. The processes employed are much more similar to that of chemical manufacture. The raw materials used are also numerous which include both organic and inorganic compounds. Some of the pharmaceutical plants do not generate any liquid effluents, while some others discharge little quantities of strong waste and others let out larger volumes.

Due to these wide variations a generalization cannot be drawn on the effluents of pharmaceutical industry. Most of the antibiotics such as penicillin, streptomycin, lysine, sulfaquinazoline, nicarbazine and vitamins such as B_1, B_2, B_{12} and many steroids (such as cortisone) are produced in the fermentation. The most troublesome waste produced in the fermentation process is the spent beer liquor. The spent beer liquor is the fermented broth remaining after the recovery of antibiotics and other valuables. It contains large amounts of organic materials, proteins and other nutrients and consequently the BOD of these effluents is abnormally high. Brown[1] reports five main pharmaceutical wastes and their characteristics are as follows:

1. Strong fermentation beers: Small in volume but having 4000 to 8000 ppm BOD.
2. Inorganic solids: Waste slurry with little BOD.

3. Washings of floor and equipment: Large percentage of total volume and BOD from 600 to 2500 ppm.
4. Chemical waste: Solution or solvents which exert a substantial BOD when diluted with other wastes.
5. Barometric condenser waste: Resulting from solids and volatile gases being mixed with condenser waste causing 60 to 120 ppm BOD.

Pollutional characteristics of the effluent from an antibiotic manufacturing unit is given in Table 29-1.

The antibiotic wastes impart objectionable odors to stream and inhibit biological population and action. If they are discharged into sewer, they must be properly diluted, otherwise they affect the sewage treatment.

Table 29-1.
Characteristics of the Effluent of an Antibiotic Manufacturing Unit*

Characteristics	General antibiotic Wastes	Special antibiotic wastes	
		Terramycin	Penicilin
BOD, ppm	1500-1900	20,000	800-13000
Suspended solids, ppm	500-1000	10	–
pH value	1-11	9.3	2-4

The characteristics of synthetic drug effluents are varied and complex, as large number of different types of

* Reproduced from N.L. Nemerow, "Industrial Water Pollution Origins, Characteristics and Treatment" – 1978, with the kind permission of the publishers. M/s.Addisson–Wesley Publishing Co, Massachusetts.

chemicals is used. The volume and composition vary from unit to unit. Approximately 1000 to 3000 litres of waste will be discharged per 100 kg of products manufactured. No specific conclusion on the characteristics of the effluents can be drawn. In general, they are either highly acidic (such as from the manufacture of organic intermediates) or highly alkaline (such as from the manufacture of sulfa drugs) and possess a high BOD and COD. Some of the effluents (as from the manufacture of sedatives) contain toxic substances like cyanides.

If the wastes are discharged into stream, they deplete the dissolved oxygen immediately. These are also corrosive due to their high acidity/alkalinity. Further, some of the substances present in them are toxic to aquatic life.

Discussion regarding radiological products is beyond the scope of this book.

TREATMENT

Antibiotic Wastes

Equalization, neutralization and clarification are the essential steps involved in the primary treatment of these wastes. Anaerobic digestion and controlled aeration are proved to be the effective secondary treatments. Activated sludge and oxidation ditch are also employed in some pharmaceutical manufacturing units. The effluent from secondary treatments may be passed on to sand filters to produce effluent of better quality.

At times, some of the antibiotic wastes are evaporated and incinerated. Residues from penicillin and other antibiotics are dried and used in stock food. It is reported that a vacuum dried mycelium from the manufacture of penicillin can be digested to produce methane while reducing the organic matter content by about 55%.

Synthetic Drug Wastes

The type of treatment largely depends on the products manufactured. Due to the varied characteristics of wastes from different sections of the plant, a careful pilot-plant study is essential. Segregation of different waste streams is a preliminary step in the treatment. Acidic wastes are neutralized with lime. Odor producing wastes are chlorinated. Cyanide bearing effluents are subjected to alkaline-chlorination. Secondary treatments include biological oxidation with acclimatized micro organisms.

PARAMETERS TO BE DETERMINED

1) Color 2) Odor 3) Temperature 4) pH value 5) Acidity/alkalinity 6) Suspended solids 7) Dissolved fixed solids 8) BOD: Substances inhibiting biological activity may be present. In these cases the BOD value has to be interpreted carefully. 9) COD 10) BOD:COD 11) Specific components such as cyanide etc.,

REFERENCES

1) J.M Brown, "Treatment of pharmaceutical Wastes". Jour. Sewage & Industrial Wastes, 23, 1017 (1951).

2) N.L.Nemerow, "Industrial Water Pollution - Origins, Characteristics and Treatment - 1978, Addison-Wesley publishing Co. Massachusetts.

3) M.N.Rao & A.K.Datta, "Waste Water Treatment" - 1979, Oxford & IBH publishing Co, Calcutta.

Chapter - 30

POULTRY

Poultry processing differs from the processing of other meats. The poultry operations include, receiving, killing, scalding, defeathering, singeing, evisceration, washing, chilling and packing.

The receiving area is usually contaminated with excreta, feathers and spilled feed. They are collected and used as manure. Cleaning of the cages produce waste of high strength which is used for irrigation purposes. The broilers are hang line on a moving chain which delivers them to the killing section where their throats are cut. The blood is collected in troughs and sent to by-products unit or rendering plant. The blood-tunnel wash water is the strongest waste. It is reported that BOD of blood of chickens is more than 90000 ppm. After killing, the chickens are conveyed to a scalding tank. This scalder waste is another major source of pollution as it contains blood, feather, manure and dirt adhering to the bird. The temperature of water is about 55 to 60°C.

The birds are then defeathered by machines, flushed with water, singed to remove pin feathers, washed and eviscerated and washed thoroughly. Washing after evisceration requires large volumes of water which are subsequently discharged as waste. (The heart, liver and gizzard are removed and processed separately which also generate significant quantities of wash water as waste). The offals, head, feet and scraps are collected separately during these operations and sent to rendering plants and are converted into tankage for feed additives or fertilizers. The dressed bird is then cut, frozen or refrigerated, weighed and packed.

CHARACTERISTICS OF EFFLUENTS

The effluent from poultry dressing operations contains varying amounts of blood, feathers, fleshings, fat, digested and undigested foods, manure and dirt. Blood and manure are the major contributors for such a high pollution load of the effluent. The characteristics of a combined effluent of a typical poultry is described below. (Table 30-1).

Table 30-1. Characteristics of a Combined Poultry Plant Effluent*

Characteristics	Result
pH Value	6.5 to 9.0
Total solids, mg/l	350 - 3200
Suspended solids, mg/l	100 - 1500
Suspended solids, % of total solids	20 - 50
Dissolved solids, mg/l	200 - 2000
Volatile solids, mg/l	250 - 2700
Volatile solids, % of Total solids	65 - 85
Total Nitrogen, mg/l	15 - 300
Total alkalinity, mg/l	40 - 350

* U.S. Department of Health Education and Welfare, Technical Report W62-3.

EFFECTS OF EFFLUENTS

Poultry effluents are highly organic in nature and may deplete oxygen and cause putrid odors. Further, they may harbour pathogens and pass on to receiving streams. Untreated poultry effluents if discharged into sewer will overload the sewage treatment systems resulting in an upset.

TREATMENT

Poultry plant wastes are amenable for biological treatment. The troublesome materials such as feathers, head, feet etc., are removed then and there in the process itself. Other suspended matter of higher size are removed by screens prior to biological treatment. Septic tanks and lagoons are the common methods in biological treatment which are reported to remove about 93% of BOD.

PARAMETERS TO BE DETERMINED

1) Color 2) Odor 3) pH value 4) Suspended solids 5) Total dissolved solids 6)Dissolved volatile solids 7) Biochemical Oxygen Demand 8) Ammonia Nitrogen 9) Organic Nitrogen 10) Oils and grease.

Chapter - 31

RUBBER INDUSTRY

There are two types of rubber namely (i) natural and (ii) synthetic. The natural rubber is obtained from the rubber tree as latex. Synthetic rubber is made by condensation of butadiene with different monomers. There are several types of synthetic rubbers. SBR (GR-S), Nitrile-NBR, Neoprene-CR, Butyl IIR, Thiokol A and Silicone R are a few to mention. They differ by their physical and chemical properties and are suited for specific purposes. Detailed discussion on the production of natural rubber and the above mentioned synthetic rubbers is beyond the scope of this book. A short description on the characters of the wastes generated during their production is given below.

Natural rubber production yields an odoriferous waste of high BOD. It is reported that this obnoxious odor exists for a stretch of several hundred miles from the point of waste discharge and also render the water unpalatable. It also imparts an undesirable taste to the receiving waters.

Production of synthetic rubber involves two stages i) production of synthetic latex and (ii) coagulation of latex either in an acid - brine solution or with alum. The produced rubber is washed several times and dried. The liquid effluent consists of the coagulation liquid (acid + brine) and a little amount of coagulated rubber and organic materials that are not polymerized. The characteristics of synthetic rubber wastes reported by Black[1] and Rostenbach[2] are given in Table 31-1 and 31-2.

The discussions made till now are regarding the wastes evolved from rubber production. Other rubber wastes arise from the production of rubber commodities. The various processes involved are washing, compounding, calendaring and curing. Large volumes of wash water are discharged as waste which contain the impurities of crude rubber.

TREATMENT

The major problem in treating rubber wastes is the removal of odor. Chlorination and sulfonation are the effective treatment methods. Phenolic substances are also removed by chlorination. Activated carbon adsorption is more effective but expensive. Biological treatments (trickling filtration and activated sludge) are equally effective and useful in reducing BOD to a greater extent. Addition of nutrients containing nitrogen and phosphates is important in biological treatment. Alternatively combined treatment with domestic sewage may be practised.

Table 31-1.
Analysis of Typical Wastes from the Synthetic Rubber Industry*

Sl.No.	Plant area	App. daily discharge mg/d	pH	Total solids ppm	Suspended solids, ppm	5 day BOD ppm	Odor concentration
1.	Butadiene waste	1.90	2.8	300	27.6	2550	16100
2.	Styrene waste Copolymer plant	4.62	6.2	150	4.5	180	690
3.	Process waste	2.34	4.3	5580	12.3	69	62
4.	Recovery and Reactor	0.39	8.0	570	23.6	492	8760
5.	Main sewer+	2.63	7.0	6530	46.0	168	930
6.	Main out fall institute++	119.40	5.4	270	15.5	81	460

+ Represents the mixture of 3 and 4.
++ Represents the mixture 1,2, and 5 plus large quantities of condenser water.

* Table 31-1 is reproduced from N.L.Nemerow "Industrial Water Pollution" - Origins, Characteristics and Treatment (1978) with the kind permission of the publishers M/s. Addison Wesley Publishing Company Inc., Massachusetts.

Table 31-2.
Wastes from Combined Operations of Three Plants Butadiene, Styrene and GR-S*

Sl. No.	Characteristics	Value of concentration
i.	pH Value	8.6
ii.	Oil, ppm	18
iii.	Settleable solids ml/litre	1.4
iv.	Odor threshold	16
v.	Flow, gpm	
	Styrene plant	190
	GR-S plant	735
	Butadiene plant	1120
	TOTAL	2045

REFERENCES

1) Block, O.R. "Study of wastes from synthetic rubber industry" Sewage Works J.18,1169 (1946)

2) Rostenbach R.E., "Status report on synthetic rubber wastes", Sewage Ind. Wastes 24,1138 (1952)

3) Maurie, M."Rubber Technology" - 1973, Van Nostrand-Reinhold.

* Table 31-2 is reproduced from N.L.Nemerow "Industrial Water Pollution" - Origins, Characteristics and Treatment (1978) with the kind permission of the publishers M/s. Addison Wesley Publishing Company Inc., Massachusetts.

Chapter - 32

SLAUGHTER HOUSE & MEAT PACKING

Three major sources of wastes of meat industry are from (1) stockyards (2) slaughter house and (3) packing house. The stockyard wastes consist primarily of animal manure with urine, straw and unconsumed food along with dirt and water used for cleaning. The stockyard wastes are strong and possess a very high BOD. Killing, dressing and some by-product processing are carried out in the slaughter house. The animals are stuck and bled on the killing floor. Blood is collected and sent for recovery. Flushing of killing floors contributes significant volumes of strong waste waters. Then hides and skins are removed, salted and piled in separate place. Viscera, paunch materials, head, feet, and bones are sent for edible rendering unit. Other bones are sent for glue manufacture. Most of the slaughter houses do possess their own inedible rendering plants where offal and other animal wastes are converted to grease and animal feed. Edible (wet) rendering generates an effluent of very bad quality.

In packing house, meat is further processed by cooking, curing, smoking and pickling. The other processes carried out in packing house are meat canning, sausage manufacture, rendering of edible and inedible fats to edible and inedible tallow respectively.

The effluent from killing floor is deep reddish brown in color, possesses a high BOD and suspended matter. (See Table

32-1). Mention has to be made that blood is highly nitrogenous and decomposes rapidly. If wet process is carried out in a rendering unit, the generated effluent will be strong as it contains excess of nitrogenous organic matter. Wash waters of the rendering plant also contribute to pollution load.

Table 32-1.
Characteristics of a Slaughter House and Meat Packing Waste*

Source	pH	BOD mg/l	Suspended solids mg/l	Organic Nitrogen mg/l
Killing floor	6.6	825	220	134
Blood and tank water	9.0	32000	3690	5400
Scalding tub	9.0	4600	8360	1290
Meat cutting	7.4	520	610	33
Gut washer	6.0	13200	15120	643
Sausage department	7.3	800	560	136
Lard department	7.3	180	180	84
By-products	6.7	2200	1380	186

EFFECTS OF EFFLUENTS

As the effluents are highly organic in nature, they deplete the oxygen of receiving waters and result in anaerobic conditions. The suspended impurities settle down to the

* Reproduced from N.L.Nemerow, "Industrial water pollution Origins, Characteristics and Treatment"- 1978 with the kind permission of publishers M/s, Addison-Wesley Publishing Co, Massachusetts.

bottom of water courses. Further, the water course gets discolored. Above all, it may transmit pathogenic organisms, (Also refer Chapter-I. "Basic Information").

TREATMENT

These effluents are amenable for biological treatment. Screening (on a fine mesh), sedimentation, chemical coagulation and clarification are the steps involved in the primary treatment.

Either trickling filtration or activated sludge process may be employed as secondary treatment. Both are equally efficient. To obtain better quality effluent in an activated sludge process, extended aeration (for 8 to 10 hrs) would be useful. To ensure safety, final effluent may be appropriately chlorinated before discharge into water courses. The treated effluents may be utilized for irrigation.

PARAMETERS TO BE DETERMINED

1) Color 2) Odor 3) pH value. 4) Total suspended solids 5) Total dissolved solids 6) Dissolved fixed solids. 7) Biochemical Oxygen Demand: Dilutions ranging from 1:199 to 1:2000 have to be made. Seeding is not necessary. Use appropriately diluted sample (10 times) for BOD dilutions. 8) Chemical Oxygen Demand 9) Ammonia Nitrogen 10) Organic Nitrogen: Use smaller volumes of sample for digestion and collect distillate in 2% boric acid solution until free from ammonia. Titrate with 0.02 N. H_2SO_4 11) Oils and grease 12) Test for pathogens.

If the effluent is disposed on to land. 13) Percent sodium and 14) Boron.

REFERENCES

1) N.L. Nemerow, "Industrial Water Pollution-Origins, Characteristics and Treatment" 1978, Addison - Wesley Publishing Company Inc., Massachusetts.

Chapter - 33

SOAPS & DETERGENTS MANUFACTURE

Soaps and detergents are exclusively used as cleaning agents in houses, laundries and industries. Soaps are prepared from edible fatty acids and tallow, while detergents are made from various chemicals. The manufacturing processes of soaps and detergents are entirely different. However mention has to be made that volume of liquid effluents generated from the manufacture of both of them are only little but are strong.

Soaps are formed by the reaction between edible oils (including tallow) and alkalis. The reaction may be carried out either in cold or hot condition depending on the facilities and final products. In some units glycerine is obtained as a by-product on hydrolysis of the fatty acids which are then allowed to react with alkali. Builders such as soda ash, phosphates and silicates of sodium, borax and talc are added for softening water and to increase the cleaning efficiency.

The effluents from soap manufacture are highly alkaline in nature and contain free fatty acids and possess a very high BOD and COD. Percent sodium usually exceeds 90%. If salting out is carried out, then the chloride content will also be much higher.

The major step in the treatment of soap manufacturing waste is the recovery of floatable fatty acids by use of trap tanks. Other steps include primary settlement and neutralization. The waste should be discharged only after secondary treatment. It is to be stated however that whatever may be the treatment given, the dissolved solids content is not at all reduced.

Detergents are a class of surface active agents and grouped as anionic, cationic and non-ionic. The widely used type is anionic which ionize in aqueous solution giving a negatively charged group and a positively charged ion. It is the active ingredient present in household and industrial cleaning agents. Typical examples are primary and secondary sodium alkyl benzene sulfonates. Cationic detergents are salts of strong quaternary organic bases which ionize to give a positively charged quaternary ammonium or pyridinium group (cation) and a negatively charged group. Cetyl pyridinium bromide is the well known cationic syndet. It is used for special purposes only. Non-ionic detergents do not ionize. Their use is on the increase. Polyglycol ethers of alkylated phenols belong to this class.

Detergents exhibit their cleaning action over wide pH ranges and they do not form insoluble curds and precipitate as soaps. They do possess wetting and emulsifying properties and are extensively used in textile, metal, leather, and rubber industries.

The most widely employed raw materials for the manufacture of syndets are alkyl benzene sulfonate, sulfuric acid and fuming sulfuric acid. Description of various processes of manufacturing of these syndets is beyond the scope of this book.

The effluents contain sulfuric acid, alkyl benzene sulfonate and other fatty alcohols and builders such as sodiumtripolyphosphate, sodium hexa metaphosphate and tetra sodium pyrophosphate. The characteristic feature of these effluents is their non-biodegradability. (Recently developed bio-degradable active component is tetra propylene benzene sulfonate). BOD values of these effluents are exceptionally high. Discharge of these effluents in a stream is highly undesirable for the following reasons.

i) Affect aeration and reoxygenation of the stream.

ii) They cause excess frothing on the surface.

iii) They are toxic to certain flora and fauna.

iv) They are not easily removable and may gain access to drinking water.

The suggested treatment is equalization, pH adjustment, chemical coagulation (if necessary), settlement and activated sludge oxidation.

REFERENCES

1) Bogan, R.H. "Biochemical oxidation of Synthetic Detergents", Proceedings of 10th Industrial Waste Conference, Purdue University, Series 80,1955.

2) Garner. W., "Detergents" - 1966, Elsevier Publishing Corporation.

3) Jones, H.R. "Detergents and Pollution Control" - 1963, Noyes Data Corporation.

Chapter - 34

SOFT DRINKS MANUFACTURE

In a soft drink manufacture plant, liquid effluent arises from washing of bottles, spillages from syrup produced, washing of floors and from water treatment plants. Bulk of the effluent is contributed by washing of bottles only. As detergents are used for washing, the combined effluent is alkaline. Suspended impurities occur due to the labels, paper and straw left in the bottle. But this problem is eliminated by direct printing on glass instead of using labels. The used straws are disposed of in the consumer point itself.

BOD of the effluent is high due to the residual drinks in the bottle. (It is to be mentioned that soft drinks possess a BOD ranging from 60000 to 85000 due to the sugars and syrup constituents). The characteristics of a carbonated beverage waste is given in Table 34-1.

Table 34-1.
Characteristics of a Carbonated-Beverage Waste*

Sl. No.	Characteristics	Amount
i)	pH value	10.8
ii)	a) Phenolphthalein alkalinity, mg/l	150
	b) Total alkalinity, mg/l	290
iiiv)	5 day BOD, mg/l	430
iv)	Suspended solids, mg/l	220
v)	Waste Volume, gal/1000 cases	10600

* Reproduced from, N.L.Nemerow, "Industrial Water Pollution Origins, Characteristics and Treatment" with the kind permission of the publishers, M/s. Addison- Wesley Publishing Company, Massachusetts.

The soft drinks bottling effluents are generally discharged into sewers as they are located in towns and cities. Separation of residual drinks reduces the BOD of the effluent to a significant extent. In some plants, wastes are utilized for irrigation after dilution.

REFERENCES

1) Porges and E.J.Struzeski, "Wastes from the soft drink bottling industry", J. Water Pollution Control Federation, 33, 167 (1961): Proceedings of 15th industrial waste conference, Purdue University, May 1960, p.331.

Chapter - 35

STEAM POWER PLANTS

The steam power plants are noted especially for their heated effluents. They may be component unit in industries or may be installed by municipalities and other authorities as a separate unit to supply electricity to public. The major process in a power plant is the production of steam, which drives turbines to generate electricity. Coal is the normally employed fuel. Oil is also used to a lesser extent. The following wastes are generated in a steam power plant.

1. Cooling water: This contributes to the major portion of effluent. Generally cooling waters are treated with phosphates and silicates and with disinfecting agents to control corrosion and to prevent algal growth and slime formation respectively. Though cooling waters are recycled, they are also discharged at frequent intervals. They do not possess any toxic substance. Temperature of cooling water is the only problem for disposal. Hence it must be cooled and then only discharged. The effects of heated effluents are described in detail in the section "Effects of Effluents".

2. Cooling Tower Blow down: This contains minerals in excess concentration along with atmospheric dust.

3. Boiler Blow down: This is concentrated saline waste and contains precipitated sludge. This waste will always be alkaline (pH about 11). It also contains antifoam agents.

4. Water Treatment Plants Wastes: It contains acids and alkalis used for regeneration of ion exchange resins for softening and demineralization.

5. Water from Power Plant Cleaning: Periodical cleaning of equipments with acids and alkalis contribute significant volume of effluents. Cleaning is carried out three or four times a year to remove scales from tubes and boiler drum surfaces. The cleaning agents commonly used are hydrochloric acid, acetic acid and detergents. Potassium bromate and ammonia are also used to a certain extent.

6. Smoke Stack Scrubber Wastes: When pulverized coal is employed as a fuel, fly ash is produced. The quantity of fly ash produced is about 10% of coal burnt. It is removed from the flue gases by use of a wet scrubber (alkaline solution), or cyclone separator or electrostatic precipitator. Fly ash is then pumped as a slurry to settling basins. The separated liquor is an important waste. Needless to say that fly ash is a troublesome solid waste. The fly ash is reported to contain carbon (1 to 60% by wt), iron, aluminium, calcium, magnesium, titanium, silica, sulfur and phosphorus.

7. Drainage: Drainage from coal storage, cinder and ash dumps are acidic in nature, but the volume is much less.

EFFECTS OF EFFLUENTS

As already indicated, temperature of the cooling water is the major problem which causes thermal pollution leading to the following effects.

i) depletion of oxygen and other gases.

ii) change in pH.

iii) increase in the rate of biochemical reactions and physical activity of aquatic animals.

iv) increase in the growth rate of bacteria, aquatic flora fauna and fish.

v) affliction of spawning and other reproductive mechanisms of fish.

vi) causes mixing of upper layer (epilimnion) and lower layer (hypolimnion) in a lake with an over all change in dissolved oxygen concentration and algal nutrients.

2. Addition of boiler compounds and its subsequent discharge as waste releases phosphates, carbonates, sulfates and certain organic compounds as tannin which cause algal bloom in a stretch of water. Further, the water will become corrosive and its hardness also increases.

3. Chlorination and addition of disinfectants and biocides such as copper sulfate and sodium penta chlorophenate to a cooling water for control of slime and algal growth imparts color, taste and odor to receiving waters. Bacterial and other biological populations of the water course are reduced considerably.

TREATMENT

1. Cooling may be reduced by utilizing waste heat from boilers for other purposes. Proper cooling and cooling water storage ponds reduce thermal pollution to a considerable extent.

2. Inplant management, improved methods such as closed circuit (evaporative) cooling, disposal of waste heat to atmosphere are some useful measures to copeup with the problem.

3. The boiler blow down may be collected in a holding tank, neutralized properly and discharged at controlled rate.

4. The water treatment plant wastes maybe passed on to equalization tanks and pH adjusted if necessary and then disposed of.

5. Waste from power plant cleaning after equalization and neutralization may be combined with the condenser cooling water, during discharge.

6. The fly ash slurry is settled in separate settling basins. The segregated water may be reused if possible. Eckenfelder[1] suggests lagooning as a method of treatment. Fly ash may be used in various ways. It can be best used as filler in light weight concrete aggregate for making bricks and used as a land filler.

REFERENCES

1) Eckenfelder W.W.Jr. "Industrial Water Pollution Control,1966, Mc Graw Hill Book Co., New York.

2) Tilden, Josephine E., "The Algae & their Life Relations"- 2nd ed - 1968, Hafner Publishers.

3) U.S. Environmental Protection Agency, "Energy Recovery from Waste" - 1972, EPA, Washington.

4) Stern A.C.(Ed), "Air Pollution" Volume I,II,III,IV, and V.

5) N. Manivasakam, "Environmental Pollution" - 1984, National Book Trust, New Delhi.

Chapter - 36

SYNTHETIC RESINS (PLASTICS) AND OTHER POLYMER MANUFACTURE

There are a wide variety of plastics and resins and a broad classification of them is given below with typical examples in the parenthesis.

1. Cellulosics (viscose rayon) 2. Vinyls (PVC) 3. Styrenes (polystyrene) 4. Polyolefins (polythene) 5. Acrylics (methyl and ethyl acrylate) 6. Polyesters and alkyd resins 7. Amino resins (urea formaldehye) 8. Phenolic resins 9. Miscellaneous resins (polyurethanes, nylon, polycarbonates etc.,) A brief on the manufacturing processes, effluents generated and treatment are as follows.

CELLULOSICS

Wood and other cellulosic materials are digested to pulp and purified. This purified pulp is the major raw material for the production of cellulosics. They range from regenerated cellulose to nitro cellulose. Manufacture of the popular cellulosic material viscose rayon, effluent generated during its manufacture and its treatment are dealt elaborately in Chapter-37. "Viscose Rayon Manufacture".

VINYLS

These are the most versatile thermoplastic resins made from compounds having a vinyl ($-CH=CH_2$) group.

The important members of this class are polyvinyl chloride, polyvinyl alcohol and polyvinyl acetate and polyvinyl acetals.

Production of monomer is the first stage in the manufacture of vinyl resins. Then they are polymerized to the required extent and are separated from the unreacted monomer either by spray-drying or by coagulation with the addition of acid. Finally, the polymer is washed, dewatered and dried (in rotary driers).

The effluent usually contains, suspending agents such as PVA, gelatin or cellulose, surface active agents and small amounts of monomer. Final polymerized product is also present in significant quantities as fine particles. In addition, reaction by-products during the production of manomers such as hydrochloric acid are also present. It is to be noted that BOD : PV ranges from 1:15 to 1:20 which indicates its resistance towards biological oxidation. The characteristics of an effluent of a PVC manufacturing unit is given in Table 36-1.

Equalization and neutralization, primary sedimentation (in multiple stage settling tanks) and activated sludge process are reported to be more effective treatments.

STYRENES

Styrene monomer is produced by the pyrolysis-dehydrogenation of ethyl benzene (which is synthesized from ethylene and benzene). Styrenes are produced by mass polymerization and suspension methods. The two major effluents arise from the production of styrenes are (i) suspension medium (reaction water) and (ii) wash water. It is

Table 36-1.
Characteristics of Effluent of a PVC Manufacturing Unit

Sl. No.	Characteristic	Result
i.	pH Value	6.5
ii.	Total Suspended Solids, mg/l	19,700
iii.	Settleable Solids, ml/l/hr	250
iv.	Total Dissolved Solids, mg/l	73,300
v.	Dissolved Fixed Solids, mg/l	54,000
vi.	Permanganate value, mg/l	76
vii.	Chlorides (as Cl), mg/l	32,000

reported that about 1500 liters of effluent is generated per tonne of styrene produced. It is to be mentioned that the waste is not strong and contain only the catalysts and suspending medium in small amounts. The suspending agents employed are PVA, polyacrylic acid, ethyl cellulose, starch, gelatin etc., Inorganic substances such as calcium carbonate, calcium phosphates and silicates may also present.

POLY OLEFINS (POLY ETHYLENES)

Polythene is the first and is still the largest in the production of polyolefin plastics. Among the total production of synthetic resins, polythenes represent about 30% as they have wide applications. Production of polyolefins generate only a little quantity of waste water that too from spillages and cleaning of equipments.

ACRYLICS

Acrylic emulsions are the well known surface coatings (paints) used in industries and houses. Acrylic resins are produced by polymerization of methyl and ethyl acrylates and methyl, ethyl and butyl methacrylates. Three processes employed (to obtain different types of finished products) are bulk, solution and emulsion polymerization. While the first two do not generate any effluent the third process liberate effluent. The concentrated waste water is obtained through washing of vats and filtration (or centrifugation) of lumps, The effluent contains both acrylic monomer and polymer, catalysts and emulsifying agents.

POLYESTER AND ALKYD RESINS

Polyester resins are complex esters formed by the reaction between a difunctional alcohol and dibasic acid (or anhydride). Unsaturated polyesters (laminates and molding compounds) are produced when any of the reactants contain unsaturation. Alkyd resins are polyester resins produced in presence of oils or fatty acids.

Effluents arise during esterification and by washing of kettles (with alkali or solvents).

AMINO RESINS

Urea - formaldehyde and melamine - formaldehyde are the two important amino resins. They are produced by condensation polymerization (40% aqueous solution of formaldehyde). The reaction is exothermic and cooling water is used to control the temperature of the reaction. After the reaction is over, the products are vacuum dehydrated.

The effluents generated are hot and contain water from the raw materials, water produced in the condensation reaction, alkalis used for cleaning the reaction kettles and cooling tower blow down.

Lagooning and thermal incineration are two methods practised at present. It is reported that foul odors are released from lagoons. Incineration may lead to air pollution problems. Finding an effective and economic method of treating these wastes is still in its infancy.

PHENOLIC RESINS

Phenolic resins are prepared from almost any phenolic body and aldehyde. Phenol formaldehyde resins constitute the greater proportion. Other phenolic resins include phenolfurfural and resorcinol-formaldehyde.

Manufacturing processes and generation of effluents are similar to that of amino resins.

It is reported that about 5000 liters of waste water is produced per tonne of resin. These effluents contain an excess concentration of phenols. (100-200 ppm) and removal of phenol is the first stage of treatment. Lagooning and incineration are the present treatment methods.

MISCELLANEOUS RESINS

This category includes, polyurethanes, nylon, polycarbonates, acetal, silicone and coumarone -indene resins. As the production of these resins is carried out in a cooled recirculated system, practically no effluent is generated.

REFERENCES

1) Golding B. "Polymers and Resins" - 1959, Van Nostrand.

2) Mark H(ed), "Encyclopedia of Polymer Science and Technology" 1966, Wiley - Interscience.

3) Sandler S.R. and W. Faro, " Polymer Synthesis" - 1974, Academic Press.

Chapter - 37

VISCOSE RAYON MANUFACTURE

Viscose rayon is regenerated cellulose. There are two types of viscose rayon, 1) Textile viscose rayon 2) High tenacity viscose rayon. Textile viscose rayon is the mostly manufactured item and is used primarily in women's apparels, upholstery and blended with wool in carpets and rugs. The high tenacity viscose rayon is used in cords for tyres, hose and belting.

MANUFACTURING PROCESS

Raw materials

The basic raw material used for the manufacture of viscose rayon is wood and cotton linters. The important chemicals needed are caustic soda, sulfuric acid, carbon-di-sulfide and zinc sulfate.

Production

The cellulosic raw materials contain both alfa cellulose and hemi cellulose. Alfa cellulose is the basic component required for rayon manufacture. It is separated from the hemi-cellulose by steeping wood or cotton in a solution of caustic soda (17 to 20% w/v). The alfa cellulose is insoluble while the hemi cellulose is soluble. The soluble portion is removed and recycled while the insoluble alkali cellulose is allowed to ripen for about 24 to 48 hours, at 75°F (This ripening or aging is allowed to get a solution after xanthation of correct viscosity for spinning). The aged alkali cellulose is converted to cellulose xanthate by treatment with carbon-di-sulfide under carefully controlled temperature and reduced pressure for two hours with churning. The cellulose xanthate is dissolved in caustic soda solution (2 to 2.5% w/v) to yield a proper viscose solution. The resulting solution known as viscose is separated by filter pressing, deaerated at high vacuum and at a temperature below room temperature to remove small air bubbles (which may either weaken the final yarn or cause breaks during spinning) and sent for extrusion.

Viscose is extruded into fine jets under pressure through thimble like spinnerette containing 750 to 2000 holes into the spinning bath containing dilute sulfuric acid (8% w/v), zinc sulfate (0.5 to 17% w/v), glucose and other chemicals which are essential to the spinning process. Sulfuric acid neutralizes free NaOH and decomposes xanthate liberating H_2S, CS_2, and S. The presence of 4% glucose prevents crystallization of salts in filaments.

The extruded yarn is washed to remove spinning liquor until free of acid with excess water. The sulfur adhering to the yarn is removed then by treatment with a very dilute solution (1%) of sodium sulfide and again washed with copious volume of water. Then the yarn is bleached with hypochlorite solution, washed with dilute acid solution and detergents, dried properly and conned. A flow chart indicating the above processes and waste streams are given in Fig.37-1.

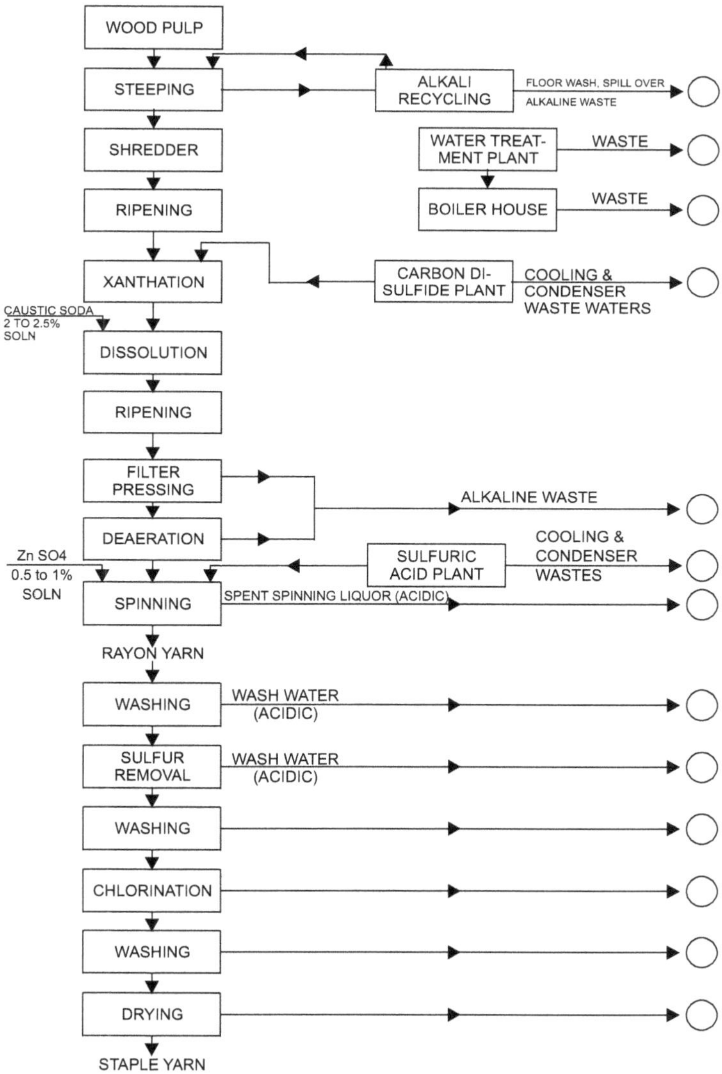

Fig 37-1. Flow Chart of Viscose Rayon Manufacturing Processes with Effluent Streams

SOURCES OF EFFLUENTS

Every rayon manufacturing industry has the following units.

1. Carbon-di-sulfide plant
2. Sulfuric acid plant
3. Rayon manufacturing plant

Among the three plants, rayon manufacturing plant contributes to the bulk of effluents. Some units would also have their own wood pulping sections which contribute larger volumes of effluents.

Carbon-di-sulfide plant effluents

Consists of only cooling and condensate waters.

Acid plant effluent

Consists only cooling and condensate waters.

Rayon manufacturing plant effluent

a) Alkaline effluents from the initial treatment of the cellulose and processing i.e., from steeping, alkali recycling, ripening, filter pressing and deaeration and carbon-di-sulfide washing for the removal of sulfides.

b) Acidic effluents arise from the spinning process and washing.

c) Effluents from bleaching and subsequent washing.

d) Other wastes include water treatment plant effluents, floor washings, leakages, spill overs etc.

VOLUME OF EFFLUENTS

The volume of effluents generated depends on the production capacity of the unit and vary from one plant to another. Further the quantity will be more if the industry has its own wood pulping unit. It is estimated that the volume of the effluent ranges from 350m^3 to 1800m^3 per tonne of the final product depending on the industry.

CHARACTERISTICS OF EFFLUENTS

In a viscose rayon industry various streams of effluents are combined and discharged as a single effluent. The effluents contain sulfuric acid, zinc sulfate, (caustic alkali which is neutralized by mixing with acidic effluents), sodium sulfate, sulfides and hemicelluloses. The color ranges from orange yellow to deep yellow. (Mention has to be made that the treated effluent will be highly colored (dark brown) than the untreated effluent and one should not get confused between treated and untreated effluents. The increase in color of the treated effluent is due to treatment with alkali-usually lime. Low pH is the most important characteristic of viscose rayon waste. The effluent usually contains high suspended solids, sulfates and less chlorides. The BOD, COD and permanganate values are always high. As the BOD is mainly due to hemicellulose and lignin compounds, (which are not easily bio-degradable) reduction of BOD is very difficult with these effluents. See Table 37-1 for typical characteristics.

Zinc is always present in appreciable concentrations which not only is toxic to certain forms of fish but also affects biological treatment.

Table 37-1.
Characteristics of Combined Effluent of a Viscose Rayon Industry (having wood pulping unit)

Sl. No.	Characteristic	Result
i.	Color	Yellow
ii.	Odor	Characteristic wood pulp odor
iii.	pH value	3.5
iv.	Total Suspended Solids, mg/l	230
v.	Total Dissolved Solids, mg/l	1230
vi.	Dissolved Fixed Solids, mg/l	780
vii.	Dissolved Oxygen, mg/l	Nil
viii.	Bio-chemical Oxygen Demand, mg/l	380
ix.	Chemical Oxygen Demand, mg/l	964
x.	Permanganate value (4 hrs), mg/l	436
xi.	Permanganate value (3 mts), mg/l	200
xii.	Chlorides (as Cl), mg/l	140
xiii.	Sulfates (as SO_4), mg/l	Nil
xiv.	Sulfides (as S), mg/l	Nil
xv.	Residual chlorine (as Cl_2), mg/l	Nil
xvi.	Zinc (as Zn), mg/l	5.5
xvii.	Tannin & Lignin compounds (as tannic acid), mg/l	110
xviii.	Oils & grease, mg/l (Petroleum ether solvent)	1.6

EFFECTS OF EFFLUENTS

The most objectionable features of the viscose rayon effluents are the extreme acidity and the immediate oxygen demand. The whole flora and fauna of the receiving stream would be destroyed if the untreated effluent is discharged in to it. Thus natural purification of stream would be upset., Further more, the acids in the effluent generate hydrogen sulfide from bottom deposits. The immediate oxygen demand depletes oxygen rapidly and anaerobic conditions result, leading to foul odors.

The color of the treated effluent poses serious problems as it cannot be removed by simple means and the color persists in the receiving streams for longer distances.

It is reported by some of the authors that the acids present in the effluent precipitate and coagulates gill secretions of fish resulting in massive deaths.

Though zinc is considered to be non toxic to human beings it is lethal to certain types of fishes. A report indicates that concentrations of zinc in soft water ranging from 0.1- 1.0 mg/l are lethal to fish. The concentration of zinc in combined rayon effluents always exceed the threshold limits. Zinc has a toxic effect towards protozoa and bacteria and it interferes in biological treatment as well as in BOD determination.

TREATMENT

Viscose rayon wastes can be treated effectively. Both primary and secondary treatments are applied.

From the alkaline wastes, caustic soda is recovered. From the spent acid baths, zinc sulfate and sodium sulfate are recovered. After recovery, the following treatments are given.

I. Primary Treatment

i) Screening

Coarser suspended particles and floating matter are removed.

ii) Equalization and Neutralization

Both the alkaline and acidic streams are allowed to combine in equalization tanks. The combined effluent will be (generally) acidic. It is neutralized by calculated doses of lime. Most of the settleable impurities settle down in equalization tank.

iii) Coagulation and Flocculation

The fine suspended impurities are removed by coagulation. The coagulants and precipitating agents used are alum and lime. Coagulated effluent is passed on to clariflocculators. The clarified effluent is subjected to secondary treatment.

II. Secondary Treatment

Though trickling filtration and activated sludge process are effective, less costlier methods such as aerated lagoon and oxidation ponds are employed. The aerated lagoon is the most effective treatment. Nutrients such as nitrogen and phosphorous must be supplied frequently.

PARAMETERS TO BE DETERMINED

Important Parameters

1) Color 2) Odor: Care must be taken while sniffing odor. 3) Temperature 4) pH value. 5) Total suspended solids 6) BOD: Dilutions ranging from 1:99, 1:199 and 1:499 have to be made. Dilution water must be seeded. As the presence of Zn results in appreciable fall of BOD value, it should be interpreted in conjunction with COD and permanganate values. 7) COD: Use 25ml sample and 25 ml or 50 ml of 0.25 N.$K_2Cr_2O_7$ 8) Permanganate value: Use 25 ml or 50 ml sample. 20 to 40 ml 0.25 N.KMnO4 may be required. 9) Zinc: Preliminary acid digestion is necessary. 50 ml sample may be taken for digestion and made upto 500 ml. Use 10 ml or 20 ml aliquot from this made up sample. 10) Sulfates 11) Sulfides: Determine immediately in the field itself.

Additional Parameters

12) Total dissolved solids 13) Dissolved fixed solids 14) Chlorides 15) Percent sodium 16) Boron 17) Residual sodium carbonate 18) Tannin & Lignin compounds 19) Oils and grease.

REFERENCES

1) R.N. Shreve and Joseph A.Brink, "Chemical Process Industries" 4th ed-1977, Mc Graw- Hill International Book Co.,

2) Indian Standards Institution, "IS:9502-1980 - Guide for Treatment and Disposal of Effluents of Viscose Rayon Industry", ISI, New Delhi.

Chapter - 38

RADIO-ACTIVE POLLUTION

Ratio-active pollution can be defined as the increase in natural back-ground radiation, emerging from the activities of man involving the use of naturally occurring or artificially produced radio-active materials.

Man-made radiations originate from the activities of man involving the use of radio active materials. Radio-nuclides have numerous applications. They are used for the production of nuclear weapons, and for the production of electricity. They are extensively used as tracers in various fields. The following pages describe various processes by which radio active contamination occurs.

Ore Processing

The primary natural materials used in nuclear processes are uranium and thorium. Uranium-238 and thorium-232 are used as artificial fissionable materials. The only element in which natural fission occurs is uranium-235. Mining and refining of these materials are the primary steps in nuclear technology. The waste waters from mining and washing processes mainly contain a small percentage of ore. The residues in the form of solids or sludges from the metallurgical units (Where the ingots are produced) contain the daughter products uranium and thorium. Of all the daughter elements, those of relatively long half-life include thorium-230, radium 226 and lead-210 (radium D) in the uranium-radium series, and radium–228 and thorium 228 in the thorium series.

The necessary isotopes of uranium and thorium are segregated from other radioactive materials by gaseous diffusion method or by some other process. Depending upon the process used, the unnecessary radio-active materials are separated in various chemical forms. Those isotopes with longer high-life will persist for much longer periods. The more soluble materials may eventually reach and contaminate waters. Of all the radio-active nuclides, radium-226 is considered to be the most harmful in a water environment due to its longer half-life, bio chemical properties, energetic radiations and its immediate daughter products. Radio-active pollution of the atmosphere also occurs through all these mining, washing, refining and separation processes, with the result of the release of radio-active gases such as radon, thoron, and subsequent adsorption of these products on particles present in the atmosphere and the formation of dusts from uranium or thorium ores.

Operation of Reactors and Nuclear Fuels

The processed nuclear fuel is introduced into the reactor, the operations of which also contribute to pollution. The sequential operations are (i) Fission process, (ii) Activation process and (iii) Thermal process. A partial transformation of the primary fuel into secondary fuel occurs in fission process with the simultaneous appearance of fission products. Fission products of the heavy metals include all the elements from about atomic number 30 (zinc) to atomic number 64 (gadolinium). The elements have excess neutrons as soon as they are formed and they rapidly change into more stable elements by radio-active decay emitting beta and gamma radiations. Activation process is one by which stable substances introduced into the channels of the reactor are activated by the neutron flux. The heat liberated during fission and activation process is converted into electricity in thermal process using coolant systems and heat exchangers. Two types of wastes are formed in these processes, (i) the fission products remaining in both the primary and secondary fuels and (ii) extraneous activation products in the coolant. Thus, the fuel elements and coolants constitute the sources of radio-active pollution. Fission is said to release radio-active gases such as xenon and krypton to the atmosphere.

The problem of disposal of radio-active wastes is more complicated. The radio-active wastes resulting from the above mentioned processes contain unwanted excess fission products and activation products. Since the wastes produced are in large quantities and have a high activity, they cause an extremely difficult public health problem. When a fuel element is dissolved, all the fission products and heavy elements are freed from confinement and pass into the liquid. The volatile materials (inert gases and halogens), pass off as vapor. When they are released into the atmosphere, they become potential contaminant of the environment as they can settle on land or can be washed of into surface waters by rain.

Nuclear Weapon Testing

Nuclear explosion tests threaten the world in such a

manner that the natural background radiation of the environment is increased largely due to these nuclear tests. The tests including fission and fusion processes employ uranium-235 and plutonium-239 as fission materials and hydrogen or lithium as fusion material. Explosion - a non-moderate chain reaction, in due course results in the appearance of a very large neutron flux which makes the surrounding matters radio active. The radio active materials primarily resulted in such an explosion are fission products such as strontium-90 caesium-137, iodine-131, and unused explosives and activation products. These activation products are formed by neutron bombardment of elements present in water or soil. Due to the greater force of explosion, the radioactive products are transformed into gases and are ejected high into the atmosphere as fine particles resulting in the atmospheric pollution. This primary pollution is followed by a secondary pollution through radio-active fall-out. These particles will subsequently be carried away by rains or otherwise settle down on soil or will be dispersed over the surface. Thus radio-active pollution occurs not only at the test sites but also at the remote corners.

When rain water containing radio nuclides falls on the ground its activity is transferred to soil by adsorption on soil particles. As the radio-activity in the soil is available to plants, it enters the food chain resulting in the possibility of eventual ingestion by human beings. When a stream acquires radioactivity from fission products through rains, the aquatic flora and fauna absorb and concentrate radio-activity. Thus once again the possibility occurs that those aquatic flora and fauna used as food by man could accumulate dangerous amounts of radio-isotopes.

The fission products are numerous out of which, strontium-89, strontium-90, yttrium-91, zirconium-95, niobium-95, ruthenium-103, ruthenium-106, caesium-137, cerium 141, cerium-144, promethium-147, and samarium-151 are the most important long-lived fission products. The short-lived fission products comprise of molybdenum-99, iodine-131, iodine-132, tellurium-142, iodine-133, iodine-135, barium-140, presodymium-143 and neody-mium-147. All these fission products vary in nature. Some of them are insoluble in water and settle out with the particulate materials and concentrate in silts and sludges. The soluble materials remain in solution and are carried farther. Some other materials enter into the biological cycle and are distributed according to the biota present.

Radio-active materials also released into the environment by irradiation with neutrons which are produced during an explosion. The nature and the quantity of the nuclides produced, entirely depend on the circumstances at which explosion occurs. It is reported that a high-altitude burst makes only the materials present in the bomb and air to be irradiated. It is also reported that a burst, close to the ground makes all the elements of the soil as potential sources of induced radio-activity. A subsurface explosion also converts materials such as sodium in water to be radio-active. Some of the important induced radio-nuclides are iron-59, carbon-14, managanese-54, iron-55, cobalt-57, cobalt-58, iron-59, cobalt-60 and zinc-65. A recent report indicates that a number of induced radio nuclides are concentrated in biological systems and this is proved by their presence in fish.

Radio Tracers

Radio tracers are extensively used in the fields of medicine, industry, agriculture and biology, to determine the course of chemical and biochemical reactions. Among them, C_{14} and I_{125} are the most important. Determination of "flow-through time" of sedimentation tanks and of "reaches" or "stretches" in rivers are achieved by the use of radio tracers. They are also used to determine the rate and direction of ground water.

It is reported that when waste waters containing radio tracers are discharged into sewers and rivers, some of the isotopes such as radio iodine and radio phosphorus accumulate in biological slimes and sludges. In rivers, radio-active material may be concentrated particularly by microscopic organisms which serve as food for fish and other aquatic life that are consumed by man. It also becomes a problem to consume water from the water supplies to which rivers containing radio-active wastes are contributing.

The effects of radio-active pollution are numerous and varied. Extensive studies have been carried out and many reports published. From the reports it is seen that radiations can cause cancers, abnormal births and mutations in man. Though these effects are common to plants and animals, such abnormalities are eliminated by the evolutionary processes of natural selection, (i.e.) the survival of the fittest. Since natural selection does not operate in man, radio-active pollution can produce great human misery. There is no threshold or safe-dose for radiation effects. Even the smallest increase in radiation above natural background radiation is reported to cause risks. Chronic exposure to radiations leads to leukemia in an individual and even affects an unborn child in the womb. Exposure to radiations causes genetic damage and affects future generations. Hence radio-active pollution is primarily a health issue of man when compared with the other environmental pollutants. Ratio-active pollution thus differs from the other pollutants in the respect that it not only affects the individuals but also future generations. Moreover, the effects due to other harmful pollutants (such as DDT, PCB etc.,) are primarily on the environment, the effects of radio-active pollution are primarily on man. Modern life and habits seem to increase the risks. The radiations from medical X-rays, color TV sets, luminous dials of clocks and watches, X-ray fluoroscopes are also contributing to the risks.

The International Commission on Radiological Protection assessed the maximum permissible weekly dose for a radiation worker as 0.3 rem*. The maximum permissible total dose for complete life-time has been fixed as 200 rem which corresponds to 0.1 rem per week or 5 rem per year for continuous exposure. In any case the total accumulated radiation shall never exceed 5 rem per year beyond age 18, nor shall the dose in any 13-week period exceed 3 rem. However, somewhat higher doses are also permitted where exposure is

---

* rem-Unit of radiations energy, roentgen-equivalent-man (rem). It is the amount of radiation that will produce an energy dissipation in man that is biologically equivalent to one roentgen of radiation of X-rays or approximately equals to 1000 ergs/g. In turn roentgen (r) is a unit of X-ray or gamma-radiation intensity. It is the amount of radiation (gamma or X-ray) that products one electrostatic unit of electricity in one cubic centimeter of dry air at N.T.P.

restricted to skin. It has been stated that strontium-90 and radium-226 should not exceed more than 3 and 10 pCi* per litre respectively. The Federal Radiation Council has fixed a limitation to a water supply that the gross beta concentration shall never exceed 1000 pCi per litre in the known absence of alpha emitters and strontium-90.

Prevention and control

Since no suitable therapy has been found out for the curing of internal contamination of radio-active pollutants, it is necessary to adopt preventive measures to reduce the artificial radio-activity and steps have to be taken such that the radio-active pollution of the environment does not reach dangerous levels. It is better to know the principles underlying the preventive measures before attempting them. The rate of decay of radio-active materials and the subsequent emission of radiations are unaltered by man and hence cannot be converted into other non-toxic materials. The preventive measures should be double edged so as to curtail the effects of both occupational exposure and exposure of the population at large. All measures should so aim that the level of radio-active pollution should not exceed the maximum permissible doses.

Radio-active wastes may be discharged into the environment, only after necessary treatment so that the radio-activity is at a lower level. Prior to discharge into sewer system or into the streams the low activity wastes have to be stored for some time to reduce the activity. In the case of high activity wastes, they cannot be discharged. The radio nuclides from

* pCi : 'Pico Curie' - unit of radio-active disintegrations per second.

those wastes are segregated by coagulation, precipitation or by ion exchange. The concentrated nuclide (in solid form) is then stored or buried. Thus, after decontamination, the waste water may be discharged. Still stronger wastes cannot be treated and they are stored indefinitely. They should never be discharged into the environment. However, the wastes may be concentrated by evaporation and the condensate with a very low activity may be discharged. The concentrated bottoms are then once again stored in special type of tanks. Now-a-days small quantities of high activity wastes are converted into solid such as concrete and then buried under ground or sea. Recent reports indicate that these wastes may also be converted, into glassy or ceramic material which is then calcined to alumina or zirconia and leached. Clay may be used to take up the leached radio nuclides through absorption or ion exchange after which it is fired. Firing causes the exchange in crystal structure of clay and the radio nuclides cannot be leached or desorbed. Disposal of the wastes at sea has also been tried. In one type, the wastes are sealed in concrete-filled steel drums and discharged into depth of 1000 fathoms* or more. Since the disposal at sea is a costlier process, underground disposal is under study.

The next step in the prevention of radio-active pollution is the limitation of the emission of radio-active pollutants. This is a fruitful measure and has to be planned before the commencement of construction.

There are several techniques and according to the nature of the industry or process these techniques have to be chosen. For example, in the chemical and metallurgical

* fathom - a measure of six feet

industries, the use of radio-active materials may be carried our under a jet of soil (or water) instead of gaseous and powder forms. In uranium mines, wet drilling may be employed in combination with under ground drainage. Closed-cycle coolant systems with gaseous coolants of very high purity may be used in the case of nuclear reactors to prevent access of extraneous activation products. Containment may also be employed to limit the radio-active emissions. This may be achieved by using hermitically sealed boxes and closed-cycle systems. If the atmospheric radio-active pollution is still high (due to long lived isotopes), then dispersal is the only possible way. Dispersal method makes the pollutants in a confined place to spread over a large area such that the pollution is weakened and the effects are less. Use of high chimneys and ventilation of the working place where radio-active contamination is high are some of the dispersion methods.

REFERENCES

1) Glasstone, S. (Ed), "The Effects of Nuclear Weapons"-1957, U.S. Govt, Printing Office, Washington, D.C.

2) Friedlander, G.J., Kennedy W. and J.N. Miller, "Nuclear and Radio-Chemistry" 2nd ed-1964, John Wiley & sons, New York.

3) Glasstone, S, "Source book on Atomic Energy", 3rd ed-1967 D. Van Nostrand Co., Inc. Princeton, N.J.

4) N Manivasakam, "Environmental Pollution"-1984, National Book Trust, New Delhi.

PART -IV

ANALYSIS OF INDUSTRIAL EFFLUENTS

Chapter - 39

INTRODUCTION

39-1 PURPOSE OF EXAMINATION

Industrial wastes are examined for the following reasons.

1. To find out the pollution load that would be exerted on receiving waters or on land or on sewers.
2. To find out the presence of toxic constituents such as cyanide, cadmium, mercury etc., and to evaluate the degree of toxicity.
3. To find out the substances that cause difficulties in treatment.
4. To find out the treatability and to design an effective treatment plant.
5. After commissioning treatment plant, to check up the effectiveness of the treatment and to monitor it.
6. Receiving waters should also be examined to assess their quality to find out the pollution in the upstream and to decide its ability to accept the pollution load and to determine the degree of self purification within a given stretch.

39-2 INFORMATION TO BE OBTAINED FROM THE INDUSTRY

The duty of sampling personnel does not end just with the collection of samples. He has to obtain various information from the industry which would be highly useful in the analysis of samples to decide the constituents to be determined, the approximate quantity of them in the effluent sample, and to have a check on the analysis, and to evolve an effective and economical treatment design.

Before collecting the samples, the sample collector should go around the factory and has to observe the processes carried out and streams of effluents. He should atleast know the basic details of the processes, and points of generation of effluents. Details of the existing treatment plant and analytical data would be highly useful. A general information questionnaire is given below. Suitable additions or alterations may be made in this questionnaire as per the industry while obtaining information.

1. Name of the industry _________
2. Place and address _________
3. Phone No _________
4. Person interviewed (Name & Designation) ________
5. Materials Manufactured/Processed (in detail) _______
6. Production/Processing capacity (per day/Per week/ per month) _________
7. Raw materials and chemicals used :

   (indicate their quantities also Per day/ Per week or Per month) Describe process wise _________
8. Detailed description of the process (along with flow charts)

9. Describe the processes in detail with particular reference to effluent generation ________

10. List of effluent streams ________

11. Periodicity and quantity of effluents generated/let out per unit time (per hour/per day) ________

12. Possible contaminants and specific mention on toxic substances such as cyanide, phenol, free ammonia etc., (along with the approximate quantities of them in the effluent) ________

13. Is there any treatment given at present ________

14. If any, full details of treatment and mode of disposal (whether into surface water or on to land or into sea) (Include the details of recovery and recycling if practiced) ________

15. Analytical data on the quality of effluents before and after treatment ________

16. Analytical data on the quality of water in the upstream and downstream (if the industry draws water from the water course or discharges its effluent into it) ________

17. Source of water (surface water / ground water/ ocean) ________

18. Use and quantity of water (in litres/day) in the industry.

    Processing: ________

    Drinking: ________

    Softened / Demineralized water: ________

    Cooling water: ________

Wash down water : ________

19. Treatment given for water ________

20. Details of the effluent generated on water treatment ________

21. Describe the treatment given to the effluent arising from boiler blow down and cooling towers (if any) ________

22. Proposed future treatment and ultimate point of disposal of effluents (including design, availability of space etc.,) ________

23. Are there other industries discharging their effluents into the same body of water. List their names and types of wastes discharged by them ________

24. Is the water course a source of potable water for municipal supply ________

25. Proximity of water course to the industry (to treatment plant) ________

26. Is it possible to discharge the waste into the municipal sewerage system ________

27. If the municipality has a treatment plant indicate type and capacity (m^3/day) ________

28. Has the municipality promulgated any regulations to the types and quality, quantities and strength of wastes to be discharged into its sewer ________

29. Municipal charges for disposal of effluent into sewer ________

30. Are the wastes of other industries in the area discharge

their effluents into the sewer. If so types and volumes of wastes _________

31. If the industry wishes to dispose of the effluent into their own land, will the land available be sufficient for it _________

32. Describe the nature of soil stratum and possibility of seepage etc., _________

33. Proximity of wells and surface streams near by the land _________

34. Types of crops cultivated _________

35. Name the legally constituted agency/authority that demands treatment (Pollution control Board / River Board Authorities / Fisheries / Municipality / Public Health Department) _________

36. The tolerance limits prescribed by the authority for discharge of effluents _________

37. Is there any treatment method proposed by the authority. If so give full details _________

38. Does the industry have adequate water for dilution of the effluents. _________

39. Is the industry given any deadline for completion of treatment plant etc., _________

40. Cess to be paid for water and disposal of effluents to the Regulatory Authority _________

41. Other Details : _________

Date :
Name & Designation of the
person obtaining information.

39-3 COLLECTION & PRESERVATION OF SAMPLES

Effluent samples are collected for various examination. Mostly the sample is subjected to physico-chemical examination. Effluents from tannery, cannery, gelatin and glue manufacture, slaughter house, poultry should also be examined for their microbiological quality. If the effluent is discharged into surface water, then two samples from the water course, one from upstream and other from downstream should also be collected and examined. Whatever may be the method of sampling, the sample must represent the conditions existing at the point taken. The samples must be of sufficient quantity to carryout all determinations. In sampling water, the sampling point should be chosen carefully so that a representative sample of the water to be tested is obtained.

Quantity of sample

Generally 2-litres of sample is sufficient for most of the physical and chemical examinations. For certain special determinations, however larger volumes of samples are necessary.

Sample containers

Either glass, plastic or polythene containers of 2.5 liter capacity may be used. The choice of containers depends on the determination to be made. It is given in the section "Preservation of Samples".

Samplers

For trade wastes and sewage, special types of samplers are not necessary. But for samples of water, sampling devices to meet the specific requirements are needed, especially for collecting samples for dissolved oxygen. (For details on special type of samplers, refer "Annual Book of ASTM Standards - 1972, Part 23).

When composite samples are to be collected, it is better to use wide mouthed bottles of capacity 200-300 ml for sub samples. They are particularly useful in transferring the contents without leaving any settleable matter.

New glass bottles should be treated either with dil. hydrochloric acid or chromic acid and then repeatedly washed with water and finally with distilled water. Subsequently they may be cleaned using detergents.

Grab Sample and Composite Sample

Liquid samples are broadly classified into two major groups, viz. the one called instantaneous, spot, snap or grab sample and the other continuous or composite sample. A grab sample is a manually collected single portion of sample of water or waste water. When the grab samples of a particular water or waste water is collected at regular intervals for a specific period say for 12 hrs or 24 hrs and mixed, then the integrated sample is known as 'composite sample'.

Collection of Industrial Effluents

The sampling and the analysis of industrial effluents require greater care and attention. Obtaining representative sample only is not the criterion and it is necessary to study during sample collection, the waste load, major sources of wastes within a plant, character of wastes, potential recovery of valuable materials at profit and the effects of discharged wastes on the receiving body of water.

Wastes discharged by industries are of great variety by nature. They are subject to rapid changes within a few minutes. In sampling industrial wastes, suitable sampling points as well as the frequency and type of sample to be collected must be decided first. Generally, sampling points are located where flow conditions encourage a homogeneous mixture.

To obtain a representative sample it is necessary to collect individual samples at uniform intervals, say 10 or 15 minutes. The degree of variation in rate of flow will determine the time interval for sampling. Care should be taken in collecting a sample containing oil. It may be helpful to collect larger volumes of such samples and withdrawing a suitable proportion using a narrow diameter sectional sampling tube. Use clean and dry glass-stoppered bottles for collecting such samples. Fill the sample bottles only partially such that when the stopper is inserted no oil is lost.

Samples for the determination of dissolved oxygen and BOD should be collected in BOD bottles by siphoning to avoid aeration (Refer Chapter 41-1 "Dissolved Oxygen").

Hot samples must be cooled to atmospheric temperature and then collected. It is always better to collect the individual streams of effluents as well as combined effluent. If treatment is given, collect samples before and after treatment.

Before collection, rinse the container with the sample being collected.

Labeling of Samples

The sample containers have to be labeled with the following information.

1. Sample number
2. Date and time of sampling
3. Industry from where the sample is collected
4. Effluent stream (in detail)
5. Name and signature of the sample collector

Information to be obtained and submitted along with the effluent sample

During sampling, various other details have to obtained from the industry and submitted along with samples. See the section "Information to be obtained from the industry" - Page 285.

Collection of Water Samples

It is essential to collect and examine water samples from the upstream and downstream, if the effluent is discharged into a water course. Downstream sampling should be made at a point where a homogeneous mixture of water and effluent could be obtained after the confluence of effluent with the water course. (In general a distance of 0.5 Km to 3 Km below the confluence is sufficient). Great care and discretion are necessary in the selection of the sampling point so as to obtain a sample that represents the real conditions and nature.

Because of a wide variety of conditions found in streams, lakes, reservoirs, and other bodies of water it is not possible to prescribe the exact point of sampling. The choice of location, depth and frequency of sampling, should be based on local conditions and upon the purpose of the investigation.

A sample taken at any point in the cross section is satisfactory in the case of stream where water is mixed to uniformity. This uniform mixing cannot be expected in large rivers and streams and three or more samples are taken at the mid point of equal cross sectional areas of such rivers and streams. These samples are then combined together to obtain a composite sample. If only a grab sample is taken, it is better to collect in the middle of stream and at mid depth.

Information to be obtained during collection of Water Samples

i. Exact location and name of place

ii. Depth below surface

iii. Rate of flow

iv. Whether the water level is above or below average

v. Recent rainfall or flood conditions

vi. Possible sources of pollution

(Trade effluents/sewer drains)

vii. Temperature of water

viii. Results of field tests

Time Interval between Collection and Analysis

In general, the shorter the time between collection and analysis, the more accurate will be the results. No specific time interval between collection and analysis can be generalized; this depends upon the nature of the sample, constituents to be determined and conditions of storage. It is advisable to keep the sample out of light (to retard the growth of algae etc.,) and low temperature (4°C). While reporting the results include the time elapsed between collection and analysis.

Make the determination of pH and of dissolved gases such as oxygen, carbondioxide, hydrogen sulfide in the field itself. In some cases, the constituents may be fixed and determined later in the laboratory.

PRESERVATION OF SAMPLES

Samples are preserved as soon as they are collected to resist changes that occur during transportation to the laboratory and in the time lapse between collection and analysis. The chemical changes include precipitation of metal cations (as hydroxides), complex formation, oxidation or reduction to higher or lower valance state. Dissolution and volatilization of some of the constituents are also important changes. Sometime, metal cations such as iron and lead may be adsorbed on the surface of the container. Hence, it is always desirable to analyze the sample as soon as possible after collection. This is especially an essential condition when the analyte concentration is in the microgram range. Biological changes are also reported to occur. Biological organisms may release the cellular materials into solution or absorb some of the ions or radicals in to their cell structure.

The preservation methods are i) pH adjustment (by the addition of acid or alkali), ii) addition of chemicals to precipitate the particular constituent or ion (eg. addition of zinc acetate to precipitate sulfide) and to retard biological growth that prevents depletion of the constituent, and iii) refrigeration and freezing to retard the volatility of substances and to arrest chemical and biological activity.

No single method of preservation is entirely satisfactory and the preservative should be chosen with due regard to the determinations that are to be made. It should be noted that the preservative added for one constituent should not affect the determination of other constituent. Keeping the sample in dark at low temperature is the best possible preservation. Collection and preservation methods are described in the appropriate texts. The following Table 39-3-1 is a summary of the preservation methods, holding time, quantity of sample to be collected and type of container.

Table 39-3-1
Recommended Methods of Preservation of Samples

	Determination	Recommended container	Required volume (ml)	Method of preservation	Holding time
I. Physical Properties					
1.	Color	P/G	100	Cool to 4° C	48 hrs.
2.	Turbidity	P/G	100	Cool to 4° C. Store in dark	24 hrs.
3.	Odor	-	Sniff the odor in the field itself	-	-
4.	Temperature	-	-	Measure immediately in the field itself	-
5.	pH value	P/G (B)	50	Measure immediately in the field itself	-
6.	Electrical Conductivity	P/G	100	Cool to 4° C	48 hrs
7.	Solids				
	a) Total	P/G	100	Cool to 4° C	7 days
	b) Dissolved	P/G	100	Cool to 4° C	7 days
	c) Suspended	P/G	100	Cool to 4° C	7 days
	d) Volatile	P/G	100	Cool to 4° C	7 days
	e) Settleable	P/G	1000	Cool to 4° C	48 hrs.

	Determination	Recommended container	Required volume (ml)	Method of preservation	Holding time
II. Inorganic (A) Non metallics					
8.	Acidity	P/G (B)	100	Cool to 4° C	2 weeks
9.	Alkalinity	P/G (B)	100	Cool to 4° C	2 weeks
10.	Bromide	P/G	100	None required	1 month
11.	Chloride	P/G	100	None required	1 month
12.	Chlorine	P/G	200	Determine in the field itself	-
13.	Cyanide	P/G	500	Add NaOH to pH > 12 Cool to 4° C (if any residual chlorine is present add ascorbic acid also)	2 weeks
14.	Fluoride	P only	300	None required	1 month
15.	Iodide	P/G	100	Cool to 4° C	24 hrs.
16.	Nitrogen				
	a) Ammonia	P/G	500	Add 0.8 ml conc. H_2SO_4 per litre and refrigerate	1 month
	b) Nitrate	P/G	50	Cool to 4° C	2 days
	c) Nitrite	P/G	50	Add 40 mg $HgCl_2$/l and refrigerate.	2 days

	Determination	Recommended container	Required volume (ml)	Method of preservation	Holding time
	d) Nitrate and Nitrite	P/G	100	Add H_2SO_4 to pH<2 Cool to 4° C	1 month
	e) Total (Kjeldahl)	P/G	500	Add H_2SO_4 to pH<2	1 month
17.	Oxygen (Dissolved)	BOD Bottle (glass)	300	Determine immediately or fix DO (See Chapter 41-1 "Dissolved Oxygen)	–
18.	Phosphate	(rinse the container with 1 + 1 HNO_3)			
	a) Ortho Phosphate (dissolved)	P/G	50	Filter in the field itself (See note 3) Cool to 4° C	2 days
	b) Hydrolyzable	P/G	50	Add H_2SO_4 to pH<2. Cool to 4^oC	1 month
	c) Total	P/G	50	Add H_2SO_4 to pH<2. Cool to 4° C	1 month
	d) Total and dissolved	P/G	50	Filter in the filed itself Add H_2SO_4 to pH<2 Cool to 4° C	24 hrs.
19.	Silica	P only	50	Cool to 4° C	1 month
20.	Sulfate	P/G	50	Cool to 4° C	1 month

	Determination	Recommended container	Required volume (ml)	Method of preservation	Holding time
21.	Sulfide	P/G	500	Determine in the field or add 2 ml 2 N zinc acetate /l plus NaOH to pH>9	1 week
22.	Sulfite	P/G	50	Determine in the field itself	–
II. Inorganic (B) Metals					
23.	All metals except chromium and mercury				
	a) Dissolved	P/G	200 to 500	Filter in the field (See note 3) Add HNO_3 to pH<2	6 months
	b) Total	P/G	200 to 500	Add HNO_3 to pH<2	6 months
24.	Chromium (VI)	P/G	200	Cool to $4^{\circ}C$	24 hrs
25.	Mercury				
	a) Dissolved	P/G	500	Filter in the field. Add HNO_3 to pH<2	1 month
	b) Total	P/G	500	Add HNO_3 to pH<2	1 month
III. Organics					
26.	BOD	P/G	300 to 1000	Cool to $4^{\circ}C$ Add H_2SO_4 to pH<2	24 hrs

	Determination	Recommended container	Required volume (ml)	Method of preservation	Holding time
27.	COD	P/G	50 to 200	Cool to 4° C	1 month
28.	Total Organic carbon	P/G	50	Add H_2SO_4 or HCl to pH<2 Cool to 4° C	1 month
29.	Phenolics	G only	500	Add H_3PO_4 to pH<4.0 and 1g. $CuSO_4 . 5H_2O$ Cool to 4°C	1 month
30.	Oils & grease	G only (wide mouth)	1000	Add H_2SO_4 to pH<2 Cool to 4° C	1 month
31.	Surfactants	P/G	250	Cool to 4° C	2 days

Notes

1. G-Glass, P-Polythene or Plastic (The bottle cap should not contain any liner) G(B) - Glass, Borosilicate.
2. Preserve the sample immediately after collection. During composite sample collection, the sub samples should be kept at 4°C until compositing and sample splitting.
3. If dissolved constituents are required, filter the samples through a 0.45 micron membrane filter and then add preservatives to the filtrate.
4. Samples should be analysed as soon as possible after collection. The holding time given is the maximum time limit upto which the sample may be hold for analysis and still considered valid.

39-4 REPORTING OF RESULTS

Analytical results are usually expressed in milligrams per litre (mg/l). The term parts per million (ppm) is a weight-to-weight ratio. The term mg/l is equivalent to parts per million when 1 litre of water, sewage or industrial waste, weighs 1 kg. If the concentrations are generally less than 1 mg/l it is convenient to express the results in terms of micrograms per litre (mg/l). This is equivalent to parts per billion*. The term 'percent' is preferred when the concentration is greater than 10,000 mg/l, 1% being equivalent to 10,000 mg/l. Sometimes results are expressed in terms of milli equivalents per litre (meq/l). It is obtained by dividing mg/l of the element or ion by its equivalent weight.

In certain determinations, such as color and turbidity reference is made to arbitrary standards. In these cases results are expressed in units without designation 0,1,2, and so on.

The format of reporting analytical results is as follows:

	Sample from	Lab No:	
	Effluent and Source		
	Date of collection		
	Date of Receipt		
	Collected by		

* Billion = 10^9

Physico-chemical measurements	GENERAL	Color	
		Odor	
		Temperature °C	
		pH value	
	SOLIDS	Total Solids, mg/l	
		Total	
		Volatile	
		Fixed	
		Suspended Solids, mg/l	
		Total	
		Volatile	
		Fixed	
		Dissolved Solids, mg/l	
		Total	
		Volatile	
		Fixed	
		Settleable Solids, ml/l/hr*	
Measures of organic pollution	OXYGEN	Dissolved Oxygen, mg/l	
		Permanganate value 4 hrs, mg/l (oxygen absorbed in 4 hrs)	
		Biochemical Oxygen Demand (5 days at 20°C) mg/l	
		Chemical Oxygen Demand, mg/l	
	NITROGEN	Ammoniacal Nitrogen, mg/l	
		Organic Nitrogen, mg/l	
		Total Kjeldahl Nitrogen, mg/l	
		Nitrite Nitrogen, mg/l	
		Nitrate Nitrogen, mg/l	

* Or for other specific period

Measures of inorganic pollution	METALS	Boron, (as B), mg/l	
		Cadmium (as Cd), mg/l	
		Chromium (as Cr),	
		Hexavalent, mg/l	
		Trivalent, mg/l	
		Copper (as Cu), mg/l	
		Lead (as Pb), mg/l	
		Manganese (as Mn), mg/l	
		Mercury (as Hg), mg/l	
		Nickel (as Ni), mg/l	
		Zinc (as Zn), mg/l	
		Percent Sodium (as Na)	
Measures of inorganic pollution	NON-METALS	Acidity / Alkalinity (as $CaCO_3$), mg/l	
		Phenolphthalein (8.3)	
		Methyl Orange (4.5)	
		Chlorides (as Cl), mg/l	
		Sulfates (as SO_4), mg/l	
		Sulfites (as SO_3), mg/l	
		Sulfides (as S), mg/l	
		Phosphates (as PO_4) mg/l	
Misc.		Oils and grease, mg/l (Specify the solvent used)	
		Results of other tests*	

* Other constituents that are not covered in this result form (e.g.) cyanide, phenols, tannin & lignin, dyes, gases such as chlorine etc.,

REFERENCES

1) H.H. Black, 'Procedures for Sampling and Measuring Industrial Wastes' - Sewage and Industrial Waste, SIWAA, 1952, 24, 45.

2) J.W. Clerk etal., "Water Supply and Pollution Control" - 2nd ed, 1971.

3) American Society for Testing and Materials, 'Annual Book of ASTM Standards,' Part-23, 1972, ASTM - Philadelphia.

4) U.S. Environmental Protection Agency, 'Methods for Chemical Analysis of Water and Wastes" - EPA-600/ 4-79-020-1983 Cincinnati-Ohio.

5) N. Manivasakam, "Physico-chemical Examination of Water, Sewage & Industrial Effluents" - 1984, Pragati Prakashan, Meerut.

Chapter - 40

GENERAL PHYSICO-CHEMICAL MEASUREMENTS

IMPORTANT NOTES

The methods described in the following pages are applicable to industrial effluents, sewage, polluted waters, raw and treated waters.

The points given below apply generally and they are not mentioned normally in individual determinations.

Reagents

The quality of all reagents should be of "Analytical Reagent" - AR grade, unless otherwise stated.

Expression of concentration of acids and alkali solutions

In addition to the terms Normality (N), Molarity (M), the concentrations are sometimes expressed in additive volumes. In additive volume (a+b), the first number 'a' indicates the volume of concentrated reagent and the second number 'b' denotes the volume of distilled water required for dilution. Thus 1+3 H_2SO_4 indicates that 1 volume of H_2SO_4 has to be diluted with 3 volumes of distilled water.

Significant Figures

The results should not be reported to a greater number

of figures than is justified by the accuracy of the method. It is sufficient to give three significant figures as maximum.

Dissolved constituents

Unless specifically stated otherwise, samples should be shaken well before taking aliquots for individual tests. If necessary, the sample may first be filtered. When the dissolved concentration is to be determined, the sample is filtered through a 0.45 micron membrane filter and the filtrate analyzed by the procedure specified.

40-1 COLOR & PHYSICAL APPERANCE

Almost all industrial effluents are colored and turbid. They impart color to the receiving waters. Among them, dye house wastes and wood pulping wastes are noteworthy. The pulp mill effluents contain lignin materials and other organic materials in dissolved form. They are highly resistant to biological attack and hence the color imparted to a water course in the upstream persists for longer distances.

Observation of color is the simplest test to determine the effectiveness of treatment of an effluent. It serves as a guide in deciding the quantity of chemicals used for the removal of color and to ensure economical treatment.

Express the color hue as observed. If any changes occur on transportation or storage observe them. Also note the other physical characteristics such as floating matter including oil and grease, solid settlement, tendency of the effluent to foam etc.,

40-2 ODOR

Describe the odor as sniffed. While smelling care must be taken as most of the effluents possess toxic and offensive odors.

40-3 TEMPERATURE

Temperature measurements are usually made with mercury filled centigrade thermometer. The reading should be made by dipping the thermometer in the sample (water or waste water). Sufficient time should be allowed before constant reading is obtained. Normally the important temperature is that at the time of sampling. The temperature should be expressed to the nearest centigrade.

40-4 SUSPENDED SOLIDS

The undissolved matter present in a water or waste water is usually referred as suspended solids. Determination of suspended solids is as important as that of BOD. It is one of the valuable parameters in judging the pollution potential of an effluent, pollution load on receiving streams and also to decide the efficacy of treatment units. It is particularly useful in determining the load on secondary biological treatment units, after the removal of settleable solids in primary settling tanks.

The suspended matter is determined by filtering or centrifuging the sample, drying the residue and determining its weight by difference.

Total Suspended Solids

(i) Filtration method

Use a gooch crucible containing asbestos mat for

filtration. Prepare asbestos mat in the gooch crucible as follows:

To 10 g. of good quality asbestos in a 250 ml. beaker add 100ml. distilled water and stir well. Decant the supernatant water containing the impurities of asbestos. Repeat this process atleast 10 times and then mix the asbestos with 1000 ml. distilled water. Pour 25 ml. of the well mixed asbestos solution into a gooch crucible of 30ml capacity with small perforations. Apply suction and dry the crucible. Place the crucible in an oven at 105° C. If 'fixed suspended solids' and 'volatile suspended solids' are to be determined, place the crucible in a muffle furnace at 550° C. Cool and weigh.

Note: 1. It is extremely important that the gooch crucibles be carefully prepared and brought to constant weight before use, otherwise considerable error will occur.

2. The thickness of the asbestos mat should be about 2 to 3 mm and should not exceed 3 mm.

Filter a suitable volume of the sample through the crucible by applying suction. The filtration should be carried out at a lower vacuum than that used in the preparation of mat. Wash the filter with distilled water to remove the soluble salts. Care should be taken to prevent loss of fine asbestos particles during filtration and washing. Dry the crucible in an oven at 103 - 105°C. The increase in weight is the total suspended solids.

Express the result as mg. total suspended solids per litre of sample.

(ii) Centrifugation method

A known volume of the well shaken sample (usually about 50-200 ml) is centrifuged under standardized conditions using special conical-end centrifuge tubes. After centrifugation, the separated liquid is poured off and the residue in the tubes is washed with distilled water, re-centrifuged, the separated water is again removed and the suspended matter in the centrifuge tube is transferred by means of a water jet to a tared platinum or silica or porcelain dish and dried at 105°C. It should be remembered that centrifugation method gives slightly lower results than that obtained by filtration method.

Note: The liquid obtained after separation from the suspended matter by filtration or centrifugation may be used for the determination of dissolved solids.

Fixed Suspended Solids

Keep the dish containing the total suspended solids in a muffle furnace at 550°C for 1 hr. Cool and weigh.

Express the result as mg fixed suspended solids per litre of sample.

Volatile Suspended Solids

mg/l Volatile Suspended Solids = mg/l Total Suspended Solids – mg/l Fixed Suspended Solids

It is a usual convention that fixed suspended solids represents the inorganic mineral matter fraction while volatile suspended solids indicate the organic matter content.

40-5 DISSOLVED SOLIDS

Determination of dissolved solids particularly of fixed dissolved solids represents the inorganic mineral matter content.

Total Dissolved Solids

Centrifuge or filter a suitable volume of the sample through a glass fibre filter paper. Evaporate the filtrate (or centrifugate) in a tared porcelain dish (The porcelain dish should have been preheated at 105°C and then at 550°C, cooled and brought to constant weight). Keep the dish at 105°C for about 1 hr, cool and weigh. The increase in weight denotes the total dissolved solids content. Express the result as mg. total dissolved solids per litre of sample.

Fixed Dissolved Solids

Ignite the residue obtained from "Total Dissolved Solids" by keeping it in a muffle furnace at 550°C for 1 hr cool and weigh.

Express the result as mg. Fixed Dissolved Solids per litre of sample.

The value represents the inorganic mineral matter content of the effluent.

Volatile Dissolved Solids

mg/l Volatile Dissolved Solids = mg/l Total Dissolved Solids – mg/l Fixed Dissolved Solids

40-6 SETTLEABLE SOLIDS

The term settleable solid is applied to solids in suspension that will settle under quiescent conditions under the influence of gravity. The coarser suspended particles with a specific gravity greater than that of water only will settle.

Determination of settleable solids of effluents and other waste waters is useful in deciding the need and design of primary settling tanks for biological treatment processes. It is also used to determine the efficacy of sedimentation tanks and to indicate the amount of sludge settling in humus tanks in the case of filter effluent.

For the determination of settleable solids an Imhoff cone is used. It is an inverted glass cone of 1 liter capacity having the apex part graduated in milliliters.

Procedure

Fill an Imhoff cone to the 1 litre mark with the thoroughly mixed sample and allow to settle. Gently dislodge the material adhering to the sides of the cone after about 45 minutes by using a glass rod or by spinning the cone. Record the volume of the sludge that has settled to the apex after one hour (time reckoned from filling the cone with the sample) or other specified time related to the works condition. Express the result as ml. Settleable Solids per litre of the sample per unit time.

40-7 pH VALUE

pH is a measure of hydrogen ion concentration or more precisely the hydrogen ion activity. pH is the most important parameter as it indicates instantaneously the acidic or alkaline condition of an effluent or water. Further all the treatment processes are based on pH values only.

The application of pH values are numerous. Noted among them are:

1. To determine the type of treatment to be given to the effluent,
2. To judge the efficacy of treatment.
3. To carry out the treatment at various desired levels.

pH can be measured electrometrically or colorimetrically. pH meters are the widely employed instruments for the electrometric measurement of pH.

pH meters are usually standardized with standard buffer solutions. The preparation of some of the buffers is given below.

pH 4* Buffer solution

Dissolve 10.12 g. anhydrous potassium hydrogen phthalate-$KHC_8H_4O_4$ in distilled water and make upto 1000 ml in a volumetric flask.

pH 7* Buffer solution

Dissolve 1.361 g. anhydrous potassium dihydrogen-phosphate, KH_2PO_4 and 1.420g anhydrous disodium hydrogen-phosphate Na_2HPO_4 (Both of which have been dried at 110°C to 130°C for 2 hr) in distilled water and make up to 1000 ml in a volumetric flask.

pH 9* Buffer solution

Dissolve 3.81 g. of sodium borate decahydrate (borax) $Na_2B_4O_7.10H_2O$ in distilled water and make upto 1000 ml.

Commercial buffer solutions are also available in tablet, powder and concentrated solution forms. These must be dissolved and made upto the stated volume with distilled water.

The points to be observed during the preparation of standard buffer solution are:

---

* at 25°C

1. Use distilled water having a specific conductance less than 2 micromhos at 25°C and pH 5.6 to 6.3.
2. The distilled water for the preparation of borax and phosphate buffer solutions should be boiled and cooled to expel the carbondioxide and to produce pH of 7.6 to 7.8.
3. Care should be taken to make complete dissolution of salts otherwise the pH may be incorrect.
4. Store the buffer solutions in polythene containers.

Measurement of pH

Follow the manufacturers' instructions to operate the pH meter and determine the pH of the sample of waste water. Before the measurement of pH, standardize the pH meter using a standard buffer solution of pH near that of the sample to be tested. Check the electrode occasionally by measuring the pH of another standard buffer solution with a different pH.

The following instructions have to be observed.

i. Wash the electrodes thoroughly with distilled water and then with the sample before the measurement of pH.

ii. Allow the system to stabilize before making the final reading.

iii.. When samples containing oil or grease are to be determined, remove them by filtration and then measure the pH. If any oily film sticks to the electrode, it can be removed with a soft tissue soaked in a suitable solvent or detergent followed by thorough washing with distilled water.

iv. If the pH value exceeds 10, sodium ions also respond. In that case use a special type of glass electrode. If no such electrode is available make suitable corrections in the measurement made with the ordinary glass electrode. Usually manufacturers of pH meter supply such correction charts, along with the instrument.

v. If glass electrodes are dried out, soak them in N/10 HCl for about 10 to 12 hrs or according to the manufacturer's instructions.

vi. Care should be taken while filling the solution in the reference electrode such that no air bubbles are entrapped in the solution or in the liquid junction.

vii. After measurement of pH, the electrodes should be thoroughly washed with distilled water.

Colorimetric method

Colorimetric methods using color glass standards or pH papers yield only approximate values. However for field use and in the case of non-availability of pH meters, these methods are applicable.

40-8 ACIDITY

The industries which discharge mineral acids are. chemical, fertilizer, pesticide, and battery manufacture, electroplating, viscose rayon manufacture, mining, iron and copper pickling etc., Organic acids are discharged from the manufacture of rayon, fermentation plants, distilleries, dyeing industries etc., Acidity data is useful in determining the dosage of lime (or other agents) required for neutralization, and to determine the size of chemical feeders, neutralization tanks and other related factors.

Acidity is usually determined by titration with standard sodium hydroxide solution (0.1 N, 0.05 N or 0.02 N). The quantity of sodium hydroxide required for the sample (pH below 4.5) to reach pH 4.5 (methyl orange end point) is a measure of "mineral acidity" and the quantity to reach pH 8.3 (phenolphthalein end point) is a measure of the total acidity.

To determine acidity, color change titrimetric method is employed for colorless and slightly colored effluents and potentiometric titration for colored effluents.

Interferences

Residual chlorine interferes by bleaching the color of the indicator. It can be removed by the addition of sodium-thiosulfate solution. A fading and impermanent end point characterizes the phenolphthalein acidity. Titrations conducted at boiling temperature yield better values, as boiling speeds up hydrolysis of metal salts ($FeCl_3$, $AlCl_3$ etc. which release mineral acidity on hydrolysis) allowing the titration to be completed more rapidly.

Reagents

1. **Methyl orange indicator solution:** Dissolve 50 mg methyl orange powder in distilled water and dilute to 100 ml.
2. **Phenolphthalein indicator solution:** See Chapter 40-9

"Alkalinity"

3. **Standard sodium hydroxide solution, 1N:** Dissolve 40g. NaOH in distilled water and make upto 1000 ml in a volumetric flask. Standardize it against standard. 1N. sulfuric acid (prepared for alkalinity determination).

4. **Standard sodium hydroxide solutions** of various strengths using appropriate volumes of 1N. sodium hydroxide.

Procedure for Colorless or Slightly Colored Effluents

A) Methyl orange acidity (Mineral acidity)

Place 50 ml or suitable aliquot of the sample (requiring not less than 10ml and not more than 25 ml of titrant) in a 250 ml conical flask. Dechlorinate the sample if it contains any chlorine by the addition of 1 drop of 0.1 N sodium thiousulfate solution (25g. $Na_2S_2O_3$. $5H_2O$/1000 ml). Add 4 drops of methyl orange indicator solution and titrate with sodium hydroxide solution of 0.1 N or 0.05 N or 0.02 N according to the acidity of the sample to faint orange color (to pH 4.5) This titre value represents mineral acidity.

B) (i) Phenolphthalein acidity (Total acidity) at room temperature

To a suitable aliquot of the dechlorinated sample in a 250 ml conical flask add 4 drops of phenolphthalein indicator and titrate with sodium hydroxide solution of appropriate normality to the appearance of faint pink color (to pH 8.3)

(ii) Phenolphthalein acidity (Total acidity) at boiling temperature

Add 10 drops of phenolphthalein indicator to a suitable volume of the sample in a conical flask and heat to boiling for 2 minutes. Titrate while hot with sodium hydroxide solution of appropriate normality to the permanent faint pink color.

Procedure for Colored Effluents

Carryout the titrations potentimetrically (using a pH meter) to the required pH values.

Calculation

$$mg/l\ Acidity\ (as\ CaCO_3) = \frac{ml.NaoH \times N.\ of\ NaOH \times 50 \times 1000}{ml.\ Sample\ for\ titration}$$

(1ml of exactly 1 N.NaOH = 50 mg $CaCO_3$)

While reporting the results, state the indicator used and the temperature at which the titration has been performed.

40-9 ALKALINITY

Effluents from mercerizing and most other textile processes, liming of skin during tanning, manufacture of soaps and detergents, bleaching and bleach liquor manufacture, edible oil refining, and boiler blow-down are highly alkaline in nature.

Effluents containing caustic alkalinity should not be discharged into water courses without treatment. Alkalinity calculations are helpful in determining the nature of alkalinity in the waste and the amount of chemicals required for neutralization. Alkalinity and pH are important parameters in determining the amenability of effluents to biological oxidation.

Most of the effluents possess caustic alkalinity

(hydroxide alkalinity) only. Quantity of carbonates and bicarbonates present in the effluents are much less. Some of the effluents contain silicates, phosphates (and sometimes borates) in significant quantities. They are also titrated as alkalinity.

For residual sodium carbonate (RSC) calculations, the carbonate and bicarbonate alkalinity values are to be taken.

Alkalinity of waste samples is determined with 1N, 0.1N, or 0.02 N. H_2SO_4 according to the amount of alkalinity present. Colorless and slightly colored effluents are directly titrated using phenolphthalein and methyl orange indicators. Electrometric titration is employed for colored effluents.

Reagents

1. **Sodium carbonate solution, 1N:** Weigh accurately 13.25 g. anhydrous sodium carbonate Na_2CO_3 (Previously dried at 140°C for 2 hrs). Dissolve it in little distilled water and make upto 250 ml in a volumetric flask.

2. **Sulfuric acid, 1N:** Place 28.0 ml conc. H_2SO_4 in a 1000 ml. volumetric flask and make upto the mark with carbondioxide free distilled water. Standardize it against 1N. sodium carbonate solution using methyl orange as the indicator and calculate the exact normality.

   $$1.0\text{ml exactly } 1N.H_2SO_4 = 50 \text{ mg } CaCO_3$$

   Dilute appropriate volumes of $1N.H_2SO_4$ to prepare 0.1N. 0.2N and 0.02N, H_2SO_4.

3. **Phenolphthalein indicator solution:** Dissolve 500 mg phenolphthalein in 50 ml ethyl or isopropyl alcohol and

add 50 ml distilled water. Add sodium hydroxide solution (0.02 N.) dropwise until a faint pink color appears.

4. **Mixed indicator solution:** (This indicator gives better color change and hence used in place of methyl orange indicator). Dissolve 20 mg methyl red and 100mg bromocresol green in 100 ml 95% ethyl or isopropyl alcohol (If the sodium salts of indicator are used, prepare the solution with distilled water instead of alcohol).

(If the component indicator substances are not available, use 0.05% alcoholic solution of methyl orange)

Procedure

I. Color change titrimetric method for colorless and slightly colored samples

If the sample contains any chlorine, dechlorinate by the addition of 1 drop 0.1N. sodium thiosulfate solution (25 g $Na_2S_2O_3.5H_2O$/1000 ml)

A) Phenolphthalein alkalinity

Place 25 ml or 50 ml or appropriate volume of the sample (requiring not more than 25 ml titrant) in a conical flask. Adjust the volume to 50 ml with distilled water. Add 6 drops of phenolphthalein indicator solution. If no pink color appears, there is no phenolphthalein alkalinity. If pink color appears, then titrate with sulfuric acid of suitable normality (1N, 0.1 N 0.2N or 0.02N) to pH 8.3 until the solution becomes colorless.

B) Total Alkalinity (Methyl Orange Alkalinity)

Add 4 drops of mixed indictor (or methyl orange) to the sample in which phenolphthalein alkalinity has been determined and titrate against sulfuric acid to pH 4.5. The color change is from emerald green to light pink. (With methyl orange, the color change is from yellow to orange red)

II. Electrometric titration for colored samples

Carryout the phenolphthalein alkalinity and total alkalinity titrations potentiometrically to pH 8.3 and 4.5. Any pH meter with a glass electrode and calomel electrode (or combination electrode) is sufficient for this purpose.

If the sample contains suspended matter, remove by filtration or centrifugation and then determine the alkalinity. If suspended mater is suspected to contain any alkalinity add a known excess of standard acid to the sample, boil off carbon-dioxide, cool and back titrate with standard alkali solution.

Calculations

mg/l Phenolphthalein alkalinity

$$= \frac{ml.\ H_2SO_4\ for\ phenolphthalin\ end\ point \times N.\ of\ H_2SO_4 \times 50 \times 1000}{ml.\ sample\ take\ for\ titration}$$

mg/l Total alkalinity

$$= \frac{ml.\ H_2SO_4\ for\ total\ alkalinity\ end\ point \times N.\ of\ H_2SO_4 \times 50 \times 1000}{ml.\ sample\ take\ for\ titration}$$

If the sample does not contain silicates, phosphates and borates in significant quantities, the following table 40-9-1 may be used to calculate hydroxide, carbonate and bicarbonate alkalinities.

Table 40-9-1
Hydroxide, Carbonate and Bicarbonate Alkalinities

	Hydroxide alkalinity (as $CaCO_3$)	Carbonate alkalinity (as $CaCO_3$)	Bicarbonate alkalinity (as $CaCO_3$)
P = 0	0	0	T
P < ½ T	0	2P	T – 2P
P = ½ T	0	2P	0
P > ½ T	2P–T	2T – 2P	0
P = T	T	0	0

Problem

The phenolphthalein and total alkalinity values of a dyeing effluent are 13,100 mg/l (as $CaCO_3$) and 16,100 mg/l (as $CaCO_3$) respectively. The waste does not contain any silicate or phosphate. Calculate hydroxide, carbonate and bicarbonate alkalinities of the sample.

	Hydroxide alkalinity (as $CaCO_3$)	Carbonate alkalinity (as $CaCO_3$)	Bicarbonate alkalinity (as $CaCO_3$)
P > ½ T	= 2 P – T	= 2(T – P)	0
	= (2 x 13600 – 16100)	= 2(16100 – 13600)	
	= 27200 – 16100	= 2(2500)	
	= 11,100 mg/l	= 5000 mg/l	

REFERENCES

1) ABCM - SAC Joint committee, "Analyst", 1956, 81, 721

2) J.F.T. Thomas & J J Lynch. "JAWWA" 1961,52,259.

3) A.I.Vogel, "A Text book of Quantitative Inorganic Analysis Including Elementary Instrumental Analysis" 3rd ed - 1964. ELBS - Longman Green & Co., Ltd., London.

Chapter - 41

MEASUREMENT OF ORGANIC POLLUTION

41-1 DISSOLVED OXYGEN

The measurement of dissolved oxygen indicates immediately the purity of water and is important for maintaining aerobic conditions in the receiving waters and in the aerobic treatment of sewage and industrial waste waters. Further DO determination is the basis for BOD test which is used to evaluate the pollutional strength of waste waters and rate of biochemical oxidation.

Sample Collection and Preservation

Samples have to be collected in BOD bottles. If volumes higher than 300 ml are required, then they have to be collected in separate 1 litre pyrex or corning bottles. Special sampling devices such as Displacement sampler (or) Kemmerer type samplers may be employed in collecting samples from depth. (For details of these samplers refer "Annual Book of ASTM Standards" -1972, part 23). Fill the bottles without entrapment of any air. As far as possible the samples should not be disturbed during collection. The liquid may also be siphoned through a glass or rubber tube into the sample bottle. The tip of the tube should be at the bottom so as to deliver to the bottom of the bottle and allowed to flow until the liquid in the bottle has been changed several times.

The temperature of the sample should be recorded as

soon as the sample is collected. It is also useful to note barometric pressure, especially, for calculating percent saturation of oxygen.

It is advisable to determine the DO in the field itself. If it is not possible, fix the DO by adding 2 ml each of manganous-sulfate solution and alkali-iodide-azide to the sample in the BOD bottle. Preserve the fixed sample by keeping at 4°C in dark. Iodometric and Electrode methods are the best available methods for the measurement of DO. Selection of the method among these two, largely depends on the interferences present and accuracy desired.

IODOMETRIC METHOD

The basic Winkler method is used to determine DO. It has been modified however to remove the most common interferences. Among the modifications, the Alsterberg's modification of adding sodium azide to remove nitrite and Rideal-Stewart's modification of adding permanganate to eliminate the interference of ferrous iron are important.

Principle

When manganous sulfate is added to the sample containing alkaline potassium- iodide, manganous hydroxide is formed, which is oxidized by the dissolved oxygen of the sample to basic manganic oxide. On addition of sulfuric acid, the basic manganic oxide liberates iodine equivalent to that of dissolved oxygen originally present in the sample. The liberated iodine is titrated with standard solution of sodium thiosulfate using starch as the indicator.

$$MnSO_4 + 2KOH \rightarrow Mn(OH)_2 + K_2SO_4$$

$$2Mn(OH)_2 + O_2(\text{Dissolved Oxygen}) \rightarrow 2MnO(OH)_2$$

(Basic Manganic oxide)
(Brown in color)

$$MnO(OH)_2 + 2H_2SO_4 \rightarrow Mn(SO_4)_2 + 3H_2O$$

Manganic sulfate

$$Mn(SO_4)_2 + 2KI \rightarrow MnSO_4 + K_2SO_4 + I_2$$

$$2Na_2S_2O_3 + I_2 \rightarrow Na_2S_4O_6 + 2NaI$$

Interferences

Nitrite is the major interfering ion. Ferrous and ferric-iron, residual chlorine, oxidizable sulfur compounds such as sulfide, sulfite, thiourea and its derivates, organic matter and suspended matter also interfere.

Nitrite is eliminated by the addition of sodium azide. Interference due to ferrous iron, nitrite and unsaturated organic compounds are eliminated by adding potassium-permanganate.

Interference by ferric compounds are suppressed by the addition of phosphoric acid, in place of sulfuric acid.

Samples containing suspended solids absorb appreciable quantities of iodine in acid solution. Suspended matter can be flocculated by the use of the alum solution and the supernatant liquid may be used for DO determination.

When sulfites, thiosulfate and polythionates are present, a preliminary oxidation with alkaline sodium hypochlorite is necessary (See procedure-Page 331).

If the sample contains any residual chlorine, it has to be removed by the addition of sodium sulfite or thiosulfate (See procedure).

Reagents

1. **Manganous sulfate solution:** Dissolve 91.0 g manganous sulfate monohydrate, $MnSO_4.H_2O$ in distilled water. Filter if necessary and dilute to 250 ml. 1 ml of this solution when treated with 50 ml. of acidified potassium iodide solution should not liberate iodine.
2. **Alkali-iodide-azide reagent:** (i) Dissolve 175g. potassium hydroxide (or 125 g. sodium hydroxide) and 37.5g. potassium iodide (or 33.7g. sodium iodide) distilled water and dilute to 250 ml.

   Note: Sodium and potassium salts may be used interchangeably.

   (ii) Dissolve 2.5g. sodium azide in 10 ml distilled water.

   (iii) Pour the azide solution to the alkali-iodide solution and mix well.

   This solution should not give color with starch solution when diluted and acidified.
3. **Conc. sulfuric acid**
4. **Phosphoric acid: 85 to 90%**
5. **Sodium thiosulfate solution:** 0.1 N: Dissolve 24.82 g. sodium thiosulfate, $Na_2S_2O_3.5H_2O$ in boiled and cooled distilled water and make up to 1000 ml in a volumetric flask. Standardize it against standard potassium dichromate solution (0.1 N) as given below :

Standardization of sodium thiosulfate using standard potassium dichromate:

Dissolve 1.2257 g. potassium dichromate (previously dried at 103°C for 1 hr) in distilled water and make upto 250 ml in a volumetric flask. This is exactly 0.1000 N solution. Place 100 ml boiled cooled (oxygen free) distilled water in a 500 ml iodine flask. Add 3 g. potassium iodide, 2.0 gm sodium bicarbonate and shake until the salts dissolve. Add 6 ml. conc. HCl. Pipet 25.0 ml. 0.1000 N potassium-dichromate solution in to the flask. Cover the flask with the stopper and keep it in the dark for 5 minutes. Rinse the watch glass and dilute the solution in the flask to 250 ml with boiled and cooled distilled water (free of oxygen). Titrate the liberated iodine with the sodium thiosulfate solution in the burette. When the solution acquires yellowish green color, add 1 ml. starch indicator solution. Now a blue shade appears. Rinse the sides of the flask and continue the titration until one drop changes the color from greenish blue to light (chrome) green. Calculate the normality of sodium thiosulfate solution.

6. **Sodium thiosulfate solution, 0.025 N:** Dilute appropriate volume of 0.1 N sodium thiosulfate solution (about 250 ml required) to 1000 ml with distilled water in a volumetric flask.

   1.0ml exactly 0.025 N sodium thiosulfate ≡ 0.2 mg oxygen

7. **Starch solution:** Dissolve 1 g. starch in a little water. Stir it with a glass rod to make it as a thin paste. Pour this paste in about 100 ml boiling distilled water and continue boiling for 2 minutes and cool.

Procedure

1. Collect the sample in a BOD bottle.
2. Add 2 ml manganous sulfate solution followed by the addition of 2 ml alkali-iodide-azide solution. The tip of

the pipet should be below the surface of the liquid.

3. Stopper the bottle without entrainment of air and mix by inverting the bottle atleast 10 times.

   Note: Now a brown precipitate of basic manganicoxide appears. Instead of brown precipitate, if a white precipitate is formed, the sample may be considered devoid of oxygen).

4. Allow the precipitate to settle completely leaving a clear supernatant liquid.

   (With saline waters the precipitate settles only with difficulty).

5. Carefully remove the stopper and add 2 ml. conc. sulfuric acid* by the sides of the bottle.

6. Stopper the bottle and mix thoroughly until dissolution is complete.

7. Measure 203 ml of the solution from the bottle (which corresponds to 200 ml of the original sample) into a conical flask of 500 ml capacity.

8. Titrate immediately with 0.025N sodium thiosulfate solution using starch as indicator to colorless end point.

Calculations

If the thiosulfate solution is exactly 0.025 N, then,

mg/l Dissolved Oygen = ml thiosulfate x 1

ml/l of DO at $0^{o}C$ and 760 mm pressure = mg/l DO x 0.70

---

* If the sample contains ferric iron, use phosphoric acid instead of sulfuric acid.

Samples containing excess nitrite (5 to 40 mg/l)

First add 0.2ml. 3% aqueous solution of sodium azide to the sample and then proceed from step No. 2.

Samples containing ferrous iron

Samples such as mine drainage waters and acid stream waters usually contain significant amounts of ferrous iron. It can be eliminated as follows:

To the sample collected in a 300 ml BOD bottle, add 0.7 ml conc. H_2SO_4 keeping the tip of the pipet below the surface of the liquid. Add sufficient amount of potassium permanganate solution (630 mg $KMnO_4$/ 100 ml) such that the liquid after mixing retains a red-violet tinge for about 20 minutes. Add carefully, potassium oxalate solution (2g $K_2C_2O_4 . H_2O$/100 ml) using a graduated 1 ml pipet. Take care to avoid large excess of potassium oxalate solution. Mix well and allow to stand in the dark. (The reaction occurs more readily in the dark). After the color of the permanganate has been discharged, add 2 ml manganous sulfate solution. Add 3 ml alkali iodide-azide reagent instead of 2 ml. Proceed from step No. 3.

Note: Since potassium permanganate and potassium oxalate solutions contribute to the volume, measure 205 ml of the final solution instead of 203 ml and titrate with 0.025 N sodium thiosulfate.

Samples containing suspended matter

Collect the sample in a 500-1000 ml bottle with usual precautions as for regular DO samples. Add 10 ml alum solution (10g. $AlK(SO_4)_2.12H_2O$ /100 ml) and add 2 ml. conc. ammonia solution. Stopper and mix well by inverting. Allow to settle for about 10 minutes. Siphon the clear supernatant into a 300 ml BOD bottle until it overflows. Stopper it carefully and proceed from step No 2.

Samples containing sulfides and excess organic matter

Electrometric method is preferable.

Samples containing sulfite, thiosulfate and polythionates and other oxidizable substances

Industrial waste waters such as paper mill wastes usually contain sulfites, thiosulfate and polythionates. Siphon two portions of 300 ml into two BOD bottles. To the sample in one bottle add 2N. sodium hypochlorite (NaOCl) solution (0.2 to 0.4 ml is enough) to oxidize the sulfur compounds to sulfates. Mix by inversion for 20 to 30 seconds. Add 1 ml potassium iodide solution (IN) and then acidify with 1 ml or more of H_2SO_4 (1+9). Mix by inversion. Neutralize the liberated iodine with 0.1N sodium sulfite solution using 0.2 ml starch indicator. Calculate the quantity of sodium hypochlorite required to neutralize the oxidizable substances and add it to the sample in another BOD bottle. Allow four minutes for completion of reaction and then proceed from step No.2 of the procedure.

Samples containing chlorine and hypochlorites

Proceed as described for samples containing sulfites but start from the addition of potassium iodide solution. Add the amount of sodium thiosulfate (or sulfite) required to neutralize chlorine or hypochorites to the sample in another BOD bottle. Proceed from step No.2 of the procedure.

ELECTRODE METHOD

Dissolved oxygen is well determined by electrode method also. It is very simple and the dissolved oxygen content can be found instantaneously. Measurements using dissolved oxygen electrodes are gaining prominence as they allow insitu measurements. The DO electrodes can be lowered to various depths is biological waste treatment tanks to monitor the DO level of any point. The rate of oxygen utilization in such a tank can also be determined by collecting the mixed liquor in a BOD bottle and dipping the electrodes in it and observing the rate of oxygen depletion.

The DO electrodes are calibrated by making measurements in water samples, the DO of which have been already determined by the Winkler method described previously. During dissolved oxygen measurements sufficient movement of the sample by electrode be maintained to prevent low readings. DO electrodes are very sensitive to temperature, and accurate measurements of temperature have to be made along with DO measurement for applying suitable correction, or instrument equipped with automatic temperature compensation may be used.

41-2 BIO-CHEMICAL OXYGEN DEMAND (BOD)

Bio-chemical Oxygen Demand (BOD) is a test of great value in the analysis of industrial effluents, sewage and grossly polluted waters. BOD is the quantity of oxygen required by bacteria and other micro organisms, during the biochemical degradation and transformation of organic matter (present in wastewater) under aerobic conditions.

Inspite of many limitations, BOD test still remains the best available single test for assessing organic pollution. It is the only test that gives a measures of the amount of biologically oxidizable organic matter. BOD is the major criterion used in stream pollution control, where organic loading must be restricted to maintain desired dissolved oxygen levels. It is highly useful in the studies of measuring the purification capacity of streams and serves regulatory authorities as a means of checking the quality of effluents discharged to such waters.

The BOD determination involves the measurement of dissolved oxygen content of the sample, before and after 5 days incubation at 20°C. If the sample does not contain any oxygen, it is supplied with oxygen and depletion is calculated as BOD.

To carryout biochemical oxidation, microbial organism (seed) have also to be provided to the sample if necessary. The following points should be read carefully before proceeding with BOD determination.

Nitrification

The BOD determination is aimed to find out the amount of biochemically oxidizable carbonaceous matter only. If nitrogenous matter (ammonia and organically bound nitrogen) are present, they will eventually be oxidized to nitrite and nitrate in the presence of nitrifying bacteria and contribute to BOD.

$$2NH_3 + 3O_2 \xrightarrow{\text{Nitrite-forming bacteria}} 2NO_2^- + 2H^+ + 2H_2O$$

$$2NO_2^- + O_2 + 2H^+ \xrightarrow{\text{Nitrate-forming bacteria}} 2NO_3^- + 2H^+$$

Fortunately, the reproductive rate of nitrifying bacteria at 20°C is much less and the required population to deplete oxygen will be attained only from 8th day. Hence nitrification does not occur to any detectable extent during the 5-day determination of industrial wastes and crude and settled sewages. However in treated effluents and river waters which have already been partially nitrified, it is doubtful whether nitrification will occur within 5 days of BOD period or after 5days. If nitrification occurs to a considerable extent within 5 days in these samples, the BOD determination has to be carried out only after suppressing nitrification.

Presence of bacteriostatic and bactericidal substances

The presence of substances that inhibit or prevent the growth of bacteria, cause errors in the BOD determination.

Acids and alkalis have a marked effect upon bacterial activity and hence before proceeding the determination, the pH of the samples should be brought to the range of 6.5 to 8.3

Free chlorine present in the sample affects the determination and the sample should be dechlorinated with sodium sulfite.

The presence of toxic metals are undesirable as they are bacteriostatic and consequently low BOD values are obtained. (The presence of 1 mg/l of copper depresses the BOD by about 33%. 1mg/l of zinc causes 17% reduction and 62.5 mg/l of zinc depresses the BOD by about 50%). It is better to conduct 4 hours permanganate test instead of BOD test with samples of industrial effluents and river water containing both organic and metallic contaminants.

Separate determinations of organic carbon and nitrogenous matter are preferable for samples of high salinity.

Presence of nutrient materials

Mineral nutrients are necessary for the growth and activities of bacteria. Carbon, nitrogen, phosphorus and sulfur are the most important elements. Traces of calcium magnesium, potassium and iron are also needed. These are supplied by the dilution water.

Dilution water

As most of the effluents contain large quantities of organic matter and do not contain dissolved oxygen, they have to be diluted and supplied with oxygen. This is done by diluting them with standard dilution water containing sufficient oxygen. As natural surface waters and tap waters are found unsuitable for direct use as dilution water, a synthetic dilution water has to be prepared.

A suitable dilution water should satisfy the following requirements: correct pH limits, proper salinity, reasonable buffer capacity, the presence of necessary mineral nutrients and the absence of bactericidal and bacteriostatic substances such as toxic metals and free chlorine.

The dilution water is usually buffered by the addition of phosphate buffer (containing a mixture of phosphates) to a pH of about 7. The buffer is added in order to maintain favorable pH conditions at all times. The sodium and potassium phosphates contained in the buffer and the calcium and magnesium salts added help to maintain the proper osmotic conditions. Further, they maintain the buffering capacity of the dilution water and are utilized by microorganisms for their growth and metabolism. The calcium and magnesium salts also contribute to the total salt content. Magnesium sulfate, ferric chloride and ammonium chloride supply to requirements for sulfur, iron and nitrogen respectively and any phosphorus required is provided by the phosphate buffer added.

Seeding with micro-organisms

It is essential to seed the dilution water when BOD determination is made on samples of industrial effluents which do not contain bacteria. The seeding material generally used is fresh and settled raw sewage. If the effluent is not readily biodegradable or toxic, acclimated seed must be used. See page 340.

Temperature

Apart from all, temperature is an important factor in the determination of BOD. As described previously, the standard temperature for the test is 20°C. The temperature should be kept constant throughout the 5 day period. Even a slight increase or decrease of 1°C causes an increase or decrease of BOD by 4.7%.

Precautions to be observed

Samples must be free from all added preservatives. The test should be performed as soon as possible after the sample has been collected. When it is impracticable to conduct the test immediately, samples can be stored at 4°C. On any account, samples should not be stored at room temperature.

Collection and Preservation of Samples

The samples have to be collected as for dissolved oxygen determination in BOD bottles with usual precautions. As mentioned above, keep the samples at $4^{O}C$ and start the test at the earliest possible but not later than 24 hrs. For samples of high BOD it is enough to collect in one BOD bottle, but for samples of low BOD (such as polluted waters) it is necessary to collect them in 2 or more BOD bottles.

DILUTION FOLLOWED BY BOD DETERMINATION TITRATION / ELECTROMETRIC METHOD

Principle

The dissolved oxygen content of the sample is determined before and after five days incubation at 20°C. The amount of oxygen depleted is calculated as BOD. Samples devoid of oxygen or containing less amount of oxygen, are diluted several times with special type of dilution water saturated with oxygen, in order to provide sufficient amount of oxygen for oxidation.

Reagents

Use good quality distilled water containing less than 10 mg/l copper and free from chlorine.

1. Reagents for dilution water

a) Calcium chloride solution: Dissolve 27.5g. anhydrous calcium chloride, $CaCl_2$ (or equivalent amount, if hydrated calcium chloride is used) in distilled water and dilute to 1000 ml.

b) Magnesium sulfate solution: Dissolve 25g. magnesium sulfate hepta hydrate, $MgSO_4 . 7H_2O$ in distilled water and dilute to 1000 ml.

c) Ferric chloride solution: Dissolve 0.25g ferric-chloride hexahydrate, $FeCl_3.6H_2O$ in distilled water and dilute to 1000 ml

d) Phosphate buffer solution: Dissolve 8.5g. potassium dihydrogen phosphate, KH_2PO_4, 21.75g dipotassium hydrogen phosphate, K_2HPO_4, 33.4g disodium-hydrogen phosphate hepta hydrate $Na_2HPO_4.7H_2O$, and 1.7g ammonium chloride, NH_4Cl, in 500 ml distilled water and make upto 1000ml. The pH of this buffer solution should be 7.2. keep this reagent bottle in a refrigerator to prevent mold growth.

Note: If biological growth is noticed in any of the above reagents during storage, discard it and prepare freshly.

2. *Dilution water*

i) A high quality organic free water must be used for dilution water. Distilled water with less than 10μg/l of copper, and free of chlorine, chloramines, caustic alkalinity and organic matter, is the preferred one. demineralized water can also be used. (A few disadvantages of demineralized water have to be mentioned. If the demineralized water is from a new demineralizer with new resin, it would contain substantial amount of organic matter that is not at all detected by conductivity meter. As there is a large surface area, bacterial growth in the columns is possible). The most practical way to produce water of low organic content is by distillation from alkaline permanganate. Store the distilled water in the BOD incubator until it reaches 20°C. Aerate required volume of the water (at 20°C) with a supply of clean compressed air. Add 1 ml each of calcium chloride, magnesium sulfate, ferric chloride and phosphate buffer solutions per litre of aerated distilled water and mix thoroughly. This is the standard dilution water. Prepare this dilution water just before use. Discard any dilution water remaining after use and clean the bottle preferably with a sterilizing agent.

ii) Seeding with micro-organisms

Seeding is the addition of small measured volume of water containing a good bacterial population to the dilution water. Seeding is not necessary for samples containing bacterial flora. (Hence sewage, sewage effluents and most river waters need not be seeded). It is essential to seed the industrial effluents which usually do not contain any bacteria. The seeding material generally preferred is fresh and settled raw sewage. Fresh and settled final effluent of good quality from an aerobic biological process is also recommended. The seed should be kept at 20°C for 24-36 hrs. before use. The recommended seed concentration is 1-2 ml per litre of dilution water.

If the waste sample to be tested contains materials not readily biodegradable by the microbes present in domestic sewage or if it contains toxic materials such as phenol, formaldehyde or similar microbic inhibitory agents, use as seed the effluent obtained from biological treatment of the industrial waste (under determination). If the waste is not treated, the receiving water below the point of discharge (3 to 8 Km) may be used as seed. In the absence of these sources an acclimated seed may be used. Acclimated seed is prepared as follows:

One to two litres of domestic (raw) sewage is aerated for about 24 hrs and allowed to settle. Then two-thirds of the supernatant liquid is siphoned off and discarded. The container is then refilled with the domestic sewage containing 10% of the effluent under determination and once again aerated for 24 hrs. This procedure is repeated with 10% increments of effluent until the fill liquid is 100% effluent. Even if this seed is ineffective, a specific waste culture should be included in the conditioning process.

Note: A satisfactory dilution water when incubated with or without seed under standard conditions should not absorb more than 0.2mg/l of oxygen.

3. *Reagents for the determination of Dissolved Oxygen:*

I. Manganous sulfate solution: See " Chapter 41-1, Dissolved Oxygen". Page No. 327

ii. Alkali-iodide-azide Reagent: See Chapter 41-1, "Dissolved Oxygen". Page No. 327

iii. Conc. sulfuric acid, or phosphoric acid, H_3PO_4 85 to 90%.

Procedure

Samples must be given pretreatment (if any required) and various dilutions of the sample are made with standard or seeded dilution water and each dilution is taken in two BOD bottles. One set is used for the immediate determination of dissolved oxygen and the other is kept for incubation at 20°C for five days. After 5 days, the DO of the incubated samples are determined and BOD values are calculated as per the equation given in calculation.

A) Pre-treatment

1) If the pH of the sample is not in the range of 6.5 to 8.5, add sufficient amount of 1 N.H_2SO_4 or 1N.NaOH to bring the pH in that range. Determine the amount of acid or alkali needed on a separate aliquot of the sample.

2) If the sample contains any chlorine, dechlorinate using freshly prepared sodium sulfite solution of 0.025N (1.575g anhydrous Na_2SO_3/1000 ml). Determine the amount of sodium sulfite required on a separate aliquot of the sample by iodometric method.

3) Some samples may be supersaturated with dissolved oxygen, especially waters containing algae. The excess oxygen above saturation level in those samples has to be removed by partly filling them in a bottle and then agitating at 20°C or by aerating at 20°C.

4) To suppress nitrification in some samples, it is necessary to add 1 ml of 0.5% solution of allyl thio-urea per litre of standard/seeded dilution water and the BOD determination can be proceeded as usual but the result should be reported as BOD, nitrification suppressed mg/l.

5) Samples that have been stored in refrigerator should be allowed to reach room temperature before dilutions are made.

6) Samples must be thoroughly mixed before dilution. Coarse suspended matter present in the samples usually causes difficulty. Those samples should be shaken vigorously just before the dilutions are made.

7) If the BOD of a polluted river water containing algae has to be determined, centrifuge the sample to remove algae and use the centrifugate for determination.

Dilution

It is recommended that a series of dilutions should be made for a sample and atleast three of the dilutions should deplete 20% to 90% of the initial dissolved oxygen. The permanganate value and COD value can be taken as a guide to the dilution. A general scheme of dilution is as follows:

Sample	Percent sample in dilution mixture	ml. sample in 1 litre dilution mixture
River waters	No dilution-100% sample	1000
Polluted river waters	50	500
	25	250
	20	200
	10	100
Biologically purified sewage effluents	50 to 10	500 to 100
Oxidized effluents	25 to 5	250 to 50
Settled sewage	5 to 1	50 to 10
Raw sewage	5 to 1	50 to 10
Industrial effluents	No limit can be suggested anyhow the following ranges may be useful	
	20%, 10%	200, 100
	4%, 2%, 1%	40, 20, 10
	0.4%, 0.2%	4, 2
	0.1%, 0.05%	1, 0.5

Note: If less than 5 ml of sample is required for dilution, dilute first a larger aliquot of the sample in a volumetric flask and use the diluted sample for BOD dilutions.

Dilution technic

i) Carefully siphon the standard dilution water (or seeded dilution water when trade wastes are to be determined), into a 1000 ml. graduated cylinder and half fill it without entrainment of air.

ii) Add the appropriate quantity of the well mixed sample into the cylinder by the sides without producing any air bubble. Make up to the 1000 ml. mark with dilution water.

iii) Mix well with a plunger type mixing rod, or any other suitable rod without entrainment of air.

iv) Siphon the mixture into two BOD bottles of 300 ml capacity. While siphoning, allow about 50 ml of mixture to waste. Fill the bottles and stopper carefully without allowing any air bubble to remain inside. Water seal both the bottles.

v) Prepare succeeding dilutions of lower concentrations in the same manner as described above.

Note: Samples may also be measured directly into BOD bottles of known capacity with a large tip volumetric pipet and diluted with dilution water.

vi) Siphon out standard dilution water/seeded dilution water into two BOD bottles and water seal them after filling completely.

Determination of Dissolved Oxygen

Use one set of sample dilutions and dilution water to find out the initial dissolved oxygen content. Keep the other set of sample dilutions and dilution water in the BOD incubator at 20°C (in dark) for 5 days. After 5 days, determine the DO content of dilution water and sample dilution, using the procedure given in the chapter 'Dissolved Oxygen' starting from the addition of manganous sulfate solution or by electrode method.

Calculations

$$BOD\ mg/l = (DO_0 - DO_5 - BC) \times \frac{100}{\text{percent sample}}$$

Where DO_0 is the initial dissolved oxygen content in mg/l
DO_5 is the dissolved oxygen content after incubation for 5 days.

BC is the difference between the dissolved oxygen contents of the blank on initial day and after 5 day incubation (blank correction).

Problem-1

The DO values of the industrial effluent of Viscose rayon manufacturing industry on initial day and after 5 days incubation were 7.8 and 4.6. Percentage dilution is 1%. DO of seeded dilution water were 7.8 and 7.6 Calculate the BOD of the sample.

$$BOD\ mg/l = (DO_0 - DO_5 - BC) \times \frac{100}{\text{percent sample}}$$

$DO_0 = 7.8$
$DO_5 = 4.6$
$BC = 7.8 - 7.6 = 0.2$

$$BOD\ mg/l = (7.8 - 4.6 - 0.2) \times \frac{100}{1}$$

$$= (3.2 - 0.2) \times 100$$

$$= 300$$

It is usual to get two or more values of BOD for a sample as many dilutions are prepared. In that case a question arises; which value is reliable and has to be taken?. The BOD value for dilution that depletes more than 2 mg/l oxygen and have residual oxygen more than 0.5 mg/l should be taken.

Problem-2

The untreated effluent of a sugar industry is analyzed to determine BOD. Dilutions made are 0.5%, 0.2% and 0.1%. DO on the initial day and after 5 day incubation are 7.6, 7.8, 7.8 and 0.3. 5.0 and 6.3 respectively. Calculate the BOD for each dilution and which BOD value is more reliable?

$$0.5\%\ BOD\ value = (7.6 - 0.3) \times \frac{100}{0.5} = 7.3 \times 200 = 1460$$

$$0.2\%\ BOD\ value = (7.8 - 5.0) \times \frac{100}{0.2} = 2.8 \times 500 = 1400$$

$$0.1\%\ BOD\ value = (7.8 - 6.3) \times \frac{100}{0.1} = 1.5 \times 1000 = 1500$$

The BOD value for 0.2% dilution is more reliable as it depletes more than 2.0 mg/l oxygen and have a residual oxygen more than 0.5 mg/l.

Immediate Dissolved Oxygen Demand

Some effluents/highly polluted waters may contain

large amounts of inorganic reducing substances such as sulfite, sulfide, ferrous iron etc., They consume the DO immediately and get oxidized. In these samples, the BOD value does not represent the oxygen demand due to organic substances only. To find out the real value of BOD, the oxygen depletion due to inorganic substances may be deducted from the calculated BOD value. For this, each dilution has to be filled in three BOD bottles. The DO of one bottle has to be found immediately. DO of another bottle has to be found out after 15 minutes. The third bottle has to be kept for incubation. Now calculate the real BOD value as follows :

DO of the dilution (titrated immediately as dilution is prepared) = IDO_0

DO of the dilution (titrated after 15 minutes) = IDO_{15}

Immediate Oxygen Demand (IDOD) = $(IDO_O - IDO_{15}) \times$ dilution factor

Real BOD value (only due to organic matter) = Calculated BOD value - IDOD

Checking the Correctness of the technique

The technique can be checked by using a standard solution containing glucose and glutamic acid. Dissolve 150 mg. glucose and 150 mg. glutamic acid (both dried at 103°C (for 1 hour) in distilled water and make upto 1000 ml. Prepare a 1 in 50 dilution using seeded dilution water and determine the BOD in the usual manner. The BOD should be approximately 200 mg/l. If the result is more then 240 mg/l. or less than 200 mg/l, then seed, dilution water or experimental technique should be suspected.

Ratio of BOD and Permanganate values

The BOD value in conjunction with permanganate value is highly useful in evaluating the amenability of the waste for biological oxidation. The BOD value indicates the relative ease with which bio-chemical oxidation takes place, while permanganate value indicates the quantity of organic impurity to be treated. In the ratio of BOD and permanganate value is less than three, the effluent will be less amenable to biochemical oxidation and if the ratio is three or more than three it is treatable by aerobic biological process.

BOD & COD values:

See introductory part of the Chapter 41-3, "Chemical Oxygen Demand (COD)".

41-3 CHEMICAL OXYGEN DEMAND (COD)

This test is highly useful to find out the pollutional strength of industrial effluents and sewage. Chemical oxygen demand as the name implies is the oxygen requirement of a sample for oxidation of organic and inorganic matter. As the oxidizable inorganic matter is usually negligible in comparison with the quantity of organic matter, COD is generally considered as the oxygen equivalent of the amount of organic matter oxidizable by potassium dichromate.

Note that COD value is not the amount of organic matter but is the amount of oxygen required for oxidation. ASTM* defines chemical oxygen demand as the amount of

* ASTM – American Society for Testing and Materials

oxygen (expressed in mg/l), consumed under specified conditions in the oxidation of organic and oxidizable inorganic matter corrected for the influence of chlorides.

Bio-chemical oxygen demand and chemical oxygen demand are two separate parameters. One should not confuse between these two.

BOD is the requirement of oxygen due to biodegradable organic matter only and is a biological phenomenon carried out at more or less natural conditions that exist in the receiving waters. In COD test, the sample is subjected to a chemical oxidation induced by chemical reagents and it is an artificial oxidation by which both biologically oxidizable and biologically inert organic matter are oxidized. Hence the COD value for a sample is always higher than BOD value.

As the inert organic materials are also oxidized, this test (COD) is of great value particularly in evaluating the pollution potential of effluents such as pulp and paper industries, rayon industries and chemical industries. Though these effluents contain considerable quantities of organic matter, the BOD of them will be much less as the materials are not biodegradable and lead to wrong conclusion that less organic matter is present and consequently to a less effective treatment system. In these cases COD reveals the real pollution potential.

The COD test is recommended as a supplement to BOD test. The major advantage of the COD test is the short time required to carryout rather than the longer 5 day procedure for BOD. If waste water remains relatively uniform over a long period, a ratio between BOD and COD may be arrived and then

BOD test substituted with COD test. In conjunction with BOD values, COD data are helpful in knowing toxic conditions and the presence of biologically resistant organic substances.

Though the COD test is advantageous, it is to be noted that it does not differentiate between biodegradable and bio-inert materials. Further, the rate at which the biologically oxidizable materials stabilize (under conditions that exist in nature) cannot be found out, whereas BOD test indicates that rate.

Sample Collection and Preservation

The sample collected for COD determination should be representative. It is best to conduct COD test as soon as possible after collection. If this is not possible, add conc. H_2SO_4 to the sample to bring pH below 2. Analyze the sample within 7 days if the acidified sample is stored at room temperature or within 28 days if the sample is kept at 4°C in a refrigerator.

Principle

The organic matter of the sample is oxidized to water, carbon-di-oxide and ammonia by refluxion with a known excess of potassium dichromate in a 50% sulfuric acid solution. The excess dichromate is titrated with a standard solution of ferrous ammonium sulfate solution.

$$6Fe_2 + Cr_2O_7^{2-} + 14H^+ \rightarrow 6Fe^{3+} + 2Cr^{3+} + 7H_2O$$

Usually silver sulfate is added as a catalyst to promote oxidation of certain classes of organic compounds such as straight chain aliphatic compounds like acetic acid, amino acids etc., Mercuric sulfate is added to eliminate the interference due to chlorides.

Note 1: Aromatic hydro carbons such as benzene and its homologs, pyridine and other heterocyclic compounds containing nitrogen, [(e.g.) pyrrole, pyrolidine, proline, nicotinic acid,] paraffins and naphthenes are not oxidized by dichromate under the experimental conditions.

Interferences & Elimination

Chloride is the most common interfering ion. During COD test, it is oxidized to gaseous chlorine and for this oxidation some of the dichromate is utilized resulting in higher COD values.

$$6Cl^- + Cr_2O_7^{2-} + 14H^+ \rightarrow 3Cl_2 + 2Cr^{3+} + 7H_2O$$

Addition of mercuric sulfate suppresses chloride interference upto 2000 mg/l by forming poorly ionised soluble mercuric chloride complex.

At concentrations above 2000 mg/l, mercuric sulfate is less effective but any how reduce interference upto 5000mg/l. In these samples it is better to evaluate a correction factor by collecting the chlorine produced during oxidation in potassium iodide solution and titrating with a standard sodium thiosulfate solution, or the sample may be diluted appropriately such that the chloride concentration of the diluted sample is within 2000 mg/l.

Nitrites interfere by oxidation to nitrate for which some dichromate would be utilised leading to high COD results. (1mg nitrite nitrogen exerts a COD of 1.1 mg). Addition of sulfamic acid eliminates this interference. But nitrite seldom occurs in effluents excepting from dye house and dye manufacture.

Ferrous iron and sulfides also consume dichromate for their oxidation leading to positive errors. If high concentration of these ions are present, a separate analysis and correction is necessary.

Apparatus

A round bottomed reflux flask of 500 ml capacity with a ground glass 24/40 (B-24) neck and anyone of the condensers - Davies double surface, or Graham, or Liebig or West with 300 mm jacket. Use of reflux condenser prevents the loss of volatile matter during digestion. (Do not apply any stop cock grease on the ground glass joint as this may lead to high COD values).

Reagents

1. **Mercuric sulfate crystals.**
2. **Sulfuric acid - Silver sulfate reagent:** Dissolve 10.1 gm. silver sulfate (Ag_2SO_4) in 1000ml conc. H_2SO_4. About 2 days will be required for dissolution.
3. **Potassium dichromate, 0.25N:** Dissolve 12.258 gm potassium dichromate $K_2Cr_2O_7$ (Previously dried at 103°C for 2 hrs) in distilled water and make upto 1000 ml in a volumetric flask.

   1.0 ml exactly 0.25N potassium dichromate $\equiv$ 2.0 mg oxygen.
4. **Potassium dichromate 0.025N:** Pipet 50ml of 0.25N

potassium dichromate solution into a 500 ml volumetric flask and make upto the mark with distilled water.

1.0 ml exactly 0.025N potassium dichromate ≡ 0.2 mg oxygen.

5. **Ferroin indicator:** Dissolve 695 mg ferrous sulfate $FeSO_4.7H_2O$, and 1.485 g. 1, 10-Phenanthroline-monohydrate together in distilled water and dilute to 100 ml.

6. **Ferrous Ammonium Sulfate 0.1N:** Dissolve 39.29 g. ferrous ammonium sulfate, $Fe(NH_4)_2(SO_4)_2.6H_2O$ in distilled water. Add 20 ml conc. sulfuric acid. Cool and make upto 1000 ml in a volumetric flask. Standardize it daily against 0.25 N potassium dichromate as follows:

   Pipet 10.0ml 0.25 N potassium dichromate solution into a 250 ml conical flask. Dilute to 100 ml and add 30 ml. conc. H_2SO_4 and cool to room temperature. Add 2 to 5 drops of ferroin indicator and titrate with ferrous ammonium sulfate solution. The end point is the color change from blue green to reddish brown. Calculate the exact normality.

7. **Ferrous ammonium sulfate, 0.01N:** Pipet 50 ml of 0.1N ferrous ammonium sulfate and make upto 500 ml in a volumetric flask. Standardize daily against 0.025N potassium dichromate as described above.

8. **Sulfamic Acid.**

Procedure

Precaution:

Do not use de-ionised water for dilution as this may lead to high results because of the organic matter content.

1) Shake the sample well, so that the contents are mixed thoroughly. (It is also recommended that if large amounts of settleable solids are present, the sample may be homogenized using a blender).

2) Place 50 ml or an aliquot in the reflux flask. Add mercuric sulfate according to the chloride concentration of the aliquot. (i.e) a ratio of 1:10 chloride: mercuric sulfate has to be maintained. For example, if the aliquot contains 10 mg chloride, add 100 mg mercuric sulfate. Mix thoroughly.

3) Measure suitable amount of sulfuric acid - silver sulfate reagent in a measuring cylinder. The amount is equal to the sum of sample volume and dichromate volume. (For example, if 50 ml aliquot is taken and 25 ml dichromate is added, then the amount of sulfuric acid-silver sulfate reagent to be added is 75 ml.) From this amount, initially transfer carefully and slowly 5 ml only to the reflux flask and mix thoroughly to dissolve the mercuric sulfate. Cool while mixing to avoid possible loss of volatile matter in the sample.

4) Pipet a pre-determined quantity of 0.25N (or 0.025 N) potassium dichromate solution into the flask and mix well. Use 0.25 N potassium dichromate solutions for samples having a COD of 50mg/l or more and use 0.025N $K_2Cr_2O_7$ solution for samples having a COD less than 50 mg/l.

5) Carefully add the remaining sulfuric acid-silver- sulfate

reagent in the measuring cylinder to the reflux flask, a little quantity at a time, cooling the flask simultaneously under tap.

Note: Always maintain the ratio of volume of sulfuric acid to sample volume + dichromate volume as 1:1. If it is smaller, the oxidizing power of the solution will decrease and it is larger, blank consumption of dichromate will be more.

6) Add a few antibumping granules (already cleaned with sulfuric acid and rinsed with distilled water) and place the flask in position. Attach the condenser and start the cooling water. Reflux for 2 hrs or for a suitable small period as required.

7) Remove the flame, allow the flask to cool and wash the condenser with distilled water into the flask.

8) Transfer the contents of the flask to a 500 ml conical flask and dilute appropriately with distilled water and cool it well. The final volume after dilution should be about 2.5 times the volume of sample + dichromate + H_2SO_4.

9) Add 2 to 3 drops of ferroin indicator and titrate with 0.1N or 0.01N ferrous ammonium sulfate solution. The end point is the sharp color change from blue green to reddish brown (Do not mind even if the blue-green color reappears again within a few minutes).

10) Conduct a blank using distilled water in place of sample. (The quantity of other reagents added is the same as that added for the sample). Note the titre value.

Calculation

$$mg/l\ COD = \frac{(Blank\ titre\ value - Sample\ titre\ value) \times Normality\ of\ ferrous\ ammonium\ sulfate \times 8 \times 1000}{ml.\ sample\ taken\ for\ determination}$$

Note : The correctness of the analysis can be checked against a standard solution of glucose. Dissolve 486.6 mg glucose in distilled water and make upto 1000ml. COD of this solution is 500 mg/l.

Procedure for samples of high nitrite content

Add 10 mg. sulfamic acid per mg nitrite nitrogen to the reflux flask before the addition of dichromate solution and proceed as described above.

Procedure for polluted waters

The general procedure is applicable to polluted waters also. Use higher aliquots of sample and 0.025N. dichromate solution.

Problem - 1

Following is a brief of COD test carried out on an untreated dairy waste. Calculate COD value of the waste.

ml. sample	mg. $HgSO_4$	ml. 0.25 N $K_2Cr_2O_7$	ml. H_2SO_4 -Ag_2SO_4	Final Volume (ml)	Titre Value
10	10	25	35	170	39.1
Blank (10 ml Distilled water)	10	25	35	170	59.2

Normality of Ferrous ammonium sulfate = 0.1037N

$$COD = \frac{(Blank\ titre\ value\ -\ Sample\ titre\ value) \times N\ of\ FAS \times 8 \times 1000}{ml.\ sample\ taken\ for\ estimation}$$

$$= \frac{(59.2 - 39.1) \times 0.1037 \times 8 \times 1000}{10} = 1667.49$$

Problem - 2

Analysis of three industrial effluents gave the following results for BOD and COD. Interpret their biodegradability.

Effluent	5 days BOD mg/l	COD mg/l
Pulp & Paper Industry	115	1490
Sugar Manufacture	820	960
Viscose Rayon manufacture	340	960

Effluent	Ratio of BOD & COD	Bio-degradability
Pulp & Paper Industry	1:13	Much less biodegradable. Refractory organics (non-biodegradable materials) are present in high concentration
Sugar Manufacture	1:1.2	Highly biodegradable. Non-bio degradable materials are almost absent.
Viscose Rayon manufacture	1:2.8	Bio-degradable but not to a significant extent as non-biodegradable materials are predominant.

41-4 PERMANGANATE VALUE (OR) OXYGEN ABSORBED (OR) TIDY'S TEST

This test determines the amount of oxygen absorbed by

the sample from potassium permanganate in acid solution. The test gives some indication of the nature of oxidizable substances present in the sample. Though the test is an empirical one, it is useful as a quick means to find out the pollution potential of waste waters and is useful in laboratories with limited facilities.

Principle

Waters and waste waters when acidified with sulfuic acid and mixed with potassium permanganate, absorb oxygen. The amount of oxygen absorbed can be calculated from the quantities of potassium permanganate added and the amount left after the reaction.

Reagents

1. **Potassium permanganate stock solution, 0.1250 N:** Dissolve 4 to 4.2 g of $KMnO_4$ in about 900 ml distilled water in a 2 litre beaker. Cover it with a watch glass and heat the solution to boiling; boil gently over a small flame for 15-30 minutes and allow the solution to cool to room temperature. Keep the solution overnight and filter through a funnel containing a plug of purified glass wool. Collect the filtrate in a 1000 ml. volumetric flask cleaned with chromic acid mixture. Make up to the mark with distilled water and standardize it against 0.1250 N. sodium oxalate solution as follows. Place the permanganate solution in an amber bottle and keep it in dark.

   Standardization of Potassium permanganate solution:

   Dissolve 2.0938g. sodium oxalate $Na_2C_2O_4$ in distilled water and make up to 1000ml. in a volumetric flask. This is exactly 0.1250 N. Place 25ml of this solution in a 500 ml conical flask and add 150ml. 2 N. Sulfuric acid (prepared by mixing 50 ml. conc. H_2SO_4 and 950 ml. distilled water). Titrate with potassium permanganate solution rapidly at room temperature to the persistence of a faint pink color. Allow to stand until the solution becomes colorless. Then heat to 55-60°C and complete the titration until a faint pink color persists for 30 seconds.

Calculate the normality of potassium permanganate solution.

1ml. of exactly 0.1250 N. $KMnO_4$ ≡ 1.0 mg. Oxygen

2. **Potassium permanganate solution, 0.01250N:** Dilute appropriate volume of 0.1250N. potassium-permanganate solution ten times such that the resulting solution is exactly 0.01250N.

   1 ml. of exactly 0.01250 N. $KMnO_4$ ≡ 0.1 mg. Oxygen.

3. **Sodium thiosulfate solution, 0.1250N:** Dissolve 31.03g. $Na_2S_2O_3 . 5H_2O$ in distilled water and make up to 1000 ml. in a volumetric flask with carbon dioxide-free (boiled and cooled) distilled water. Standardize it against standard 0.1250 N potassium permanganate solution iodometrically using potassium iodide, and starch as indicator.

   1 ml. of exactly 0.1250 N.sodium thiosulfate ≡ 1.0 mg. oxygen.

4. **Sodium thiosulfate solution, 0.0125N:** Dilute appropriate volume of 0.1250N solution ten times with carbondioxide-free distilled water such that the resulting solution is exactly 0.0125N.

   1ml. of exactly 0.01250 N. sodium thiosulfate ≡ 0.1 mg. oxygen

5. **Sulfuric acid, 1+3:** Add carefully 250 ml conc. H_2SO_4 into 750 ml. distilled water with constant stirring and cool.

6. **Potassium iodide solution, 10%:** Dissolve 10g potassium iodide, KI in 100ml. distilled water. Keep it in an amber bottle. (When it becomes colored, reject it).

7. **Starch indicator solution:** See Chapter 41-1, "Dissolved Oxygen" Page 328.

Procedure

A) Industrial Effluents

Rinse two 500ml stoppered glass bottles first with sulfuric acid and then with distilled water. Place 100ml or 50 ml or suitable aliquots of the well mixed sample in one bottle (aliquot volume depends on the type of the effluent). Add 10 ml sulfuric acid followed by 10 ml of 0.1250 N potassium-permanganate solution. The volume of potassium-permanganate added may be such that half of it remains after 4 hrs. Accordingly 20ml, or 30 ml or more volume may be added depending upon the nature of the sample. If the $KMnO_4$ volume is more than 10 ml, add 10 ml sulfuric acid per each 10 ml increase of $KMnO_4$. Prepare a blank also in the other bottle using 10 ml sulfuric acid and same volume of potassium permanganate solution used for sample. Keep the sample and blank at 27°C in dark for 4 hrs. At the end of 4 hrs, add 2.5 ml. potassium iodide solution per 10 ml $KMnO_4$ added and titrate with 0.1250 N sodium thiosulfate solution using starch as indicator.

Calculations

If sodium thiosulfate solution is exactly 0.1250N,

mg/l Oxygen absorbed (Permanganate value or Tidy's value)

$$= \frac{(\text{ml } 0.1250\text{N. thiosulfate for blank} - \text{ml } 0.1250\text{N. thiosulfate for sample}) \times 1 \times 1000}{\text{ml. sample taken for determination}}$$

Note : Chlorinated effluents should be properly dechlorinated with sodium sulfite prior to determination. The permanganate value obtained above indicates an empirical sum of oxygen requirement of both the organic and inorganic matter. The amount of inorganic matter alone can be obtained as described below:

Oxygen absorbed in 3 minutes:

Conduct the experiment as described above, but exactly after 3 minutes (from the addition of $KMnO_4$) add potassium-iodide solution and titrate with 0.1250N sodium thiosulfate.

Inorganic substances such as nitrite, sulfide, sulfite and ferrous iron react with potassium permanganate and oxidized within 3 minutes. Thus the 3 minutes permanganate value is regarded as the oxygen requirement by inorganic matter only.

B) Polluted waters, Raw waters and Treated Waters

Adopt the same procedure as for effluents, but use suitable higher volumes of the sample and lesser volumes of potassium permanganate solution. Use 0.01250N. $KMnO_4$ solution. Titrate with 0.01250 N. sodium thiosulfate solution.

Calculations

If sodium thiosulfate solution is exactly 0.01250N,

mg/l Oxygen absorbed =

$$= \frac{\text{(ml 0.01250N. thiosulfate for blank – ml. 0.01250N. thiosulfate for sample} \times 0.1 \times 1000}{\text{ml. sample taken for determination}}$$

41-5 AMMONIA NITROGEN

Ammonia is present in excess concentrations in the effluents of coke manufacture and petroleum industry. Production of nitrogenous fertilizers also release ammonia in their effluents.

Aerobic or anaerobic decomposition of nitrogenous organic mater evolves ammonia. It is determined by distillation followed by colorimetric/ titrimetric procedure.

DISTILLATION - NESSLERISATION/ TITRIMETRIC METHOD

Principle

The sample after pH adjustment to 7.4 with phosphate buffer is distilled and the distillate containing ammonia is treated with Nessler reagent and the yellow brown color produced is compared with that of standards.

If the sample is known to contain excess ammonia (more than 5 mg/l) the distillate is collected in boric acid solution and titrated with standard sulfuric acid.

Interferences

Sulfide interferes by producing turbidity in the distillate on addition of Nessler reagent but can be eliminated by adding lead carbonate to sample before distillation. Volatile organic compounds such as formaldehyde interfere and they are eliminated by boiling at low pH after which the normal distillation is followed. Aliphatic and aromatic amines, organic chloramines, ketones, aldehydes and alcohols interfere by producing a yellowish or greenish off color or turbidity following the addition of Nessler reagent.

Reagents

1. **Ammonia-free distilled water:** Prepare ammonia-free distilled water by passing distilled water through a strong cation exchanger or add 0.1 ml conc. H_2SO_4 per litre of distilled water and redistill.

2. **Phosphate buffer solution:** Dissolve 14.3 g. of potassium dihydrogen phosphate KH_2PO_4 and 68.8 g. dipotassium hydrogen phosphate K_2HPO_4, in little ammonia-free distilled water and dilute to 1000ml.

3. **Sodium sulfite solution, app N/70:** Dissolve 0.2g. sodium sulfite, Na_2SO_3 in ammonia-free distilled water and dilute to 200 ml. Prepare freshly.

4. **Nessler reagent:** Dissolve 100g. mercuric iodide HgI_2 and 70g. potassium iodide, KI, in about 100ml ammonia - free distilled water. Add this solution slowly with constant stirring to a cool solution of 160 g NaOH in 500 ml distilled water. Dilute to 1 litre. Keep it in a rubber stoppered pyrex bottle in dark.

   (or)

   Dissolve 35g. potassium iodide (KI) and 12.5 mercuric

chloride, $HgCl_2$ in about 700ml. water and add a saturated solution of mercuric chloride with stirring until a slight permanent red precipitate is formed. (about 40 to 50 ml would be required). Then mix with a cooled solution containing 120 g. sodium hydroxide, NaOH in 150 ml water. After the mixture has cooled down, transfer to a 1 litre volumetric flask, add a further 1 ml saturated mercuric chloride solution and shake well. Finally dilute to the 1 litre mark and shake again.

5. **Ammonia stock solution:** Dissolve 3.819 g. of ammonium chloride, NH_4Cl (dried at 100°C) in ammonia free distilled water and make up to 1 litre in a volumetric flask.

   1.0ml = 1.0 mg N or 1.22 mg NH_3

6. **Ammonia intermediate solution:** Pipet 10.0 ml ammonia stock solution into a 1000 ml volumetric flask and make upto the mark with ammonia - free distilled water.

   1ml = 0.01 mg N or 0.0122 mg NH_3

7. **Ammonia working solution:** Place 125.0 ml intermediate ammonia solution and make upto 500ml in a volumetric flask.

   1ml = 0.0025 mg N or 0.003 mg NH_3

Procedure

Use a pyrex distillation flask of capacity 1000 to 1500 ml attached to a vertical condenser as shown in Fig. 41-5-1.

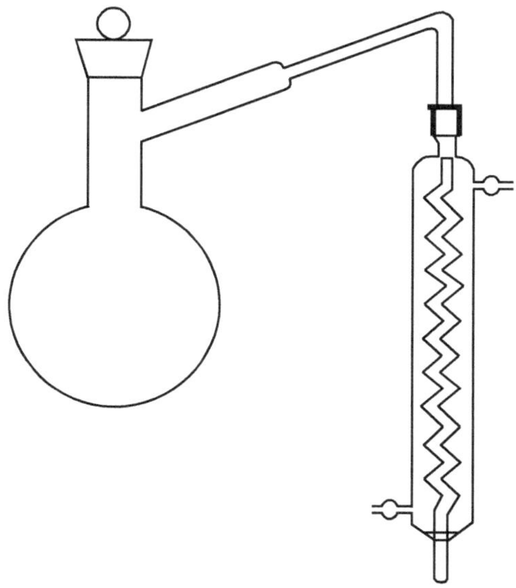

Fig 41-5-1. Distillation apparatus for Ammonia-Nitrogen, Organic Nitrogen, Phenol and Selenium Determinations

Preliminary Treatment

Place appropriate volume of the sample in a beaker. If the sample is untreated effluent, take 100ml, 50 ml or 25 ml according to the strength. If it is a treated effluent, take 100 ml or 250 ml. For polluted waters take 100 ml or more.

Dechlorinate the sample if it contains any chlorine with sodium sulfite solution. (1 ml N/70 sodium sulfite removes 0.5 mg. chlorine)

If necessary, neutralize the sample to approximately pH 7 with 1N. NaOH solution or 1N. H_2SO_4 Transfer the contents of the beaker to the distillation flask.

(For the removal of other interfering substances, see the

section 'Interferences')

Distillation

Add 10 ml phosphate buffer solution. (If the sample contains calcium above 250 mg/l, add an additional 10 ml buffer solution for each 250 mg in the sample and adjust to pH 7.4 with acid or base. Immediately start distillation without delay. (See note). Distill at a rate of 6-10 ml/ minute and collect the distillate in a volumetric flask until the last distillate shows no ammonia by testing with Nessler's reagent. Compensate the loss of water during distillation by adding ammonia-free distilled water to the distillation flask.

Note : The whole distillation assembly should be made free from ammonia prior to the sample distillation, using 500 ml ammonia free distilled water and 10 ml phosphate buffer solution with a few glass beads to prevent bumping.

Color comparison

Pipet into a series of 100 ml. Nessler tubes, 2.0, 4.0 6.0, 8.0, 10.0, 12.0 and 15.0 ml standard ammonium solution (1 ml = 0.0025 mg N). Include a Nessler tube as blank. Pipet in a 100ml Nessler tube an aliquot of the distillate. (If the sample is unpolluted water, transfer completely 100ml of the distillate). Make up the contents of all the Nessler tubes to 100ml with ammonia-free distilled water. Add 2.0 ml Nessler reagent to each tube and mix thoroughly. Compare visually the color of the sample with that of the standard after 10 minutes. If photometric measurement is made, measure the absorbance or transmittance in a spectrophotometer at 400 to 500 nm using a light path of 1 cm or longer or filter photometer equipped with a violet filter.

Calculation

mg/l Ammonia nitrogen (as N) =

$$= \frac{\dfrac{\text{matching std. in mg}}{\text{vol. of aliquot of the distillate in ml.}} \times \text{Total vol. of the distillate in ml.}}{\text{vol. of sample (in ml) taken for distillation}} \times 1000$$

If the sample is known to contain more than 5 mg/l ammonia nitrogen, collect the distillate in boric acid solution and titrate with 0.02N. H_2SO_4 using mixed indicator as described in "Total Kjeldahl Nitrogen"

41-6 ORGANIC NITROGEN

Organic nitrogen is the nitrogen present in the nitrogenous organic matter. It is the difference between total nitrogen and ammonia nitrogen. Hence deduction of ammonia nitrogen from total nitrogen gives the amount of organic nitrogen. It may also be determined by digestion of the sample after removal of free ammonia with subsequent distillation and titration using 0.02 N. H_2SO_4 as described in "Total Kjeldahl Nitrogen".

Express the result as mg. organic nitrogen (as N) per litre of the sample.

41-7 TOTAL KJELDAHL NITROGEN

Total kjeldahl nitrogen is the sum of ammonia nitrogen and organic nitrogen. This does not include nitrite nitrogen and nitrate nitrogen. The classical kjeldahl method is used to determine the total nitrogen content.

KJELDAHL METHOD

Principle

The nitrogen of the organic matter is converted in to ammonium sulfate when treated with sulfuric acid in presence of copper sulfate catalyst. An excess of alkali is then added (to liberate the ammonia from ammonium sulfate) and distilled and the distillate is either treated with Nessler reagent or titrated with standard sulfuric acid after absorption in boric acid solution.

$$2NH_3 + 2H_3BO_3 \rightarrow 2NH_4^+ + 2H_2BO_3^-$$

$$H_2SO_4 \rightarrow 2H^+ + SO_4^{2-}$$

$$2NH_4 + 2H_2BO_3^- + 2H^+ + SO_4^{2-} \rightarrow (NH_4)_2SO_4 + 2H_3BO_3$$

Reagents

1. **Sulfuric acid, conc.**
2. **Copper sulfate solution, 10%:** Dissolve 10g. Coppersulfate $CuSO_4$, $5H_2O$ in 100 ml distilled water.
3. **Potassium sulfate, crystals**
4. **Phenolphthalein indicator solution:** See Chapter 40-9 'Alkalinity', Page 319.
5. **Sodium hydroxide solution, 50% :** Dissolve 100g NaOH in 200 ml distilled water. Place it in a rubber stoppered bottle.
6. **Boric acid solution, 2%:** Dissolve 10 g. boric acid H_3BO_3 in ammonia-free distilled water and dilute to 500 ml.
7. **Mixed indicator solution:** Dissolve 200 mg. methyl-red

indictor in 100 ml 95% ethyl or isopropyl alcohol. Dissolve 100 mg. methylene blue in 50 ml 95% ethyl or isopropyl alcohol. Mix the two solutions. Prepare monthly.

8. **Standard sulfuric acid solution, 0.02N:** See Chapter 40-9, "Alkalinity" Page 319.

Procedure

Digestion: Place 100 ml or an appropriate volume of the sample in a kjeldahl flask. Add 10 ml conc. H_2SO_4 and 1 ml. copper sulfate solution. If the organic matter is hard to destroy, add 20 ml conc. H_2SO_4 and 5 g. potassium sulfate. Add a few glass beads and boil under hood, until the solution becomes clear. Then digest for additional 30 mts. Allow to cool.

Distillation: Transfer the contents of the flask carefully in to a distillation flask and dilute to about 300 ml. Make the solution in the flask alkaline with sodium hydroxide using phenolphthalein indicator. Start the distillation after immersing the tip of the condenser in 50 ml boric acid solution in a conical flask. Collect about 200 ml of the distillate.

Titration: Add 0.5 ml mixed indicator solution to the distillate. Titrate against 0.02N. H_2SO_4. End point is the color change from pale green to lavender.

Conduct a blank also, starting from the digestion step to final titration.

Calculation

mg/l Total Kjeldahl Nitrogen (as N) =

$$= \frac{(ml\ 0.02N.H_2SO_4\ for\ sample - ml.\ 0.02NH_2SO_4\ for\ blank) \times 0.28 \times 1000}{ml.\ sample\ taken\ for\ determination}$$

REFERENCES

1) Edwin W. Taylor, "The Examination of waters & Water supplies". Sixth edition, 1949. J.A. Churchill Ltd. London.

2) T.R. Camp, "Water and its Impurities", 1953, Reinhold publishing Corporation, Newyork.

3) D.F. Boltz, Colorimetric Determination of Non-Metals - 1958, Interscience Publishers, Newyork.

4) Louis Klein "River pollution, Vol. 1-Chemical Analysis", 1959, Butterworths, London.

5) A.I.Vogel, "A Text book of Quantitative Inorganic Analysis including Elementary Instrumental Analysis", 3rd edition - 1964 - ELBS and Longmans Greenman & Co. Ltd.

6) Louis Klein & Bolton, "Sewage Treatment - Basic Principles and Methods," 1971, Butterworths, London.

7) Department of Environment, "Analysis of Raw, potable and waste waters", 1972. Her Majesty's Stationery Office, London.

8) American Society for Testing and Materials, "Annual Book of ASTM Standards," Part-23, 1972, ASTM , Philadelphia.

9) American Public Health Association, "Standard Methods for the Examination of Water and Waste Water" - 14th ed, 1976, APHA, Washington.

10) N. Manivasakam, "Physico - Chemical Examination of Water, Sewage and Industrial Effluents" - 1984, Pragati Prakashan, Meerut.

Chapter - 42

INORGANIC CONSTITUENTS - METALS

42-1 PRELIMINARY TREATMENT OF SAMPLES

The industrial effluents usually contain organic matter (which forms complexes with metals easily), suspended particulate matter, color and other interfering substances. They must be eliminated before commencing the determination for metals. The usual practice is to subject the samples to acid digestion. This preliminary acid digestion not only destroys organic matter but also brings all metallic compounds in suspension to solution. Interferences due to nitrite, cyanide, sulfide, sulfite, thiosulfate and thiocyanate are also eliminated during digestion.

For the determination of calcium, magnesium, sodium-potassium, lithium and strontium, samples are either dry ashed or subjected to nitric acid digestion. For the determination of heavy metals (aluminium, barium, beryllium, cadmium, chromium, copper, iron, lead, manganese, mercury, nickel, selenium, silver, vanadium and zinc) either nitric acid-sulfuric acid digestion, or nitric acid-perchloric acid digestion (if the organic matter is hard to destroy) is employed. Certain metals however require special treatments which are described in appropriate texts.

Preliminary Treatment for Ca, Mg, Na, K, Li, & Sr.

A) Dry ashing: 1. Evaporate a suitable aliquot of the sample in a porcelain or silica dish on a water bath. Ignite the contents of the dish by keeping in muffle furnace at 550° to 600°C until a white ash is obtained.

2. Dissolve the ash in a minimum amount of conc. HCl and warm distilled water. Mix thoroughly, filter and neutralize with conc. ammonia solution and make up to a known volume.

B) Digestion with Nitric acid: Acidify a suitable volume of the sample taken in a 250 ml conical flask with conc. HNO_3 and evaporate to dryness on a water bath. Add about 25 ml conc. HNO_3 to the residue in the flask and heat to near boiling. Evaporate the acid to a small volume. Care must be taken to avoid spattering if any unoxidised matter remains. It will be indicated by brown fumes. Complete the ashing process by repeated additions of small volumes of HNO_3 or HNO_3 with little amount of hydrogen peroxide. The final ash should be white unless such elements, as iron and copper are present. Proceed from step No. 2 of, "Dry Ashing".

II. Preliminary Treatment for Al, Ba, Be, Cd, Cr, Cu, Pb, Fe, Mn, Hg, Ni, Se, Ag, V & Zn.

A) Digestion with Nitric and Sulfuric acids: Place a suitable volume of the well mixed sample in an evaporating dish (porcelain or silica). Acidify to methyl orange with conc. H_2SO_4 and add 5 ml conc. HNO_3 and 2 ml 30% hydrogen peroxide to reduce chromate by placing over a water bath or hot plate to about 10 ml. Transfer to a 125 ml conical flask using 5 ml conc.

* If necessary cover the evaporating dish with watch glass to avoid loss of material by spattering.

HNO_3. Add 10 ml conc. H_2SO_4 and a few glass beads. Evaporate on a hot plate until white dense fumes characteristic of SO_3 just appear in the flask. (If the solution is still not clear, add a further 10 ml conc. HNO_3 and evaporate to fumes of SO_3). Cool to room temperature, add 50 ml distilled water, boil to dissolve the solids if any present and filter through a sintered glass crucible or porcelain crucible. Transfer the filtrate to a 100 ml volumetric flask and make up to the mark with distilled water. The resulting solution is 3N in H_2SO_4. Use aliquots of this made up solution for the determination of metals.

Note: If lead is present in the sample, add 50 ml ammonium acetate solution (200 g/300 ml water) to the conical flask in which digestion was carried out and heat to incipient boiling. Pour through the crucible in which the digestate was filtered. Lead present ad $PbSO_4$ remaining on the filter will be dissolved and carried out into solution. Transfer the filtrate to a 100 ml. volumetric flask and make upto the mark with distilled water.

Determine lead from this solution also.

B) Digestion with Nitric and Perchloric acids:
(See caution, page 373)

If it is difficult to oxidize the organic matter by digestion with nitric and sulfuric acids, digestion may be carried out using nitric and perchloric acids.

Measure a suitable volume of the well mixed sample in an evaporating dish and acidity to methyl orange with conc. HNO_3. Add a further 5 ml. of conc. HNO_3 and evaporate to about 10 ml by placing over a water bath or hot plate. Transfer it to a conical flask of 125ml capacity and cool. Add 5 ml. of conc. HNO_3 and 10 ml of perchloric acid $HClO_4$ (70%). Add a few glass beads or boiling chips and heat gently by placing on a hot plate until white dense fumes of $HClO_4$ just appear. (If even at this stage, the solution is not clear, add further 10 ml. of conc. HNO_3. Place a watch glass over the mouth of the conical flask and keep the solution just barely boiling until it clears).

Allow to cool and dilute to 50 ml with redistilled water. Boil to expel oxides of nitrogen and chlorine present if any. Filter through a sintered glass crucible or porcelain filter crucible and make up the filtrate to 100 ml with distilled water. The solution now obtained is 0.8 N. in $HClO_4$. Use aliquots of this made up solution for the determination of heavy metals.

Caution: Digestion with nitric acid and perchloric acids should be done carefully as the mixture of organic matter and perchloric acid may explode violently. Observe the following precautions.

1. Before the addition of perchloric acid ensure that the sample is pretreated with nitric acid.
2. Do not add perchloric acid to hot solutions.
3. Fuming should be done only under hoods.

42-2 CADMIUM

Cadmium salts are usually found in wastes from electro plating industries, textile printing, pigment works, lead mines and chemical industries.

The following dithizone method is applicable to all types of industrial effluents as well as for polluted, raw and potable waters. (When the other metallic concentrations are exceedingly high, a different procedure has to be followed, Refer "Standard Methods"-14th edition, 1976)

DITHIZONE COLORIMETRIC METHOD

Principle

Cadmium reacts with dithizone to form a red complex and is measured colorimetrically.

Interferences

Lead and zinc interfere above concentrations 6 mg/l and 3 mg/l respectively. The interference by copper and nickel are suppressed by complexing them with cyanide.

Reagents

1. **Cadmium stock solution:** Dissolve 100mg. pure cadmium metal in 50 ml. 10% nitric acid in a 250 ml. beaker. Boil to expel the oxides of nitrogen and make up to 1000 ml. in a volumetric flask.

   1.0 ml = 100 μg Cd.

2. **Cadmium working solution:** Pipet 10.0 ml cadmium stock solution into a 1000 ml volumetric flask and make upto the mark with 1% nitric acid.

   1.00 ml = 1 μg Cd.

3. **Thymol blue indicator solution:** Dissolve 400 mg thymol sulphonephthalein sodium salt in 100 ml. distilled water.

4. **Sodium hydroxide solution, 6N :** Dissolve 24g of NaOH in distilled water and dilute to 100 ml.

5. **Tartaric acid solution, 2%:** Dissolve 20g. tartaric acid $H_2C_4O_6$ in distilled water and make up to 1000 ml. Keep it

in a refrigerator. The solution must be cold when used.

6. **Chloroform**

7. **Hydroxylamine-hydrochloride solution:** Dissolve 20 g. hydroxyl amine hydrochloride, NH_2OH. HCl in distilled water and make up to 100ml.

8. **Dithizone stock solution:** Dissolve 80 mg. well powdered dithizone in 1000ml. chloroform. Place it in an amber bottle and keep it in refrigerator. Use the solution while still cold.

9. **Dithizone working solution:** Dilute 50 ml dithizone stock to 500 ml with chloroform.

10. **Sodium hydroxide - potassium cyanide solution I:** Dissolve 40 g. NaOH and 1.0g. potassium cyanide, KCN in 100 ml distilled water. Store it in a polythene bottle.

11. **Sodium hydroxide-potassium cyanide solution II:** Dissolve 40g. NaOH and 0.05g. potassium cyanide, KCN in 100 ml distilled water. Store it in a polythene bottle.

12. **Hydrochloric acid, conc.**

13. **Sodium potassium tartrate solution:** Dissolve 25g. sodium potassium tartrate $C_4H_4O_6$ KNa.$4H_2O$ in 100 ml distilled water.

Procedure

1. In to a series of separating funnels, measure appropriate volumes of cadmium working solution covering the range upto 20μg Cd. Include a separating funnel

containing distilled water as the blank.

2. Digest appropriate volume of the effluent sample as described in the "Preliminary Treatment of the Samples" and place a suitable aliquot (containing not more than 20μg Cd) of the digested sample in a separating funnel.

3. Adjust the pH of the sample, standards and blank to 2.8 using thymol blue indicator, and dilute to 25 ml. with distilled water.

   (If a raw or potable or polluted water containing less than 10 μg Cd has to be analyzed, place 200 ml of the sample in an evaporating dish. Add 0.5 ml conc. HCl and evaporate to 20 ml. Add a few drops of thymol blue indicator solution dropwise to the appearance of yellow color. The pH of the sample now will be about 2.8. Place it in a separating funnel and dilute to 25 ml. with distilled water).

4. To the sample, standards and blanks in the separating funnel, add the following reagents in order with mixing after each addition:

   i) 1 ml. sodium potassium tartrate solution,
   ii) 5 ml. sodium hydroxide - potassium cyanide solution I,
   iii) 1 ml. hydroxylamine hydrochloride solution, and
   iv) 15 ml. dithizone stock solution.

Shake well* and draw the chloroform layer into a second separating funnel containing 25 ml tartaric acid solution. Add 10 ml chloroform to the first separating funnel and shake well. Drain the chloroform layer again into the

---

* Release the vapor pressure by opening the stopper rather than opening the stop cock.

second separating funnel. Do not permit even a trace of the aqueous layer into the second separating funnel.

Note 1: Complete the extraction operations as quickly as possible since cadmium dithizonate decomposes on prolonged contact with strong alkali saturated with chloroform.

Note 2: If lead and zinc are present in appreciable concentrations, the pink color of their dithizonates persists for several primary extractions. It should not be mistaken that cadmium has not been completely extracted.

Note 3: Absence of orange color of excess dithizone in the aqueous phase indicates that the volume of aliquot taken is too high. The determination has to be repeated then with smaller volumes of the sample.

5. Shake the separating funnel containing the combined chloroform extracts in tartaric acid solution for about two minutes. Allow the phases to separate and discard the lower chloroform layer. Wash the contents of the funnel by shaking with 5 ml. chloroform. Discard the chloroform layer making the separation as fine as possible. Gently blow air into the funnel to evaporate the floating drops of chloroform. All the cadmium now will be in the tartaric acid solution and is free from interfering metallic ions.

6. Add the following reagents in order with mixing after each addition.

   i). 0.25 ml hydroxylamine hydrochloride solution,

   ii). 15.0 ml dithizone working solution, and

   iii). 5 ml sodium hydroxide - potassium cyanide solution II

7. Shake well, allow the phases to separate and filter the chloroform layer through a small plug of cotton wool in a funnel.

8. Measure the absorbance using a spectrophotometer at 518 nm or a filter photometer with a suitable green filter, taking water as the reference. Prepare a calibration curve. Find out microgram Cd equivalent to the net optical density of the sample. (Visual comparison can also be made using Nessler tubes, and chloroform for dilution)

Express the result as mg/l Cadmium (as Cd).

42-3 CALCIUM + MAGNESIUM

Calcium and magnesium content of effluent samples are not usually determined but they are determined if percent sodium is to be calculated for disposal of the effluent on land for irrigation.

EDTA TITRIMETRIC METHOD

Principle

Calcium and magnesium ions react with EDTA to form soluble complexes and the completion of reaction is indicated by the color change of a suitable indicator such as Eriochrome-Black T.

Interferences

Several metal ions interfere in this determination. They may cause, fade or indistinct end points. However these interferences are eliminated by ashing the sample, or by the addition of sodium sulfide inhibitor before titration.

Reagents

1. **Calcium standard solution:** Weigh accurately 1.000 gm. of pure calcium carbonate (previously dried at 105°C) and place in a 250 ml conical flask using 50 ml. distilled water. Place a funnel in the neck of the flask. Add 20.5 ml 1N. HCl. Warm until the solution is complete. Cool and transfer to a 1,000 ml. volumetric flask and make upto the mark with distilled water.

   $$1.0 \text{ ml} = 1.0 \text{ mg } CaCO_3$$

2. **Standard EDTA titrant, 0.02 N (0.01 M):** Weigh accurately 3.723 g. AR grade disodium ethylene diamine tetra acetate dihydrate, [also called (ethylene dinitrilo) - tetra acetic acid disodium salt] EDTA, $Na_2H_2 C_{10}H_{12}O_8N_2.2H_2O$ and dissolve in distilled water and make up to 1000ml. Check the titer by standardizing against standard calcium solution as in the 'procedure for clear samples'. Store it preferably in a polythene bottle or in a pyrex bottle.

   $$1.0 \text{ ml of exactly } 0.02 \text{ N. EDTA} \equiv 1.0 \text{ mg } CaCO_3$$

3. **Ammonia - ammonium chloride buffer:**
   i) Dissolve 16.9g. ammonium chloride, NH_4Cl in 143 ml. conc ammonia solution. (Sp. gr 0.90)
   ii) Dissolve 1.179g. di-sodium salt of EDTA and 780mg. $MgSO_4.7H_2O$ (or 644 mg $MgCl_2.6H_2O$)

   in 50 ml. distilled water.

   Add solution, (ii) to (i) and dilute to 250 ml with distilled water. Keep in a tightly stoppered polythene container.

4. **Sodium sulfide inhibitor:** Dissolve 5.0 gm $Na_2S.9H_2O$ or 3.7 g $Na_2S.5H_2O$ in 100 ml distilled water. This inhibitor deteriorates on oxidation, and air must be excluded. Keep the bottle tightly fitted with a rubber stopper.

5. **Eriochrome Black T indicator:** Alcoholic solutions of this dye are not so stable and either one of the following preparations may be used.
   i). Mix 0.5 to 1.0g. of the dye in 100g. triethanol amine.
   ii) Mix 0.5g. of the dye with 100g sodium chloride to obtain a dry powder mixture.

Procedure

Evaporate an aliquot of the sample to dryness on a steam bath. Heat it in a muffle furnace to 550°C until the organic matter is oxidized. Dissolve the residue in 20 ml 1N. HCl and neutralize to pH 7 with 1N. NaOH and dilute to 50 ml. with distilled water.

Add 1 ml. of buffer solution per 50 ml volume. The pH of the titer should be 10.0±0.1. Add 1 ml. sodium sulfide inhibitor. Add 1 drop of indicator solution or an appropriate amount of dry powder.

Titrate with the standard EDTA solution slowly, until a reddish tinge appears and add the last few drops within 3-5 seconds. At the end point, the solution will be blue. The whole titration should be completed within 5 minutes after the addition of buffer.

Calculation

If the EDTA titrant is exactly 0.02 N,

$$mg/l \text{ calcium} + \text{Magnesium (as } CaCO_3) = \frac{ml.EDTA \text{ titrant} \times 1 \times 1000}{ml. \text{ sample for estimation}}$$

42-4 CALCIUM

Calcium is determined on effluent samples to calculate percent sodium value.

EDTA - TITRIMETRIC METHOD

Principle

The pH of the sample is made sufficiently high (12-13) to precipitate magnesium as hydroxide and calcium only is allowed to react with EDTA in the presence of selective indicator.

Interferences

Under the conditions of the test, the following concentrations of ions cause no interference.

Aluminium 5 mg/l	Iron (Fe^{2+}) 20 mg/l	Tin 5 mg/l
Copper 2 mg/l	Lead 5 mg/l	Zinc 5 mg/l
Iron (Fe^{3+}) 20 mg/l	Manganese 10 mg/l	

Gravimetric procedure may be employed when the concentration of the above interfering substances exceeds the limit and when orthophosphates are present which precipitate calcium at the pH of the test.

Reagents

1. **Sodium hydroxide, 1N:** Dissolve 40 g. NaOH in about 100 ml distilled water and make upto 1000 ml.

2. **Indicators:** Any one of the following two indicators may be used.

   a) Murexide (Ammonium purpurate) indicator: It is available commercially as tablets. The end-point with this indicator is from pink to purple.

   b) Patton and Reeder's indicator: The indicator is 2-hydroxy-1-(2-hydroxy-4sulpho-naphthyl azo)-3-naphthoic acid. The end point with this indicator is from wine red to pure blue. Prepare the indicator by grinding in a mortar 1g. of the dye with 100 g. sodium sulfate.

3. **Standard EDTA titrant, 0.02 N:** See Chapter 42-3 "Calcium + Magnesium" (Page 379).

Procedure

1. Place a suitable aliquot of the sample (ashed and leached with HCl and neutralized with NaOH) as described in Chapter 42-3 "Calcium + Magnesium".

2. Add 2.0 ml sodium hydroxide solution (per 50 ml volume) to produce a pH of 12 -13 and mix well by swirling the flask.

3. Add 1 tablet of murexide indicator or 0.1-0.2g of Patton and Reeder's indicator.

4. Titrate immediately with EDTA to the proper end point. With murexide, the color change is from pink to purple.

With Patton-Reeder's indicator color changes from wine red to blue.

Calculation

If the EDTA titrant is exactly 0.02 N,

$$mg/l\ calcium\ (as\ CaCO_3) = \frac{ml.EDTA\ titrant \times 1 \times 1000}{ml.\ sample\ for\ taken\ estimation}$$

42-5 MAGNESIUM

The industrial and commercial uses of magnesium salts are numerous and the occurrence of magnesium in industrial effluents is common, but the determination is not usually made on effluents excepting for the calculation of percent sodium.

Magnesium is usually calculated as the difference between total hardness and calcium hardness.

$$mg/l\ Magnesium\ (as\ CaCO_3) = \underset{[Calcium\ +\ Magnesium\ (as\ CaCO_3)]}{mg/l\ Total\ hardness} - \underset{(as\ CaCO_3)]}{mg/l\ Calcium\ hardness}$$

42-6 CHROMIUM

Chromium compounds are used extensively in industrial processes such as metal pickling, electroplating, aluminium anodizing, leather tanning and in the manufacture of paints, dyes paper, explosives and ceramics. In addition, chromates are frequently added to cooling water for corrosion control. Chromium exists in hexavalent and trivalent forms. (Chromates and dichromates are hexavalent chromium compounds. Tannery spent liquors contain mostly trivalent chromium). Since chromium exists in two forms, two separate determinations are necessary. The following s-diphenyl carbazide method is more selective and precise.

s-DIPHENYL CARBAZIDE COLORIMETRIC METHOD FOR HEXAVALENT CHROMIUM (CHROMIUM PRESENT AS CHROMATES & DICHROMATES)

Principle

Under acidic conditions, hexavalent chromium reacts with s-diphenyl carbazide to form a reddish violet colored complex which can be determined either visually or photometrically.

Interferences

Iron above 1 mg/l interferes but can be eliminated by preliminary treatment with cupferron*. Other common ions generally present do not interfere.

Reagents

The distilled water used for the preparation of reagents and dilutions should be free from chromium and organic matter.

1. **Chromium stock solution:** Dissolve 283 mg dried potassium dichromate, $K_2Cr_2O_7$ in distilled water and

---

* Acidify a suitable aliquot of the sample with HCl or H_2SO_4 such that it contains 5 to 10% acid. Precipitate iron with 6% aqueous solution of cupferron under cold condition. Filter and use filtrate for Cr determination.

make upto the mark with distilled water.

1.0 ml = 100 μg Cr

2. **Chromium working solution:** Pipet 10.0 ml chromium stock solution into a 1000 ml volumetric flask and make upto the mark with distilled water. Prepare freshly.

   1.0 ml = 1.0 μg Cr.

3. **s-Diphenyl carbazide reagent :** Dissolve 200 mg. s-diphenyl carbazide in 100 ml. 95% ethyl alcohol. Add 400 ml 1+9 sulfuric acid into it. Place it in an amber bottle. Keep it in a refrigerator. (The reagent is stable for several months if refrigerated). Discard the solution when it becomes appreciably brown.

Procedure

1. Place appropriate volumes of chromium working solution in 50 ml. volumetric flasks (or Nessler tubes) covering the range upto 20μg. Cr. Dilute to 50 ml with distilled water. Include a flask (or Nessler tube) containing 50 ml. distilled water as the blank.

2. Place 50 ml or an aliquot (containing not more than 20μg. Cr) of the neutralized sample in a 50 ml. volumetric flask (or Nessler tube). Dilute to 50 ml with distilled water if necessary. (If the sample is turbid, centrifuge or filter through a sintered glass crucible* and then place in the volumetric flask or Nessler tube).

3. To the blank, standards, and sample add 2.5 ml

* Not to be filtered through filter paper since it may absorb some chromium.

s-diphenyl carbazide reagent and mix well, and set aside for 5 minutes.

4. Measure the optical densities of the blank, standards and sample using a spectrophotometer at a wavelength of 540 nm (or filter photometer equipped with a suitable green filter) taking water as the reference.

   Prepare a calibration curve. Find out the microgram Cr equivalent to the observed optical density.

   (If Nessler tubes are used, visual comparison can be made)

   Express the result as mg/l Chromium (as Cr).

s-DIPHENYL CARBAZIDE COLORIMETRIC METHOD FOR TOTAL CHROMIUM (HEXAVALENT CHROMIUM + TRIVALENT CHROMIUM)

This method is applicable to potable waters, treated waters and waste waters,

Principle

Chromium is oxidised by permanganate to chromate and then treated with diphenyl carbazide to obtain the violet color complex which is measured colorimetrically.

Interferences

Mercury and vanadium interfere but they occur only in negligible concentrations. Iron interferes when its concentration exceeds 1 mg/l but can be eliminated by preliminary treatment with cupferron.

Reagents

In addition to the reagents required for the determination of hexavalent chromium, prepare the following reagents.

1. **Potassium permanganate solution, approximately 0.1 N:** Dissolve 316 mg. potassium permanganate in little distilled water and dilute to 100 ml.

2. **Sodium azide solution:** Dissolve 500 mg. sodium azide, NaN_3 in 100 ml chromium-free distilled water.

Procedure

A) Samples containing no organic matter

Place appropriate volume of the sample containing not more than 20μg Cr in a 100 ml. beaker. Add 10 ml. 10% sulfuric acid. Add potassium permanganate solution dropwise until the solution becomes pink in color. Add about 30 ml. water and place on a water bath for 20 minutes. Add sodium azide solution dropwise until the solution becomes colorless and continue heating for one minute. Cool and filter the solution if necessary. Transfer to a 50 ml volumetric flask (or to 50 ml Nessler tube) and add 2.5 ml diphenyl carbazide solution and make up to the mark with chromium-free distilled water. Mix well and compare the colors visually or photometrically. Find out the chromium content per litre of the sample.

B) Samples containing organic matter

Digest the sample with nitric and sulfuric acids as described in "Preliminary treatment of samples" (chapter 42-1). Neutralize an appropriate aliquot of the acid digested sample with conc. ammonia solution and proceed as described for samples containing no organic matter. The value obtained indicates total chromium content.

mg/l. Trivalent Chromium = Total Chromium – Hexavalent Chromium

TITRIMETRIC METHODS

When the chromium concentration is exceedingly high as in electroplating wastes, metal pickling wastes, chrome tanning wastes etc., the following titrimetric method is preferable to colorimetric methods. It is applicable to all samples (including cooling water) containing 100 to 1500 mg/l chromium.

As in colorimetric method, two separate determinations - one for hexavalent and the other for total chromium are necessary.

AVAILABLE CHROMATE (Hexavalent Chromium)

Principle

When a sample containing chromate is acidified, chromates are converted to dichromate. This dichromate can be determined by titration with standard sodium thiosulfate or standard ferrous ammonium sulfate.

$$CrO_4^{2-} + 2H^+ \rightarrow Cr_2O_7^{2-} + H_2O$$

Interferences

Iron, copper and nitrite interfere. Iron is made innocuous by the addition of ammonium bifluoride and nitrite by the addition of sulfamic acid. If copper is present, the solution has to be made alkaline and filtered. However copper does not occur in such concentrations to interfere with the test.

Procedure

A) Titration with Thiosulfate

1. Place 100 ml. of the sample in a 500 ml conical flask.
2. Add 2 ml 10N. H_2SO_4, 1 g. ammonium bifluoride and 1g. sulfamic acid with mixing after each addition.
3. Add 2g. potassium iodide, swirl and keep it in dark for 5 minutes for completion of the reaction.
4. Titrate with standard 0.1 N sodium thiosulfate solution using starch as indicator.

Calculation

Available Chromium (as Cr) =

$$\frac{\text{ml. thio} \times \text{Normality of thio} \times 17.332 \times 1000}{\text{ml. sample taken}}$$

Available Chromium (as chromate CrO_4^{2-}*) =*

$$\frac{\text{ml. thio} \times \text{Normality of thio} \times 38.66 \times 1000}{\text{ml. sample taken}}$$

B) Titration with Ferrous Ammonium Sulfate

After step No.2 in procedure (A), add 2 to 3 drops of ferroin indicator (See page 352 for its preparation) and titrate with 0.1 N ferrous ammonium sulfate. Calculations are the same as in (A), substitute ferrous ammonium sulfate in place of thio.

TOTAL CHROMIUM

Principle

Trivalent chromium is oxidized to dichromate on boiling with excess persulfate in the presence of silver nitrate catalyst. The excess persulfate remaining after oxidation is completely decomposed by boiling. The resultant solution with the dichromate can either be directly titrated with ferrous ammonium sulfate or by the addition of known excess of standard ferrous ammonium sulfate and back titration of the excess ferrous ammonium sulfate with standard dichromate.

$$2Cr^{3+} + 3S_2O_8^{2-} + 7H_2O \xrightarrow{AgNO_3} Cr_2O_7^{2-} + 6HSO_4^- + 8H^+$$

$$2S_2O_8^{2-} + 2H_2O \rightarrow O_2\uparrow + 4HSO_4^-$$

Procedure

1. Place 100 ml. of the sample or suitable aliquots diluted to 100 ml. in a 500 ml conical flask.

2. Add 4 ml. silver nitrate solution (1.7g. $AgNO_3$/100 ml) and about 1 to 2g. potassium or ammonium persulfate and boil the liquid gently for 20 mts.

3. Cool and add 20 ml. of 10N. sulfuric acid. Mix thoroughly and follow any one of the titrations given below.

A) Direct titration with Ferrous Ammonium Sulfate (FAS)

Add 3 drops of ferroin indicator and titrate with 0.1 N. ferrous ammonium sulfate until the color changes to reddish brown. Do not mind even if the blue-green color reappears again within a few minutes. Find out the normality of FAS as described in COD (Page 352).

Calculation

$$mg/l\ Total\ Chromium\ (as\ Cr) = \frac{ml.\ FAS \times N.\ of\ FAS \times 17.332 \times 1000}{ml.\ sample\ taken}$$

B) Back titration

To the solution in the conical flask add a known excess of 0.1N FAS and 0.5 ml N-Phenyl anthranilic acid indicator (prepared by dissolving 100 mg in 4.8 ml of 0.1 N.NaOH and dilution to 100 ml with distilled water) Titrate with standard 0.1N potassium dichromate until the color changes from green to violet red.

Calculation

$$mg/l\ Total\ Chromium\ (as\ Cr) = \frac{\left[ml.\ FAS - \left(\frac{Dichromate\ Titre\ value \times N.\ of\ dichromate}{Normality\ of\ FAS}\right)\right] \times N.\ of\ FAS \times 17.332 \times 1000}{ml.\ sample\ taken\ for\ determination}$$

42-7 COPPER

Copper occurs in copper plating wastes, copper pickling liquors and cuprammonium rayon wastes. Copper compounds are toxic to fish and other aquatic life and micro organisms.

NEOCUPROINE COLORIMETRIC METHOD

Principle

Copper is reduced with hydroxyl amine hydrochloride and the pH is adjusted to 4.0 to 6.0. The cuprous ions react with neocuproin to form a yellow colored complex which is extracted with chloroform and measured colorimetrically.

Interferences

None are likely. Possible interferences from organic matter, sulfide, cyanide and chromium are eliminated by preliminary acid digestion of the sample.

Reagents

Use copper-free distilled water (glass distilled water or deionised distilled water) for the preparation of solutions and dilutions

1. **Conc. Hydrochloric acid**
2. **Hydroxylamine-hydrochloride solution, 20%:** Dissolve 50g. hydroxyl amine hydrochloride, NH_2OH. HCl in re-distilled water and dilute to 250 ml.
3. **Neo cuproine solution:** Dissolve 200 mg. neocuproine (2, 9 dimethyl-1,10 phenanthroline hemihydrate) in 200 ml methyl alcohol. Keep it in a refrigerator. The reagent is stable under ordinary storage conditions for a month or more.
4. **Sodium citrate solution, 25%:** Dissolve 125g. sodium-citrate dihydrate, $Na_3C_6H_5O_7.2H_2O$ in re-distilled water and dilute to 500 ml. Transfer to a 1000ml separating funnel, add 10 ml hydroxyl amine hydrochloride solution, and 10 ml neocuproin solution. Extract with 50ml chloroform to remove copper impurities and discard the chloroform layer.

5. **Copper stock solution:** Place 100.0 mg polished electrolytic copper in a conical flask. Add 10 ml redistilled water and 3 ml conc. nitric acid. After the reaction has subsided, warm gently for complete dissolution and then boil to expel oxides of nitrogen. (Take care to avoid loss of copper by bumping etc.,) Cool and add about 50 ml re-distilled water. Transfer quantitatively to a 1000 ml volumetric flask and make upto the mark with re-distilled water.

   1.0 ml = 100 μg Cu

6. **Copper working solution:** Pipet 50.0 ml of stock solution and make upto 1000 ml. in a volumetric flask.

   1.0 ml = 5 μg Cu

Procedure

1. Into a series of 125 ml separating funnels, pipet 1.0, 2.0, 4.0,..... 10.0 ml of copper working solution. Dilute to 50 ml. with re-distilled water. Include a separating funnel containing 50 ml re-distilled water as the blank.

2. Place an appropriate aliquot of the sample (already digested with sulfuric and nitric acids as given in "Preliminary Treatment of Samples" and pH adjusted to 4-6 with conc. ammonia) Dilute to 50 ml with re-distilled water if necessary.

3. Add 0.1 ml conc. HCl to each separating funnel and mix well.

4. Add the following reagents in the order with shaking after each addition.

i). 5 ml. hydroxyl amine hydrochloride solution
ii). 10 ml sodium citrate solution, and
iii). 10 ml neocuproin solution.

5. Add 20 ml chloroform and shake the funnel vigorously. Allow the phases to separate and drain the chloroform layer into 50 ml. Nessler tubes if visual comparison is made or into 50 ml. volumetric flasks if photometric comparison is made. Extract with another 20 ml portion of chloroform and combine with the first chloroform extract. Make upto 50 ml with isopropyl alcohol.

6. Compare the color of the sample with that of the standards visually or measure the absorbance of standards and samples using blank as the reference at 457 nm using a light path of 1 cm. Prepare a calibration curve using a semilog graph paper. Find out the microgram equivalent of the sample to the observed optical density.

Express the result as mg/l Copper (as Cu).

42-8 IRON

Iron pickling effluents and mine waters contain excessive amounts of iron. Iron usually precipitates as hydroxide and forms heavy sludges in receiving streams as well as in treatment plants.

Though iron exists in two valence states (Fe^{2+} and Fe^{3+}) the amount of total iron only is of important consideration. One colorimetric method for samples of smaller concentrations and one titrimetric method for higher concentration are described here.

COLORIMETRIC THIOCYANATE METHOD FOR TOTAL IRON

This method is applicable to samples of industrial effluents, domestic waste, raw and treated waters.

Principle

Ferric iron combines with thiocyanate ions to form red colored ferric thiocyanate which is measured colorimetrically.

Interferences

Chromium (in higher concentrations), zinc, copper, cobalt, mercury, phosphates and molybdate ions interfere.

Reagents

1. **Conc. Hydrochloric acid**
2. **Dilute Hydrochloric acid, 1+1**
3. **Potassium permanganate solution, approximately 0.2 N:** Dissolve. 600 mg. $KMnO_4$ in 80 ml distilled water and dilute to 100 ml. This is about 0.2 N.
4. **Iron stock solution:** Add 20 ml conc. H_2SO_4 to 50 ml distilled water. Dissolve 0.7022g. of ferrous ammonium-sulfate crystals $FeSO_4\ (NH_4)_2SO_4.6H_2O$ in the above acid sollution. Warm it gently. Add potassium-permanganate solution dropwise until a slight pink color persists. Make upto 1000 ml in a volumetric flask with distilled water.

   1 ml = 0.1 mg. Fe

5. **Iron standard solution:** Dilute 50 ml. stock solution to 500 ml. with distilled water.

$$1 \text{ ml} = 0.01 \text{ mg. Fe.}$$

6. **Thiocyanate solution:** Dissolve 10 g. potassium thiocyanate, KCNS, or ammonium thiocyanate, NH_4CNS in 80 ml distilled water and dilute to 100 ml.

Procedure

A) For relatively clear samples

1. Place 100 ml or suitable aliquot of the well shaken sample in a 250 ml. beaker and add 5 ml 1+1 HCl. Reduce the volume to 40 ml by placing on a bot plate.

2. Cool and add potassium permanganate solution drop by drop until a pink colour persists for atleast 5 minutes. Transfer to a 50 ml Nessler tube and make upto the mark.

3. Pipet 1.0ml, 2.0ml, 3.0ml, 4.0ml, 5.0ml, 6.0ml, of iron standard solution into 50 ml Nessler tubes. Add 1ml dil. HCl and 2 drops of potassium permanganate solution Mix well and make upto the mark with distilled water.

4. Add to the standards and sample, 1 ml thiocyanate solution and mix well. Compare the color of the sample with that of standards and find out the mg. equivalent of the iron present in the sample as follows:

Calculation

$$\text{mg/l Iron (as Fe)} = \frac{\text{Matching std} \times 0.01 \times 1000}{\text{ml. sample taken for determination}}$$

B) For samples having higher concentration of iron and/or organic matter

Measure 20 ml or 10 ml or suitable aliquot of the sample in an evaporating dish and add 10 ml (iron free) conc. HCl. Evaporate to dryness on a hot plate and ignite by heating until a grayish white ash is obtained. (Note: Do not heat to a red glow).

Cool the ash and add 5 ml. dil. HCl and transfer to a 100 ml. volumetric flask and make upto the mark with distilled water. Pipet aliquots from this made up solution into 250 ml beakers and proceed as for clear samples.

C) Iron (Total) by Titrimetric method

Treat the sample as given in the preceding section procedure 'B'.

Place 25 ml of the treated sample in a 500 ml conical flask. Add 20 ml conc. HCl. Bring the solution to near boiling (70° to 90°C). Reduce the Fe^{3+} to Fe^{2+} by adding stannous chloride solution, (prepared by dissolving 30 g. $SnCl_2 . 2H_2O$ in 100 ml conc. HCl and diluting to 200 ml. with distilled water) drop by drop until the yellow color disappears. Care must be taken not to add excess stannous chloride. Cool the solution rapidly under tap to about 20°C (with protection from air). Then rapidly add 10 ml of 5% mercuric chloride solution and mix thoroughly to remove the slight excess of stannous chloride. A slight silky white precipitate of mercurous chloride should be obtained at this stage (If a heavy or grey or black precipitate is formed it indicates the presence of excess stannous chloride. In this case the reduction should be repeated).

Allow to stand for 5 minutes. Add 200 ml. 2N. sulfuric acid and 0.5 ml. N-phenyl- anthranilic acid indicator (prepared by dissolving 100 mg. in 4.8 ml. of. 0.1N. NaOH and diluting to 100 ml with distilled water). Titrate with standard potassium dichromate solution of appropriate normality. The color change is from green to violet red. From the titre value calculate the iron content, using the following equivalent.

1 ml. of exactly 1 N. $K_2Cr_2O_7$ = 55.85 mg. Fe

42-9 LEAD

Lead paint wastes, mine waters, battery manufacture wastes and electroplating wastes are the principal wastes containing lead in significant quantities. The following dithizone colorimetric method is more precise.

DITHIZONE COLORIMETRIC METHOD

This method is applicable to all types of effluents, domestic waste waters, polluted waters, raw and treated waters.

Principle

Lead is made to react with dithizone at a pH of about 11.5 to form lead dithizonate which is soluble in chloroform. In the presence of an alkaline cyanide solution, the free green colored dithizone is not extracted by chloroform. The interference from bismuth and tin is eliminated by preliminary treatment of the sample with dithizone at a pH of about 2.3.

Interferences

Interference by metals which form dithizonates is

eliminated by extraction at controlled pH. Organic matter when present will interfere and may be removed by the preliminary acid digestion of the sample. If organic matter is absent, digestion is not necessary. It is important to note that glassware may contribute lead in the estimation. Hence absolute cleaning of glassware with 1+1 HNO_3 followed by rinsing with lead-free water is necessary. If possible, separating funnels for lead estimation may be reserved. It is better to collect the samples in polythene containers.

Reagents

Use lead-free distilled water for solutions and dilutions. Lead free water may be prepared by distilling distilled water in an all glass apparatus. It may also be prepared by passing the distilled water through a mixed bed of ion exchange resins.

1. **Ammonia solution, approximately 0.5N:** Dilute 3.5 ml ammonia solution to 100 ml with water.

2. **Dithizone stock solution, 0.1% (in chloroform):** Dissolve 100 mg. dithizone in 100 ml chloroform. Keep it in an amber bottle and store in a refrigerator.

3. **Dithizone working solution (in water):** Transfer 12 ml. of stock dithizone solution into a 100 ml separating funnel. Add 20 ml. 0.5N ammonia solution and shake well. Allow the phases to separate and reject the lower chloroform layer. Filter the aqueous layer through a wetted filter paper (to remove droplets of chloroform) in to a 50 ml amber bottle.

4. **Sodium hexametaphosphate solution 10%:** Dissolve 10g. sodium hexametaphophate in 100 ml distilled water. Remove traces of lead by extraction with stock dithizone solution (0.1% chloroform solution) after adjusting the pH to 9 with conc. ammonia solution. Acidify the aqueous layer in the separating funnel using 1+1 HCl and then extract with chloroform to remove free dithizone until the chloroform layer becomes colorless. Reject the chloroform layer. Adjust the pH of the aqueous solution to 9.5

5. **Hydroxyl - amine hydrochloride solution 1%:** Dissolve 1g. of $NH_2.OH.HCl$ in 100 ml. distilled water.

6. **Alkaline cyanide solution:** Prepare a mixture of solution containing 340 ml. ammonia (sp.gr. 0.88) and 680 ml. water. Dissolve 3.0 g sodium sulfite $Na_2SO_37H_2O$ in the mixture. Add 30 ml 1% potassium cyanide solution and mix well. Keep it in a tightly stoppered bottle.

   (**Caution:** Do not use mouth pipets to deliver volumes of cyanide solution)

7. **Lead stock solution:** Dissolve 1.599 g. lead nitrate. $Pb(NO_3)_2$ in small amount of water. Add 10 ml conc. nitric acid and make up to 1000 ml.

   1ml = 1.0 mg. Pb.

8. **Lead intermediate solution:** Dilute 10 ml. of the above lead stock solution to 1000 ml. with distilled water (prepare freshly).

   1.0ml = 10 µg. Pb.

9. **Lead working solution:** Dilute 10 ml. of the lead intermediate solution to 100 ml. with distilled water (prepare freshly)

   1.0ml = 1.0 μg. Pb

Procedure

Standards

1. Place 50 ml. of lead-free distilled water into a series of short stemmed separating funnels. Pipet 0.0, 1.0, 2.0, 3.0, 4.0 .... 10.0 ml of lead working solution in to them.
2. Add the following reagents in order with shaking after each addition.
   i. 1.0 ml sodium hexametaphosfate solution
   ii. 1.0 ml hydroxylamine hydrochloride solution,
   iii. 30 ml. alkaline cyanide solution
   iv. 0.5 ml. dithizone working solution, and
   v. 10 ml. chloroform.
3. Shake the funnel vigorously and allow the layers to separate. Dry the stem of the funnel with filter paper strips. Draw the chloroform layer in the optical cell.
4. Measure the optical densities using a spectrophotometer at 510 nm or filter photometer with a green filter using a 10 mm cell. Plot a calibration curve.

Sample

If organic matter is present, it must first be destroyed by digestion with sulfuric and nitric acids. (See "Preliminary acid digestion of the sample", Chapter 42-1). If tin and bismuth are known to be present, then a preliminary extraction at a controlled pH is necessary.

The following is the procedure for samples free of organic matter, tin and bismuth.

1. Measure 50 ml or suitable aliquot of the original sample or acid digested sample (neutralized to pH about 7) containing not more than 20μg lead in a separating funnel. Adjust the volume to 50 ml. with lead-free distilled water.
2. Then proceed from step No.2 of the procedure for standards.
3. Measure the optical density of lead dithizonate in chloroform and find the number of micrograms equivalent to the optical density from the calibration graph.

Express the result as mg/l Lead (as Pb)

Removal of tin and bismuth

If the sample contains tin and/or bismuth eliminate by the following procedure.

Adjust the pH of the sample to 2-2.5 using 1+1 HCl. This aliquot is taken in a separating funnel to which 20 ml. of 0.1% dithizone in $CHCl_3$ has to be added. Shake vigorously. Discard $CHCl_3$ layer. Extract with 5 ml. portions of 0.1% dithizone in $CHCl_3$ until the last portion remains green. Wash the aqueous phase with 5 ml. chloroform and discard the chloroform layer. The aqueous phase is now free from tin and bismuth. Adjust the pH to about 7 (and dilute with lead free water if necessary). Then proceed from step No.2 of the procedure for standards.

42-10 MANGANESE

Manganese compounds present in chemical waste waters and in drainage from coal and ash tips. They react with the natural alkalinity of receiving waters in the presence of air to form the characteristic brown precipitate manganic hydroxide.

Among the methods available for the determination of manganese, the persulfate method is precise, easy to carryout and applicable to all types of effluents and waters.

PERSULFATE COLORIMETRIC METHOD

Principle

The soluble manganese compounds are oxidized to permanganate by persulfate in acid solution. The resulting color of the permanganate solution is measured either visually or photometrically.

$$2MnSO_4 + 5(NH_4)_2S_2O_8 + 8H_2O + 10HNO_3 \rightarrow \underset{\textit{Permanganic acid}}{2HMnO_4} + 12H_2SO_4 + 10NH_4NO_3$$

Interferences

Chloride and ferric ions interfere. Chloride as NaCl up to 100 mg. is eliminated by the addition of mercuric sulfate. Ferric iron is eliminated by the addition of phosphoric acid which forms soluble complexes.

Reagents

1. **Manganese stock solution:** Dissolve 1.000 g. high purity manganese metal in 100 ml 7% sulfuric acid. Cool and make upto 1000 ml in a volumetric flask. Pipet 50 ml of this solution into a 1000 ml volumetric flask and make up to the mark with distilled water.

   1.0ml = 0.05 mg. Mn.

   (Alternatively, dissolve 144 mg. $KMnO_4$ in about 100 ml distilled water. Add 25 ml dil. H_2SO_4 (20%), reduce the permanganic acid by dropwise addition of H_2O_2 (20 volume), and transfer to a 1000 ml volumetric flask and make up to the mark with distilled water).

   1ml = 0.05 mg Mn

2. **Manganese working standard solution:** Pipet 100 ml of the stock solution and make up to 500 ml with distilled water.

   1.0 ml = 0.01 mg. Mn

3. **Special reagent:** Add 400 ml conc. HNO_3 to 200 ml distilled water in a 2 liter beaker. Dissolve 75g. mercuric sulfate $HgSO_4$ in it. Add 200 ml 85% phosphoric acid, H_3PO_4, and 35 mg silver nitrate, $AgNO_3$ and dilute the cooled solution to 1000 ml.

4. **Ammonium persulfate, $(NH_4)_2S_2O_8$, crystals.**

Procedure

Clear samples

1. Place 100 ml. or suitable aliquot of the sample in a 250 ml beaker. (Adjust the pH to about 7 if the sample is

alkaline, with dil.sulfuric acid).

2. Place appropriate volumes of manganese standard solution in 250 ml beakers. Dilute each to 100 ml or to the volume equivalent to that of the sample taken. Include a beaker containing distilled water as the blank.

3. To the blank, standards and sample, add 5 ml. special reagent.

4. Heat the beakers on a hot plate and when the contents are reduced to 40 ml. add 1g. ammonium persulfate while boiling. Continue the boiling for one minute*. Then cool the beakers under tap.

5. Transfer to 50 ml. Nessler tubes. Dilute to the mark with distilled water and compare visually. Photometric measurements may also be made by using either spectrophotometer at 545 nm or filter photometer with a suitable green filter. Prepare a calibration curve and find out the mg manganese equivalent to the observed optical density.

Express the result as mg/l Manganese (as Mn).

Samples containing suspended, matter, turbidity and organic matter

To a suitable aliquot of the sample in an evaporating dish, add 5 ml. conc. HCl and evaporate to about 25 ml. Digest at the boiling point to bring dissolution of the particulate materials and once again evaporate to dryness. Cool and add 5

* Too long a boiling and slow cooling results in the loss of permanganate color. Hence should be cooled immediately.

ml conc. H_2SO_4 and 10 ml. conc. HNO_3 and digest as described in "Preliminary treatment of samples" (Chapter 42-1). Use suitable aliquots of the digested sample for estimation. Carryout the determination as for clear samples starting form step No.1.

42-11 MERCURY

Mercury enters the biosphere as a waste product from a number of industrial activities. A major source is the production of chlorine where mercury is used as an electrode in the electrolysis of brine. Pulp and paper, oil refining, and plastic industries, and battery manufacturing industries also contribute to mercury pollution. Mercury also enters into water courses through the use of mercurial fungicides.

To determine trace amounts of mercury, cold vapor flameless atomic absorption method is the best choice. The following dithizone colorimetric method though tedious is also precise and mercury concentration down to 1.0μg can be measured.

DITHIZONE COLORIMETRIC METHOD

This method is applicable to all types of effluents, polluted waters, raw and treated waters.

Principle

Mercury ions are allowed to react with dithizone (in chloroform) at a pH of about 1 to form an orange yellow colored complex and the color is measured using a spectrophotometer.

Interferences

Copper is the major interference and is eliminated during the procedure by keeping the copper dithizonate in organic phase. The other interfering elements, silver, gold, palladium, platinum (divalent) are usually absent. Because of the volatility of mercury and its compounds, wet digestion is necessary.

Reagents

Note: Use mercury free water (redistilled or deionised water) for the preparation of all solutions and dilutions.

1. **Mercury stock solution:** Dissolve 135.4 mg mercuric chloride, $HgCl_2$, in about 750 ml. water. Add 1.5 ml conc. HNO_3 and make upto the mark with distilled water.

   $$1.0 \text{ ml} = 100 \mu\text{g. Hg}$$

2. **Mercury working solution:** Pipet 10.0 ml. mercury stock solution into a 1000 ml volumetric flask and make up to the mark with distilled water.

   $$1 \text{ ml} = 1 \mu\text{g Hg}$$

   Prepare just before use.

3. **Potassium permanganate solution, 5%:** Dissolve 5 g. potassium permanganate, $KMnO_4$ in 100 ml distilled water.

4. **Sulfuric acid, conc.**

5. **Potassium persulfate solution, 5%:** Dissolve 5g. potassium persulfate, $K_2S_2O_8$ in 100 ml. distilled water.

6. **Hydroxyl-amine hydrochloride, 50%:** Dissolve 50 g. hydroxylamine hydro chloride, NH_2OH. HCl in 100 ml distilled water.

7. **Dithizone solution:** Dissolve 6 mg. dithizone in 1000 ml chloroform.

8. **Sulfuric acid, 0.25 N:** Dilute 6.9 ml conc H_2SO_4 to 1000 ml. with distilled water.

9. **Potassium bromide solution, 40%:** Dissolve 40 g. potassium bromide, KBr in 100 ml. distilled water.

10. **Phosphate-Carbonate buffer:** Dissolve 150 g. disodium-hydrogen phosphate Na_2HPO_4. $12H_2O$ and 38 g. potassium carbonate (anhydrous) in 1000 ml distilled water. Purify the buffer as follows: Place the buffer in a 1500 ml separating funnel and extract with 10 ml. portions of dithizone until the last extract remains green (bluish green). Remove the excess dithizone from the buffer by extraction with chloroform.

11. **Sodium sulfate, Na_2SO_4, anhydrous.**

Procedure

1. Place 0.0, 1.0, 2.0, 4.0 .... 10.0 ml of mercury working solution into a series of 1000 ml reflux flasks. Dilute each to 500 ml. or to the volume similar to that of sample taken for determination.

2. Place 500 ml or suitable volume of the sample in a reflux flask.

3. To the blank, standards and sample, add 1ml. potassium permanganate solution, and 10 ml. conc. H_2SO_4.
4. Mix well and boil (If the permanganate is decolorised, add more $KMnO_4$).
5. After 30 to 45 minutes, add carefully 5 ml. potassium-persulfate solution and allow to cool for 30 minutes.
6. Discharge the pink color by the addition of one or more drops of hydroxyl-amine hydrochloride solution.
7. Transfer the contents of each flask to individual 1000 ml. separating funnels.
8. Extract the contents of each separating funnel with 25 ml portions of dithizone, until the last extract remains green (bluish green) as that of the original dithizone solution.
9. Collect the chloroform extracts in individual 250 ml. separating funnels.
10. Wash the combined chloroform extracts by treatment with 50 ml. of 0.25 N. H_2SO_4.
11. Draw the chloroform layer into another series of 250 ml. separating funnels.
12. Add 50 ml. of 0.25N H_2SO_4 and 10 ml. potassium-bromide solution to each funnel and shake well. Now the mercury dithizonate will be in aqueous phase and other dithizonates (particularly of copper) will be in organic phase.
13. Discard the lower organic layer.

14. Wash the aqueous phase by shaking with a small volume of chloroform and discard the lower chloroform layer. To each separating funnel add 20 ml. phosphate-carbonate buffer and 10 ml dithizone solution. Shake well and allow the phases to separate.

15. Pass the organic layer containing mercury dithizonate through a layer of sodium sulfate (in a funnel) into a cuvette and measure the absorbance at 490 nm.

16. Prepare a calibration curve on a linear graph paper plotting absorbance (optical density) versus microgram of mercury.

17. From the calibration curve find out the number of micrograms of mercury equivalent to the optical density of the sample.

Express the result as mg/l Mercury (as Hg).

42-12 NICKEL

Nickel occurs in electroplating wastes. It is toxic to fish and other aquatic life. It is best determined by the following dimethyl glyoxime method.

DIMETHYL GLYOXIME COLORIMETRIC METHOD

Principle

Nickel reacts with dimethyl glyoxime in the presence of an alkaline oxidizing agent to form a characteristic red color complex which can either be measured visually or photometrically.

Interferences

Iron, manganese and copper interfere. If they are present in concentration thrice that of nickel, their interference is suppressed by the addition of sodium citrate. If they are present in excess concentrations, nickel dimethyl glyoxime complex alone is separated by extraction with chloroform and proceeded further. (A separate procedure is given).

Reagents

1. **Nickel stock solution:** Dissolve 447.9 mg nickelsulfate, $NiSO_4.6H_2O$ in distilled water and make upto 1000 ml. in a volumetric flask.

   1.0ml = 100 μg Ni

2. **Nickel working solution:** Pipet 10.0 ml nickel stock solution into a 100 ml volumetric flask and make upto the mark with distilled water.

   1.00 ml= 10 μg Ni

3. **Hydrochloric acid, 0.5N (app):** Dilute 43 conc. HCl to 1000 ml with distilled water.

4. **Sodium citrate solution:** Dissolve 125 g. sodium citrate, $Na_3C_6H_5O_7.2H_2O$ in 500 ml distilled water.

5. **Iodine solution, 0.05N (app):** Dissolve 20 g. potassium-iodide, Kl in 50 ml. distilled water. Dissolve 6.4 g. iodine in this iodide solution, and dilute to 1000 ml.

6. **Dimethyl glyoxime solution, 0.5%** Dissolve 1 g. dimethyl glyoxime in 100 ml conc. ammonia solution. Add 100 ml. distilled water, Filter if necessary.

7. **Additional reagents for the removal of interferences:**
   - (i) Dilute ammonia solution: Dilute 10 ml. conc. ammonia solution to 500 ml with distilled water.
   - (ii) Chloroform.

Procedure

1. Place appropriate volumes of nickel working solution covering the range up to 100μg in to a series of 50 ml volumetric flasks (or 50 ml. Nessler tubes). Include a 50 ml volumetric flask (or 50 ml Nessler tube) as the blank.

2. Place a suitable aliquot of the neutralized (acid digested) sample containing not more than 100μg nickel in a 50 ml. volumetric flask (or 50 ml. Nessler tube).

3. To the blank, standards and sample add 20 ml, 0.5 N. HCl.

4. Then add the following reagents in order with mixing after each addition :
   - i). 10 ml. sodium citrate solution,
   - ii). 2 ml. iodine solution, and
   - iii). 4 ml. dimethyl glyoxime solution.

5. Make upto 50 ml. with distilled water and allow to stand for 20 minutes.

6. Compare the color visually or photometrically using a spectrophotometer at 470 nm or filter photometer with a suitable blue filter taking water as the reference. Prepare a calibration curve and find out the microgram nickel equivalent to the observed optical density.

Express the result as mg/l Nickel (as Ni).

Sample containing excessive amounts of Iron, Manganese and Copper

Place an aliquot of the original sample or acid digested sample containing not more than 100μg Ni, in a separating funnel. Add 10 ml sodium citrate solution and make slightly alkaline using dilute ammonia solution. Add 4 ml dimethyl glyoxime solution and shake well for one minute.

Extract the nickel dimethyl glyoxime complex with three 10 ml. portions of chloroform. Collect every time the chloroform layer into another 100 ml. separating funnel. Add 20 ml. dilute ammonia solution to the combined chloroform extracts and shake well. Draw the chloroform layer into another separating funnel and extract with two 10 ml. portions of 0.5 N. HCl. Heat combined aqueous extracts gently to remove traces of chloroform. Cool and transfer to a 50 ml. volumetric flask (or 50 ml. Nessler tube). Proceed from step No. 4. of the above procedure.

42-13 SODIUM

Tannery effluents, textile dyeing effluents, edible oil refinery effluents, wastes from soap manufacture contain abnormal quantities of sodium. Sodium is best determined by flame photometric method.

FLAME PHOTOMETRIC METHOD

The flame photometric method is by far the simplest and most rapid method for the determination of sodium. Even very small amounts of sodium in large proportion of water can be determined by this method.

Reagents

1. **Deionized distilled water:** Prepare by passing distilled water through a mixed bed of ion exchange resins. Use only deionised distilled water for the preparation of solutions and dilutions.

2. **Sodium stock solution:** Dissolve 2.5422 g. sodium chloride, NaCl (Previously dried at 140°C for 1 hour) in little deionized distilled water and make up to 1000 ml. in a volumetric flask.

   1.0ml= 1.0 mg Na

   Make dilutions as required for use.

Note: Store all reagents in polythene containers only.

Procedure

Follow the manufacturer's instructions for operating the instrument. The following points will however be useful.

1. The reproducibility and accuracy of the instrument should be checked constantly. For this, a standard may be included for test every time the instrument is operated.

2. Sample should be free from suspended impurities. If the sample contains any suspended matter, filter it.

Prepare the sample by dry ashing as described in "Preliminary Treatment of Samples" (Chapter 42-1) and then determine sodium content.

Express the result as mg/l Sodium (as Na).

PERCENT SODIUM

Though effluents are mainly discharged into water courses, the increasing tendency is to use the effluent for irrigation. Among the parameters that limit the use of the effluent for irrigation, percent sodium is important. In agricultural practice, sodium concentration is expressed in terms of 'percent sodium', which is defined as the percentage of sodium concentration to the total cationic concentration. It is calculated as follows:

$$\text{Percent sodium (as Na)} = \frac{A}{A + B + C + D} \times 100$$

Where A = Sodium content (as Na) in meq/l

B = Calcium content (as Ca) in meq/l

C = Magnesium content (as Mg) in meq/l

D = Potassium content (as K) in meq/l

Conversion of mg/l to meq/l

$$1 \text{ milli equivalent / litre} = \frac{mg/l}{\text{equivalent. wt}}$$

Sodium 1 mg/l	=	0.04349 meq/l
Calcium 1 mg/l	=	0.04990 meq/l
Magnesium 1 mg/l	=	0.08226 meq/l
Potassium 1 mg/l	=	0.02558 meq/l

CALCULATION OF PERCENT SODIUM - EXAMPLE

A tannery effluent contains calcium 510 mg/l (as $CaCO_3$), magnesium 980 mg/l (as $CaCO_3$), sodium 3960 mg/l (as Na), and potassium 56 mg/l (as K). Calculate percent sodium.

Since the values of calcium hardness is expressed in terms of $CaCO_3$ convert it to Ca by multiplying with 0.40 and likewise the value of magnesium hardness (as $CaCO_3$) to Mg by multiplying with 0.24.

Then convert all the mg/l values to meq/l values.

Sodium (meq/l)	=	3960 x 0.04349	=	172.20
Calcium (meq/l)	=	(510x0.4) x 0.04990	=	10.18
Magnesium (meq/l)	=	(980 x 0.24) x 0.08226	=	19.35
Potassium (meq/l)	=	56 x 0.02558	=	1.43
				203.16

$$\textit{Percent sodium (as Na)} = \frac{172.20}{203.16} \times 100 = 84.7$$

42-14 POTASSIUM

Potassium is found in appreciable concentrations in distillery effluents (spent wash) and coconut coir processing effluents. Potassium is best determined by flame photometric method.

DETERMINATION OF POTASSIUM

1. Prepare potassium stock solution by dissolving 1.907 g. potassium chloride, KCl dired at 110°C in little de-ionized distilled water and make up to 1000 ml in a volumetric flask.

1.0 ml= 1.0 mg K

Make dilutions as required for use with deionized distilled water.

2. Prepare the sample by dry ashing as described in "Preliminary treatment of samples" (Chapter 42-1)

3. Determine potassium content of the sample using flame photometer with usual precautions (Refer Sodium).

Express the result as mg/l Potassium (as K).

42-15 ZINC

The major industrial wastes contributing to zinc are, galvanizing wastes, zinc plating wastes, viscose rayon wastes, dye manufacturing and dye house wastes and rubber factory wastes. Industries manufacturing paints and pigments, pharmaceuticals, insecticides and cosmetics also contribute to zinc. Zinc is also found in zinc-mine wastes. It is best determined by dithizone method using bis-2-hydroxy ethyl dithiocarbamate, which suppresses most of the interferences.

DITHIZONE COLORIMETRIC METHOD

Principle

Zinc reacts with dithizone to form a co-ordinate compound which when extracted with carbon tetra chloride is red in color and is used to measure the zinc content.

Interferences

Copper, cadmium and lead interfere when they present in excess concentration and must be removed by precipitation with hydrogen sulfide. A separate procedure for such samples is given. Another important factor in the determination of zinc is the purity of reagents and glassware. The reagents should be free from zinc. As glassware contribute to zinc, they must be thoroughly cleaned with sodium citrate solution before use.

Reagents

Prepare all reagents using zinc-free distilled water which can be prepared by redistilling the distilled water in an all pyrex distillation apparatus.

1. **Sodium citrate solution:** Dissolve 50 g. sodiumcitrate, $Na_3C_6H_6O_7.2H_2O$ in 450 ml zinc-free distilled water. Purify by dithizone extraction.

2. **Zinc stock solution:** Dissolve 100 mg. zinc metal (30 mesh) in about 1 ml 1+1 HCl and make upto 1000 ml. in a volumetric flask using zinc-free distilled water.

   1.0ml= 100µg Zn

3. **Zinc working solution:** Pipet 10.0 ml zinc stock solution into a 1000 ml volumetric flask and make upto the mark with zinc-free distilled water.

   1.0ml = 1 µg Zn

4. **Methyl red indictor solution:** Dissolve 100 mg. methyl-red (sodium salt) in distilled water and dilute to 1000 ml.

5. **Conc. ammonia solution (0.91 to 0.88)**

6. **Potassium cyanide solution, 5%:** Dissolve 5g. potassium cyanide, KCN in 100 ml zinc-free distilled water.

7. **Acetic acid, conc.**

8. **Carbon tetra chloride.**

9. **Bis (2-hydorxy ethyl) dithio carbamate solution:** Mix 4.0 g. diethanol amine and 1 ml. carbon-di-sulfide with 40 ml methyl alcohol. Prepare freshly.

10. **Dithizone solution, 0.005%:** Dissolve 5 mg. dithizone thoroughly in a little carbon- tetrachloride and dilute to 100 ml with CCl_4.

11. **Sodium sulfide solution I:** Dissolve 3.0 g $Na_2S.9H_2O$ or 1.65 g. $Na_2S.3H_2O$ in 100 ml zinc-free distilled water.

12. **Sodium sulfide solution II:** Dilute 4.0 ml of sodium sulfide solution I to 100 ml with zinc-free distilled water. Prepare just before use.

Procedure

1. Pipet 1.0, 2.0 .... 10.0 ml of zinc working solution in to a series of 125 ml separating funnels and dilute to 20 ml with zinc-free distilled water.

2. Include a separating funnel containing 20 ml. zinc-free distilled water as the blank.

3. Place in a separating funnel an aliquot of the sample containing not more than 10μg Zn and dilute to 20 ml with zinc-free distilled water.

4. To the blank, standards and sample, add 2 drops of methyl red indicator solution and 2.0 ml sodium citrate solution. If the color is not yellow at this point, add conc.

ammonia solution dropwise to the separating funnels, until the indicator just turns yellow.

5. Add 1 ml potassium cyanide solution and then conc. acetic acid dropwise until the solution becomes neutral peach in color.

6. Add 5 ml. carbon tetrachloride to each separating funnel to extract the methyl red and discard CCl_4 layer.

7. Add 1 ml bis-2-hydroxy ethyl dithio carbamate solution. Mix well by shaking, and then add 10 ml dithizone solution. Shake well and draw off the CCl_4 layer into another separating funnel. Extract the aqueous solution in the first separating funnel with 5 ml portions of dithizone solution until the last CCl_4 extract remains green. Combine all the CCl_4 extracts in the second separating funnel.

8. To the combined extracts in the second separating funnel, add 10 ml. sodium- sulfide solution II. Shake well and discard the aqueous layer. Wash the CCl_4 layer with further 10 ml portions of sodium sulfide solution II until the unreacted dithizone has completely been removed as shown by the color of the aqueous layer either colorless or very pale yellow.

9. Draw off the CCl_4 layer into a 50 ml. volumetric flask after removing the water adhering to the stem of the funnel using a cotton swab. Make up to the mark with CCl_4.

10. Determine the absorbance of the blank, standards, and sample at 535 nm. Plot a calibration curve. Find out the

microgram equivalent of zinc to the optical density.

Express the result as mg/l Zinc (as Zn).

Samples containing excess amounts of Copper, Lead and Cadmium

If the aliquot taken for the estimation contains more than 2 mg. of copper, lead and cadmium, either separately or jointly, they should be precipitated as sulfides.

1. Place the aliquot of the acid digested sample in a 100 ml beaker and adjust the volume to 20 ml (either by evaporation or by dilution).
2. Bring the acidity to 0.5N by adding either 6N. HNO_3 or ammonia solution as necessary.
3. Pass H_2S to the cold solution for five minutes, and filter through a sintered crucible. Wash the precipitate with hot water and combine this wash water with the filtrate.
4. Remove H_2S from filtrate by boiling. Cool and transfer to a 125 ml. separating funnel. Proceed from step No.4 of the procedure described above.

REFERENCES

1) F.D. Snell & C.T Snell, "Colorimetric Methods of Analysis" 3rd ed, 1949, D.Van Nostrand Co, Princeton, N.J.

2) E.B. Sandell, "Colorimetric Determination of Traces of Metals"- 3rd ed, 1959, Interscience Publishers, New York.

3) J.A. Dean, "Flame Photometry". 1950, Mc Graw Hill Publishing Co., New York.

4) A.I. Vogel, "Quantitative Inorganic Analysis Including Elementary Instrumental Analysis" - 3rd ed, 1964, THe ELBS & Longman,Green & Co. Ltd., London.

5) American Society for Testing and Materials, "Annual Book of ASTM Standards", Part 23, 1972, ASTM Philadelphia.

Chapter - 43

INORGANIC CONSTITUENTS NON-METALLICS

43-1 CHLORIDE

Chloride is the common anion found in industrial effluents. Tanneries, textile processing and dye manufacture effluents contribute significant quantities of chlorides. It is best determined by argentometric titration (Silver nitrate titrant) (Mohr's method).

SILVER NITRATE TITRIMETRIC METHOD

Principle

Silver nitrate reacts wtih chloride ion to form silver chloride. The completion of reaction is indicated by the red color produced by the reaction of silver nitrate with potassium-chromate solution which is added as an indicator.

Interferences

Bromide, iodide, and cyanide will titrate as chloride. Orthophosphate in excess of 25 mg/l interfere. Iron in excess of 10 mg/l masks the end point. Sulfide, sulfite, thiosulfate and thiocyanate ions also interfere. However, these interferences can be eliminated by the addition of hydrogen peroxide.

Reagents

Use chloride-free distilled water for the preparation of all reagents.

1. **Standard silver nitrate titrant, 0.0282N:** Dissolve 4.791 g. silver nitrate, $AgNO_3$ in distilled water and make upto 1000 ml. in a volumetric flask. Standardize it against 0.0282N sodium chloride solution as described in the 'procedure for clear samples'. Store it in an amber bottle.

   1.00 ml of exactly 0.0282N. $AgNO_3$ ■ 1.0 mg Cl

2. **Standard sodium chloride titrant, 0.0282N:** Dissolve 1.648 g. of sodium chloride, NaCl and make upto 1000 ml. in a volumetric flask.

   1.00ml = 1.0 mg Cl

3. **Potassium chromate indicator solution:** Dissovle 25 g. potassium chromate K_2CrO_4 in 100 ml distilled water. Add silver nitrate solution dropwise until a slight red precipitate is formed. Allow to stand for 12 hours. Filter and dilute the filtrate to 500 ml.

4. **Special reagents for removal of interferences:**

   i) **Aluminium hydroxide suspension:** Dissolve 100 g. potassium aluminium sulfate $K_2SO_4 \cdot Al_2(SO_4)_3 \cdot 24H_2O$ in 1000ml distilled water. Warm to 60°C and add slowly with stirring 55 ml. conc. ammonia solution. Allow the precipitate to settle for about one hour. Wash by decantation with distilled water to make the precipitate free from chloride. Check it by treating a portion of the decantate every time with silver nitrate solution.

After the precipitate is free from chloride, dilute it to 1000 ml. with distilled water.

ii) **Sodium hydroxide, IN:** Dissolve 40g. sodiumhydroxide, NaOH in distilled water, and dilute to 1000 ml.

iii) **Sulfuric acid, 1N :** Place 27.7 ml conc. H_2SO_4 in 500 ml distilled water in a 1000 ml volumetric flask. Make up to the mark with distilled water.

iv) **Hydrogen peroxide:** 30 volume.

Procedure

(A) For Clear Samples

Place 100 ml. of the sample or an aliquot containing not more than 10 mg. chloride in a porcelain basin of about 500 ml capacity. (If the pH of the sample is in the range of 7 to 8 it can directly be titrated. If the pH is not within this range, adjust the pH to be within this range using sulfuric acid or sodium hydroxide solution). Dilute to 100 ml with distilled water. Add 1 ml. potassium chromate indicator solution. Titrate against standard silver nitrate solution with constant stirring until a slightest perceptible reddish coloration persists. Conduct a blank by placing 100 ml chloride-free distilled water instead of sample. A blank of 0.2 to 0.3 ml is usual for the method.

Calculation

mg/l Chloride (as Cl^-)

$$= \frac{(\text{ml. } AgNO_3 \text{ titrant} - \text{ml. } AgNO_3 \text{ required for blank}) \times 1 \times 1000}{\text{ml. sample taken for determination}}$$

(B) For Colored and Turbid Samples

If the sample is highly colored or turbid, add 3 ml aluminium hydroxide to a measured volume of the sample in a beaker. Stir well and allow to settle. Filter and wash the precipitate with chloride-free distilled water. Combine the filtrate and washings (bring the pH in the range 7-8 if necessary) and titrate as described above.

(c) For samples containing cyanide, sulfide, thiosulfate and thiocyanate.

Measure a suitable volume (not exceeding 100 ml) of the sample in a casserole. Add 1 ml hydrogen peroxide (30 volume) and stir well for 1 minute. (Adjust the pH to 7-8 if necessary). Dilute to 100 ml. Add 1 ml. potassium chromate indicator and titrate as described above.

Note 1: Use always a uniform sample volume, preferably 100 ml so that ionic concentrations need to indicate the end point is constant.

Note 2: The pH of the samples must be in the range of 7 to 8. At high pH Ag+ is precipitated as Ag OH. At low pH, CrO_4^{2-} is converted to $Cr_2O_7^{2-}$

Note 3: 1 ml of the potassium chromate indicator is used to provide proper concentration of $Cr_2O_4^{2-}$. Excess indicator leads to the formation of Ag_2CrO_4 soon and less indicator delays it.

43-2 FLUORIDE

Phosphatic fertilizer industry is considered to be the major industry that discharges excess amounts (2000 to mg/l 3000 mg/l) of fluorides. Effluent from the production of aluminium from bauxite also contains fluorides.

As the color, turbidity and other constituents of effluents interfere with the colorimetric determination of fluorides, a preliminary distillation is necessary.

PRELIMINARY DISTILLATION

Sample is subjected to distillation with H_2SO_4 and the fluorides are distilled as hydrofluoric acid.

Apparatus

The apparatus for distillation is as shown in Fig. 43-2-1.

Procedure

Place 400 ml. distilled water in a 1 litre boiling flask. Carefully add 200 ml conc. H_2SO_4 and swirl the flask to make the contents homogeneous Add glass beads and connect the flask as shown in the figure. Make sure that all joints are tightly fitted. Heat the flask slowly at first and then rapidly. (The distillate must be cool whatever rapid the heating may be). As the temperature reaches 180°C, stop the heating and allow it to cool. Discard the distillate. The process serves to remove fluorides (if any) in the glassware or in the reagents and to adjust the acid : water ratio for subsequent distillation. After cooling to 120°C, add 300 ml. sample, mix thoroughly and distill until the temperature reaches 180°C. (Do not allow the temperature to exceed 180°C as it may cause sulfate carry over). Collect the distillate in a 300 ml. volumetric flask.

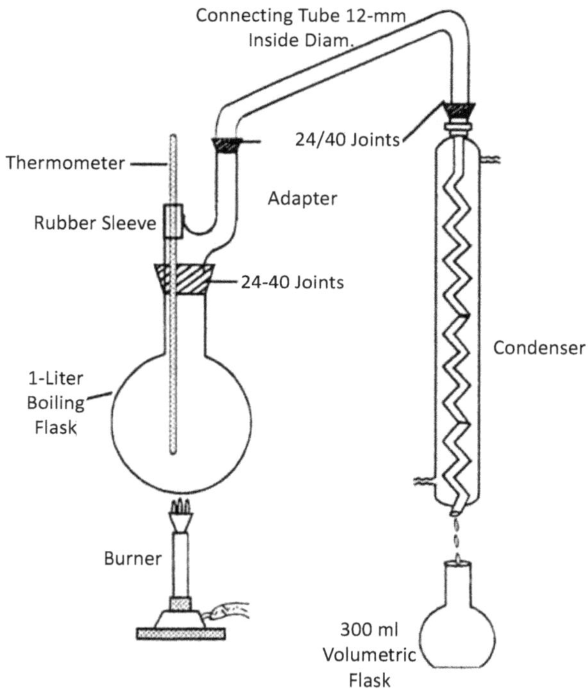

Fig 43-2-1. Direct Distillation Apparatus for Fluoride

If the sample contains high amounts of fluoride, continue the distillation with 300 ml. distilled water after the sample has been distilled. Combine both the distillates for fluoride determination.

If the chloride content of the sample exceeds 2000 mg/l add silver sulfate crystals to the distilling flask at a rate of 5 mg per milligram of chloride.

Note: The sulfuric acid solution in the distilling flask can repeatedly be used until the accumulation of contaminants from samples affect the fluoride recovery.

ZIRCONIUM - ALIZARIN COLORIMETRIC METHOD

This method is applicable to effluents and domestic waste waters (after distillation), raw waters and treated waters.

Principle

Fluoride decolorises the zirconium alizarin complex and decolorization is proportional to fluoride concentration.

$$\underset{\text{(Reddish color)}}{(\textit{Zr. alizarin lake})} + 6F^- \rightarrow \underset{\text{(Yellow)}}{\textit{Alizarin}} + (ZrF_6)^{2-}$$

Interferences

Aluminium in excess of 0.1 mg/l interferes. Phosphate as P_2O_5 in excess of 0.5 mg/l interferes. Residual chlorine interferes but can be eliminated by the addition of sodium arsenite. Color and turbidity also interfere but eliminated by preliminary distillation.

Reagents

1. **Fluoride stock solution:** Dissolve 221.0 mg anhydrous sodium fluoride, NaF, in distilled water and make upto 1000 ml. in a volumetric flask.

   1.00 ml = 0.1 mg F

2. **Fluoride standard solution:** Dilute 100 ml. stock solution to 1000 ml in a volumetric flask with distilled water.

   1.00ml = 0.01 mg F

3. **Acid zirconium-alizarin solution:**

a) Alizarin solution: Dissolve 700 mg alizarin red-S. (sodium alizarin sulfonate) in 100 ml. distilled water.

b) Zirconyl chloride solution: Dissolve 450 mg. zirconyl chloride, $ZrOCl_2.8H_2O$ in 100 ml distilled water.

c) Sulfuric acid solution: Add carefully with stirring 70 ml conc. H_2SO_4 to 700 ml water in a beaker and cool.

Pour slowly solution (a) to solution (b) and then add solution (c) and make upto 1000 ml. Use it after 24 hrs. Store in dark.

4. **Sodium arsenite solution:** Dissolve 1.0 g sodium arseni-te, in 100 ml. distilled water. Keep it in a tightly stoppered bottle.

Procedure

1. Place 50 ml. or a suitable aliquot of the distillate in a 50 ml Nessler tube. Adjust the pH to about 7 and dilute to 50 ml. with distilled water.

   (If the sample is unpolluted water, place 50 ml. or suitable aliquot of the sample diulted to 50 ml. If the sample contains any suspended matter, filter through Whatman No. 42 or equivalent filter paper. If the sample contains residual chorine, add 0.1 ml sodium arsenite solution to 50 ml of the sample in the Nessler tube and mix well).

2. Pipet 0.0, 1.0, 2.0, 4.0, 6.0, 8.0, 10.0, 12.0 ml. of standard

fluoride solution into 50 ml Nessler tubes and make upto the mark with distilled water.

3. Add 1.0 ml. acid zirconium-alizarin solution to the standards and sample. Mix well immediately and compare the color after 20 minutes.

Express the result as mg/l Fluoride (as F^-).

ELECTRODE METHOD

Fluoride ion-selective electrode is very useful for the rapid determination of fluoride in water samples particularly when contaminated with color, turbidity and other interfering substances. The fluoride-sensitive electrode is of the solid state type, consists of lanthanum fluoride crystal. In combination with a reference electrode (normally calomel electrode) it forms a cell. An expanded scale pH meter or specific ion meter may be employed for measuring the potential. For use of the electrode, suppliers' instructions should be followed. Generally the solution under test in which electrodes are dipped should be thoroughly agitated for about 3 minutes to attain a steady reading. To eliminate the interferences, an ionic stength adjustment buffer should be added to the test solution. It may be purchased from the suppliers and added to the test solution in the recommended proportion. The preparation of one of such buffers is given below:

To 500 ml. distilled water in a 1 litre beaker, add 57ml conc (glacial) acetic acid, 58 g. sodium chloride and 12 g. sodium citrate dihydrate, $Na_3C_6H_5O_7.2H_2O$. Stir well to dissolve the contents and cool to room temperature. Adjust the pH of the solution between 5.0 and 5.5 using 6N. sodium hydroxide solution. (About 125 ml is required). Cool to room temperature. Transfer to a 1 litre volumetric flask and make upto the mark with distilled water.

To 50 ml. of the sample or an aliquot diluted to 50 ml, add 50.0 ml. of this ionic strength adjustment buffer. Fluoride standards diluted to 50 ml. should also be added with 50 ml. of this buffer.

43-3 NITRATE

Synthetic fertilizer wastes contain enormous quantities of nitrate. Nitrates are the end products of the aerobic stabilization of organic nitrogen. Among the methods available, the following brucine sulfate method is sensitive and subject to less interference. The concentration range that can be best measured by this method is 0.1 to 2 mg/l NO_3 - N*. For determining higher concentrations, sample must be diluted.

BRUCINE SULFATE COLORIMETRIC METHOD

Principle

Nitrate ion reacts with brucine in strong sulfuric acid solution to form a yellow color which is measured photometrically.

Interferences

All strong oxidizing and reducing agents interfere. Chlorine interference upto 5 mg/l is eliminated by the addition of sodium arsenite. Ferrous iron, ferric iron and tetravalent manganese interfere when present in concentration more than 1

* As per ASTM (American Society for Testing and Materials), the measurable range extends upto 8 mg/l nitrogen.

mg/l. The interference by nitrite is eliminated by the use of sulfanilic acid. High concentrations of organic matter present in effluents interfere but eliminated by pretreatment with aluminium hydroxide suspension.

Reagents

1. **Nitrate stock solution:** Dissolve 722 mg. potassium-nitrate KNO_3 (previously dried at 105°C for 24 hr) in distilled water and make upto 1000 ml. in a volumetric flask.

   1.0 ml = 0.1 mg. Nitrate nitrogen (N)

2. **Nitrate standard solution:** Pipet 100 ml. nitrate stock solution in to a 1000 ml. volumetric flask and make upto the mark with distilled water.

   1.0 ml. = 0.01 mg. Nitrate nitrogen (N)

3. **Brucine-Sulfanilic acid solution:** Dissolve 1 g. brucine sulfate and 100 mg. sulfanilic acid in 70 ml. hot distilled water. Add 3 ml. conc. HCl, cool and dilute to 100 ml with distilled water. This solution is stable for several months.

   Caution: This solution is very toxic and take care to avoid ingestion.

4. **Sulfuric acid solution:** Add carefully 500 ml conc. H_2SO_4 to 75 ml. distilled water. Cool to room temperature.

5. **Sodium arsenite solution:** Dissolve 1.83 g. sodium arsenite, $NaAsO_2$ in 100 ml. distilled water.

   Caution: Toxic, take care to avoid ingestion.

6. **Aluminium hydroxide suspension:** For preparation, see Chapter 43-1 "Chloride".

Procedure

1. Into a series of 50 ml. beakers, pipet 0.5, 1.0, 2.0, 2.5, 3.0, 4.0, 4.5 and 5.0 ml. nitrate standard solution. Dilute each to 5 ml. with distilled water. Include a beaker containing 5 ml. distilled water as the blank.

2. Place 2 ml. or a suitable aliquot of the sample (not exceeding 5 ml) containing not more than 10 mg/l nitrate nitrogen in a 50 ml beaker. Adjust the volume to 5 ml.

3. (If the sample contains any residual chlorine, add 0.1 ml. sodium arsentite solution to a 50 ml. portion of the sample for each 0.05 mg/l chlorine. Add one drop in excess and mix well. Take aliquots of this arsenite treated sample for nitrate nitrogen determination. If the sample contains excess organic matter, or turbid, add 3 ml. $Al(OH)_3$ suspension, mix well, allow to settle and filter the supernatant. Discard the first portion of the filtrate. Use the filtered sample for estimation).

4. Add 1 ml. brucine-sulfanilic acid solution to blank, standards and sample and mix well.

5. Into a second series of 50 ml. beakers, place 10 ml. sulfuric acid solution.

6. Transfer the contents of the first series of 50 ml beakers containing blank, standards and sample to each of second series of 50 ml. beakers containing sulfuric acid

solution and mix well. Pour from one beaker to the other 4 to 6 times to ensure complete mixing.

7. Keep the beakers in the dark for 10+1 minutes.

8. While the color is developing, measure 10 ml. distilled water to each of the empty beakers. After 10 minutes, add the water to the respective beakers containing blank, standards and sample.

9. Allow to cool in the dark for 20-30 minutes.

10. Measure the absorbance of the standards and sample after setting the blank at 100% transmittance at a wave length of 410 nm. If spectrophotometer is not available, use a filter photometer equipped with a violet filter having a maximum transmittance between 400 and 425 nm.

11. Prepare a calibration curve and find out the mg equivalent of nitrate nitrogen in the sample.

    Note: Run a series of standards with each set of sample and prepare the calibration curve every time.

Express the result as mg/l Nitrate nitrogen (as N).

Calculation

$$\text{mg/l Nitrate N} = \frac{\text{mg. Nitrate Nitrogen} \times 1000}{\text{ml. sample taken for estimation}}$$

$$\text{mg/l } NO_3 = \text{mg/l Nitrate N} \times 4.43$$

43-4 NITRITE

Though nitrite determination is important for natural waters, it is seldom determined on industrial effluents. However it is useful in determining whether sufficient available nitrogen is present for aerobic biological treatment. This data is helpful for suppression of nitrification in BOD test. Dye manufacturing effluent is the only effluent that contains excess amounts of nitrites. Nitrite treated cooling waters also contribute to nitrites. Nitrite is best determined by the following diazotisation method.

DIAZOTISATION (COLORIMETRIC) METHOD

This method is applicable to industrial effluents, domestic waters, raw and treated waters.

Principle

Sulfanilic acid is diazotized by nitrite and the diazo compound is coupled with naphthyl amine hydrochloride to form a reddish purple azo dye at a pH of 2.0 to 2.5.

Interferences

Strong reducing agents, strong oxidizing agents and amines interfere. Color and turbidity interfere, but the interference may be eliminated or suitable corrections may be applied.

Reagents

Use nitrite-free distilled water for the preparation of reagents.

1. **EDTA solution:** Dissovle 500 mg. disodium ethylene diamine tetra acetate, dihydrate, $Na_2H_2C_{10}H_{12}O_8N_2.2H_2O$ in distilled water and dilute to 100 ml.

2. **Sulfanilic acid solution:** Dissolve completely 600 mg. sulfanilic acid in 70 ml. hot distilled water. Cool and add 20 ml. conc. HCl. Dilute to 100 ml with distilled water.

3. **α-Naphthylamine hydrochloride solution:** To 50 ml. distilled water in a beaker add 1 ml conc. HCl. Dissolve in it 600 mg. α-naphthylamine hydrochloride. Dilute to 100 ml. with distilled water. Store in a refrigerator.

4. **Sodium acetate buffer solution, 2M:** Dissolve 16.4g. sodium acetate $NaC_2H_3O_2$ or 27.2g sodium acetate trihydrate $NaC_2H_3O_2.3H_2O$ in distilled water and dilute to 100 ml.

5. **Nitrite stock solution:** Dissolve exactly 493 mg. sodium-nitrite, $NaNO_2$, in distilled water and make upto 1000 ml. in a volumetric flask.

   1.00 ml = 100 µg Nitrite nitrogen (N)

6. **Nitrite intermediate solution:** Pipet 10.0 ml. stock nitrite solution and make up to 100 ml. Prepare daily.

   1.00ml = 10 µg Nitrite nitrogen (N)

7. **Nitrite working solution:** Pipet 5.0 ml. nitrite intermediate solution into a 100 ml. volumetric flask and make up to the mark.

   1.00 ml = 0.5 µg nitrite nitrogen (N)

   Prepare this solution freshly.

8. **Aluminium hydroxide suspension:** See Chapter 43-1, "Chloride".

Procedure

A. Colorless and Clear Samples

1. Pipet 0.1, 0.2, 0.4, 0.6, 0.8, 1.0, 1.5, 2.0, 2.5, 3.0, 4.0, 5.0ml nitrite working solution into 50 ml Nessler tubes and make upto the mark with distilled water. Include a Nessler tube as blank and place 50 ml. distilled water in it.

2. Transfer 50.0 ml of the clear sample (neutralized to pH 7) or an aliquot diluted to 50 ml. in a Nessler tube. Add 1.0 ml EDTA solution and stir well.

3. Add 1.0 ml sulfanilic acid to blank, standards and sample and mix thoroughly. After 10 minutes add 1.0 ml α-naphthylamine hydrochloride solution and 1.0 ml sodium acetate buffer solution and mix thoroughly.

4. After 10 minutes, compare the color visually or measure the absorbance using a spectrophotometer or a filter photometer at 520 nm. Prepare a calibration curve and find out mg nitrite nitrogen, equivalent to the optical density.

Express the result as mg nitrite nitrogen as N per litre of sample. If it has to be expressed in terms of nitrite, then mg/l NO_2 = mg/l nitrite nitrogen x 3.29.

B. Colored and Turbid samples

If the sample is colored or contains suspended solids,

place 100 ml. sample in a 250 ml beaker. Add 2 ml aluminium-hydroxide suspension and stir well. Cover it with a watch glass and allow it to stand for a few minutes. Decant the supernatant and filter through a filter paper into a 50 ml. Nessler tube and proceed as described above, starting from step No 3.

C. Chlorinated samples

Chlorinated samples may contain nitrogen trichloride which may impart a false red color. This effect can be minimized by adding α-naphthyl amine hydrochloride solution first and then sulfanilic acid solution.

43-5 PHOSPHATE

Effluents from phosphatic fertilizer industries, boiler blow down and cooling tower waste waters contain excess concentration of phosphates. Phosphates occur in natural waters and trade wastes in various forms. They are commonly classified as ortho phosphates, condensed phosphates (meta, pyro and polyphosphates), and organically bound phosphates. These various forms of phosphorous may occur in soluble form or in particulate form.

According to "Standard Methods" the phosphates which respond to the colorimetric tests without preliminary hydrolysis or oxidative digestion of the sample are considered as "Orthophosphates". The phosphate fraction which is obtained along with orthophosphates on mild acid hydrolysis is known as condensed phosphates. This fraction includes, meta, pyro and poly phosphates. The phosphate fraction obtained on acid digestion along with ortho and acid hydrolysable phosphate is known as the 'organic' or 'organically bound' phosphates.

DETERMINATION OF ORTHOPHOSPHATES BY STANNOUS CHLORIDE COLORIMETRIC METHOD

Principle

Ammonium molybdate reacts with phosphate to form molybdophosphoric acid which is reduced to a blue colored complex, 'molybdenum blue' by the addition of stannous-chloride.

Interferences

Arsenate, fluoride, sulfide, thiosulfate, thiocyanate and excess molybdate cause negative interference. Sulfides may be removed by oxidation with bromine water.

Reagents

1. **Phenolphthalein indicator solution:** See "Alkalinity" (Chapter 40-9)
2. **Sulfuric acid-nitric acid solution:** Carefully add 75 ml. conc. H_2SO_4 to about 150 ml. distilled water and cool. Add 1 ml. conc. HNO_3 and dilute to 250 ml with distilled water.
3. **Ammonium molybdate, solution:** (a) Dissolve 25g. ammonium molybdate, $(NH_4)_6MO_7O_{24}.4H_2O$ in about 200 ml. distilled water. (b) Add carefully 280 ml. conc. H_2SO_4 to 400 ml distilled water and cool.

   Add the molybdate solution to the diluted acid and

dilute the mixture to 1000 ml.

4. **Stannous chloride solution:** Dissolve 2.5 g fresh stannous chloride, $SnCl_2.2H_2O$ in 100 ml. glycerol and heat on a water bath. Mix by stirring with a glass rod. This reagent is stable and requires no special storage.

5. **Phosphate stock solution:** Dissolve 439 mg. potassium-dihydrogen phosphate, KH_2PO_4 in distilled water and make up to 1000 ml in a volumetric flask. Add two drops of toluene as preservative.

   1.0 ml= 100 μg P

6. **Phosphate standard solution:** Pipet 10.0 ml. phosphate stock solution into a 1000 ml. volumetric flask and make upto the mark with distilled water. Prepare freshly.

   1.0 ml = 1 μg P

Procedure

Important Note: Do not use synthetic detergents containing phosphates for cleaning of the glassware. Use only acid solutions.

1. Place 100 ml. or a suitable aliquot of the sample containing not more than 20 μg P in a 100 ml. Nessler tube. (The sample should be free from color and turbidity). Add 1 drop of phenolphthalein indicator. If any pink color appears, destroy it by adding one or 2 drops of sulfuric-nitric acid solution. If more than 5 drops are required, take a smaller aliquot, add phenolphthalein, discharge the pink color and dilute to 100 ml.

2. Into a series of 100 ml. Nessler tubes, pipet appropriate

volumes of phosphate working solution covering the range upto 20 μg P. Dilute to 100 ml. Include a Nessler tube containing 100 ml. distilled water as the blank.

3. To the blank, standards and sample, add 4.0 ml. ammonium molybdate solution and 0.5 ml. stannous-chloride solution, mixing after each addition.

4. After 10 minutes but before 12 minutes, measure the color using a spectrophotometer at 690 nm.

5. Prepare a calibration curve and find out the number of micrograms of P equivalent to the observed optical density of the sample.

Express the result as mg/l Phosphate (as P).

If it has to be expressed in term of PO_4 multiply by 3.066

INORGANIC PHOSPHATES (ORTHOPHOSPHATES + CONDENSED PHOSPHATES)

To 100 ml. sample or an aliquot diluted to 100 ml. add 1 drop of phenolphthalein indictor. If any pink color appears, discharge by the addition of sulfuric-nitric acid solution. Then add 1 ml excess of sulfuric acid - nitric acid solution and boil gently for about 90 minutes. Distilled water may be added during boiling to keep the volume between 25 and 50 ml. Cool and neutralize with 6N. sodium hydroxide solution (240 g. NaOH/1000 ml) using phenolphthalein indicator. Make up to 100 ml.

Determine the phosphates by the procedure as

described for orthophosphates.

The result obtained is the sum of total inorganic phosphates (i.e) the sum of condensed (meta-, pyro-, and poly) phosphates and orthophosphates.

If the concentration of condensed phosphates only is needed subtract the ortho phosphate value from the total inorganic phosphate value.

TOTAL PHOSPHATES (ORTHOPHOSPHATES+CONDENSED PHOSPHATES+ORGANIC PHOSPHATES)

Destroy the organic matter first by sulfuric acid-nitric acid digestion as described below.

Sulfuric acid-nitric acid digestion

Place a suitable aliquot of the well mixed sample in a micro-kjeldahl flask and add 1 ml. conc. H_2SO_4 and 5 ml. conc. HNO_3. Digest the sample to a volume of 2 ml. and continue the digestion until the solution becomes colorless. Cool and add about 20 ml. distilled water and neutralize with 1N. sodium-hydroxide solution (40g. NaOH/1000 ml.) using phenolphthalein indicator. Transfer to a 100 ml. volumetric flask (filtering if necessary) and make up to the mark with distilled water.

Determine the phosphate content of this made up solution by the procedure as described for orthophosphates.

The result obtained indicates the concentration of total phosphates. If the concentration of organic phosphates is required, subtract the inorganic phosphate from total phosphate.

43-6 SULFATE

Sulfate estimation is useful in determining the magnitude of problems that arise from reduction of sulfates to hydrogen sulfide during anaerobic reduction/digestion of industrial effluents. The most precise method to determine sulfates is gravimetric method.

BARIUM SULFATE GRAVIMETRIC METHOD

This method is applicable to all types of industrial effluents, domestic waste waters, polluted waters, raw and treated waters.

Principle

Sulfate is precipitated as barium sulfate in the presence of hydrochloric acid. The precipitated barium sulfate is filtered, dried, ignited and weighed as $BaSO_4$. (The addition of HCl to the sample prevents the formation of barium carbonate with highly alkaline wastes (and waters). The precipitating agent barium chloride is added in excess to produce sufficient common ion to precipitate sulfate ion as completely as possible. Digestion of the sample after precipitation (at 90°C) is carried out to convert the colloidal precipitate to crystalline form.

Interferences

The gravimetric estimation of sulfates suffers from many interferences. Suspended matter, silica, nitrate, sulfide, and barium chloride precipitant lead to high results. Alkali metal sulfates yield low results. Chromium and iron also cause low results. Strict adherence to the procedure minimizes the interferences.

Reagents

1. **Methyl red indicator:** Dissolve 100 mg. methyl red sodium salt in 100 ml distilled water.
2. **Hydrochloric acid 1+1**
3. **Barium chloride 5%:** Dissolve 25 g. bariumchloride-dihydrate $BaCl_2.2H_2O$ in 500 ml distilled water. (If any turbidity appears on standing, reject the solution)
4. **Silver nitrate - Nitric acid reagent** (to test residual chloride) Dissolve 8.5 g silver nitrate and 0.5 ml. conc. HNO_3 in 500 ml. distilled water.

Procedure

A) Colorless and Clear Samples

1. Measure in to a 400 ml beaker, appropriate volume of clear sample containing 8 to 60 mg sulfate, (usually 200 ml). Dilute the sample to 200 ml. with distilled water, it necessary. (If the sulfate content of the sample is very low, take higher volumes of the sample and evaporate on a water bath to 200 ml).
2. Adjust the pH of the sample to 4.5-5.0 with HCl using methyl red indicator (to orange color). Then add additional 2 ml. HCl.
3. Boil the solution for about one minute and add 10 ml. hot barium chloride solution slowly from a pipet with

constant stirring.

4. Keep the beaker on a water bath to digest the precipitate at 80-90°C for atleast 2 hrs. or preferably over night.

5. Filter the contents of the beaker through an ashless filter paper (Whatman No. 42 or equivalent). Wash the precipitate with hot water until the washings give no opalescence with silver nitrate-nitric acid reagent.

6. Place the filter paper with precipitate in a previously weighed platinum or silica crucible and heat in a low flame with its lid loosely covered. Gradually increase the heat until the paper chars. Do not allow the filter paper to flame as it may cause loss. After charring is complete ignite at 800°C for 1 hr. (with free access of air). Cool in a desiccator and weigh.

Note: If silica crucible is used, pre-ignite it at 800°C for 1 hour, cool in a desiccator and weigh.

Calculation

$$mg/l \text{ sulfate as } (SO_4) = \frac{mg.\ BaSO_4}{ml.\ \text{sample taken for estimation}} \times 411.5$$

B) For colored and turbid samples

Place a suitable volume of the sample in a 400 ml beaker. Adjust the pH to 4.5 - 5.0. Add 2 ml. conc. HCl and digest for about 15 minutes. Filter through Whatman No. 42 filter paper. Adjust the volume to 200 ml. Proceed from step No. 3 as for colorless samples.

Alternatively, completely ashed sample may be taken, leached with dil. HCl and made upto 200 ml. Proceed from step No 2.

43-7 SULFIDE

Effluents from tannery, paper mill, dye house, oil refinery and edible oil refinery contain significant quantities of sulfide. Viscose rayon wastes, septic sewage and ammoniacal gas liquor also contain sulfide. Two methods, viz., colorimetric and titrimetric are described here. The colorimetric procedure is simple to carryout, rapid, sensitive and applicable to sulfide concentrations of 0.2 - 2.0 mg/l. Titrimetric procedure, though is applicable to samples containing more than 1 mg/l sulfide, it is generally used for samples of higher concentrations and containing more interferences.

Sample collection and preservation

Sample should be collected with least aeration as sulfide is volatilized and oxygen destroys sulfide. Samples have to be preserved by the addition of 2 ml. of zinc acetate solution (2N) per litre sample. Adjust the pH of the sample > 9 by the addition of sodium hydroxide. The sample should contain representative proportion of suspended solids.

COLORIMETRIC METHYLENE BLUE METHOD*

Principle

This colorimetric method determines total sulfides

* This method is not applicable to colored waste waters such as from dye house. In such cases electrode method is preferable.

which include dissolved H_2S, HS^-, S_2^- ions and acid soluble metallic sulfides (iron, manganese, zinc and lead sulfides). Acid insoluble metallic sulfides such as copper sulfide are not determined by this method.

This method is based on the reaction between p-amino-dimethyl aniline, ferric chloride and sulfide ions, resulting in the formation of methylene blue.

Interferences

Some strong reducing agents prevents the color formation or reduce the intensity. Sulfites and thiosulfate interfere but are tolerable upto 10 mg/l each. Their interference upto 40 mg/l of SO_3^{2-} or $S_2O_3^{2-}$ can be eliminated by increasing the amount of ferric chloride solution to 6 drops instead of 3 drops. Higher concentrations of sulfide itself prevent color production but dilution of the sample eliminates this problem.

Reagents

1. **Sulfuric acid 1+1:** Add carefully 500 ml conc. H_2SO_4 to 500 ml. distilled water with constant stirring, and cool.

2. **Amine-sulfuric acid stock solution:** Add carefully 50 ml of conc. H_2SO_4 to 20 ml. distilled water and cool. Dissolve in this solution 27 g. p-amino dimethyl aniline oxalate (also called N, N-dimethyl p-phenylene diamine) and dilute to 100ml. with distilled water. Store in an amber bottle.

3. **Amine-sulfuric acid working solution:** Dilute 12.5 ml. of amine-sulfuric acid stock solution to 500 ml. with 1+1

sulfuric acid. Store in an amber bottle.

4. **Ferric chloride solution:** Dissolve 100 g. ferric chloride, $FeCl_3.6H_2O$. in 40 ml. distilled water.

5. **Diammonium hydrogen phosphate solution:** Dissolve 100 g. diammonium hydrogen phosphate, $(NH_4)_2\ HPO_4$ in 200 ml. distilled water.

6. **Methylene blue solution I:** Dissolve 1.0 g. methylene-blue (USP grade) in distilled water and make up to 1000 ml in a volumetric flask. Standardize this solution with standard sulfide solution as described below.

Weigh 12 to 15 g. pure sodium sulfide (white) and wash several times with distilled water to remove the adhering black sulfide and dissolve the remaining in distilled water and make upto 1000 ml. in a volumetric flask.

To 100 ml distilled water in an iodine flask add a suitable aliquot (5 ml or 10 ml) of sulfide solution below the surface of water. Add a known excess of 0.025N. iodine solution (1 ml = 0.4 mg sulfide). Add 5 ml. conc. HCl, stopper and shake well and backtitrate with 0.025N. sodium thiosulfate using starch indicator.

mg. sulfide present in the aliquot = [ml. iodine(0.025N) - ml. thio (0.025N)]x0.400

From the above sulfide solution, prepare a standard sulfide solution containing 1 mg. sulfide per ml. To about 980 ml. distilled water in a 1000 ml. volumetric flask, add 2 ml, of this standard sulfide solution such that the sulfide concentration of the resulting solution is 2 mg/l. Use this solution for standardization of methylene blue solution by the colorimetric procedure described below. Use 2 drops of methyleneblue and compare the color with that of the sulfide solution. If the color of methylene blue is more than that of sulfide solution, it has to be diluted. If the intensity is less, small increments of dye (MB) has to be added such that the color of both the sulfide solution and MB are the same. Now the methylene blue solution I is standardized.

0.05ml. (1 drop) of MB solution I ≡ 1.0 mg/l. sulfide

7. **Methylene blue solution II:** Pipet 10 ml. methylene blue solution in to a 100 ml volumetric flask and make upto the mark with distilled water.

   0.05ml. (1 drop) of MB solution II ≡ 0.1 mg/l sulfide.

Procedure

1. Place 7.5 ml. of the sample into two 10 ml. comparator tubes. (125 mm long and 15 mm out side diameter). Mark the tubes as A and B.

2. To tube A add 0.5 ml. amine-sulfuric acid working solution and 0.15 ml. (3 drops) ferric chloride solution. Mix immediately by inverting the tube only once. If any sulfide is present, a blue color will appear. Allow three minutes for the complete color development.

3. To tube B add 0.5 ml. 1+1 sulfuric acid and 0.15 ml ferric chloride solution.

4. After three minutes, add. 1.6 ml diammonium hydrogen phosphate solution to tube A and B to remove the color of the ferric chloride. Allow to stand for about 10 minutes.

5. If the sample seems to contain more than 5 mg/l sulfide, add methylene blue solution I, drop by drop to tube B until the color matches with that developed in tube A. If

the sulfide concentration of the sample is less than 5 mg/l, add methylene blue solution II dropwise to tube B.

Note: If the sulfide concentration of the sample exceeds 20 mg/l dilute the sample ten times and use the diluted sample for determination.

Calculation

(i) If the sample volume is 7.5 ml. and methylene blue solution I is used, then

mg/l sulfide = No. of drops of methylene blue solution I x 1.

(ii) If the sample volume is 7.5 ml. and methylene blue solution II is used, then

mg/l sulfide = No. of drops of methylene blue solution II x 0.1

TITRIMETRIC METHOD

This procedure measures total sulfides excepting acid insoluble metallic sulfides.

Principle

Sulfide in the sample is made to evolve by acidification and passing a stream of CO_2. The evolved sulfide is absorbed in zinc acetate solution which is later acidified, excess iodine added and back titrated with standard thiosulfate solution.

Reagents

1. **Zinc acetate solution, 2N:** Dissolve 220 g. zinc acetate, $Zn(C_2H_8O_2)_2.2H_2O$ in 500 ml. distilled water and make upto 1000 ml.

2. **Sulfuric acid, conc.**

3. **Iodine solution, 0.025N:** Dissolve 25 g. potassium-iodide in a little water and add 3.175 g. iodine. After all the iodine is dissolved completely, make upto 1000 ml. in a volumetric flask. Standardize against 0.025 N. sodium thiosulfate solution, using starch as indicator.

   1ml of exactly 0.025N. iodine ≡ 0.400 mg. sulfide.

4. **Hydrochloric acid, conc.**

5. **Sodium thiosulfate solution, 0.025N:** See "Dissolved Oxygen" (Chapter 41-1 Page 328).

6. **Starch solution:** See "Dissolved Oxygen" (Chapter 41-1, Page 328)

Procedure

1. Place a suitable volume (250 to 500 ml) of the well mixed sample in a 1 litre wide mouth bottle fitted with two holed rubber stopper carrying a fritted - glass diffuser tube and outlet tube. (Before placing the sample, the oxygen in the bottle may be displaced by passing a stream of CO_2)

2. Connect the outlet tube to a series of 125 ml. conical flasks (two flasks) each containing 5 ml zinc acetate solution plus 95 ml. distilled water.

3. Acidify the sample in the bottle with 10 ml. conc. H_2SO_4. Pass CO_2 through the sample for 1 hr (or for an appropriate period above which no sulfide is evolved

detected by suitable tests).

4. To each of the flasks containing zinc acetate solution, add a known excess of 0.025N. iodine solution and 5 ml. conc. HCl. Stopper and shake well.

5. Back titrate with 0.025N. sodium thiosulfate solution using starch indicator.

Calculation

mg/l Sulfide (as S)

$$= \frac{[ml.\ iodine\ (0.025N) - ml.\ thiosulfate\ (0.025N)] \times 0.400 \times 1000}{ml.\ sample\ taken\ for\ determination}$$

43-8 SULFITE

Pulping of wood and other cellulosic materials by sulfite process releases sulfites in the effluent. Boiler blowdown may contain sulfites as it is used for scavenging oxygen in the boiler feed. It is best determined by the following iodometric method.

IODOMETRIC METHOD

Principle

Sulfite is oxidized under acidic conditions to sulfate when treated with iodate-iodide solution. The completion of reaction is indicated by the appearance of blue color due to the starch added.

$$SO_3^{2-} + I_2 + H_2O \rightarrow SO_4^{2-} + 2HI$$

Sulfide, ferrous iron and organic matter interfere seriously. Nitrate if present will oxidize sulfite during acidification.

Reagents

1. **Sulfuric acid, 1+1:** Dilute 100 ml. conc. H_2SO_4 to 200 ml. with distilled water.
2. **Potassium iodate-iodide solution, 0.0125 N :** Dissolve 445.8 mg potassium iodate, KIO_3 (previously dried at 120°C for several hrs) in 200 ml. distilled water. Add 4.35 g. potassium iodide, KI and 310 mg. sodium bicarbonate, $NaHCO_3$ and make up to 1000 ml. in a volumetric flask.

   1.00ml 0.0125N. potassium iodate-iodide solution ≡ 500 µg SO_3

3. **Starch indicator:** See "Dissolved oxygen" (Chapter 41-1, - Page 328)

Procedure

Precaution

Conduct the determination as soon as possible. Collect the sample allowing only a little contact with air. It is advisable that the temperature of the sample should not exceed 20°C. If hot samples are collected, cool it in cooling apparatus as described in "Standard Methods For the Examination of Water and Waste Water" published by APHA".

Titration

Place 1 ml. 1+1 H_2SO_4 in a 250 ml. conical flask. Measure 50 ml. of effluent / polluted water using a measuring cylinder and transfer to the flask. Add 1 ml. starch indicator and titrate with potassium iodate-iodide solution until a faint permanent blue color develops.

Calculation

$$\text{mg/l Sulfite (as } SO_3) = \frac{\text{ml. potassium iodate - iodide x 0.5 x 1000}}{\text{ml. sample taken for determination}}$$

If sulfite has to be expressed in terms of sodium sulfite, then,

$$\text{mg / l } Na_2SO_3 = \text{mg/l } SO_3 \times 1.57$$

REFERENCES

43-2 FLUORIDE

Preliminary Distillation:

1) American public Health Association, "Standard Methods for the Examination of Water and Waste Water" - 14th ed. 1976, APHA, Washington.

Zirconium Alizarin Method:

2) Lim C.K, "Analyst", 1962, 87, 197

3) Department of Environment, "Analysis of Raw, Potable and Waste Waters" -1972, Her Majesty's Stationery Office, London.

Electrode method

4) Frant, M.S and J.W. Ross, Jr. "Anal. Chem"-1961, 46-1169

5) Patterson S. J etal., "Jour Society for Water Treatment and Examination" - 1969, 18, 182.

43-3 NITRATE

1) American Society for Testing and Materials "Annual Book of ASTM Standards" - 1972, ASTM, Philadelphia.

43-4 NITRITE & 43–5 PHOSPHATE

1) Boltz D.F. (Ed), Colorimetric Determination of Non metals," - 1958, Interscience Publishers, New York.

43-7 SULFIDE

1) R.Pomeroy" Sewage works Jour" - 1941, 13, 498

43-8 SULFITE

1) Konrich and L.W. Haase, "Water Pollution Abs" No 15, 1942.

Chapter - 44

MISCELLANEOUS DETERMINATION

44-1 ARSENIC

SILVER DIETHYL DITHIO CARBAMATE METHOD

Principle

Inorganic arsenic is reduced by zinc in acid solution to arsine which is absorbed in silver diethyl dithiocarbamate solution to form a red complex which is measured colorimetrically at 535 nm.

Interferences

Copper, chromium, cobalt, nickel and silver ions interfere. But these metals are normally present in very small quantities and hence are insignificant. Antimony present in five times excess of arsenic concentration interferes by forming a red complex stibine.

Apparatus

Modified Gutzeit apparatus (devised by NEERI, India) as shown in the figure 44-1-1.

Reagents

1. **Zinc :** 20 to 30 mesh, arsenic - free.
2. **Potassium iodide solution, 15%:** Dissolve 15 g. KI in 100 ml. distilled water. Store in an amber bottle.

3. **Lead acetate solution, 10%:** Dissolve 10 g Lead acetate. $Pb(C_2H_3O_2)_2.3H_2O$, in 100 ml distilled water.

4. **Silver diethyl dithiocarbamate solution, 0.5%:** Dissolve 0.5g. silver diethyldithiocarbamate in 100 ml. pyridine.

5. **Hydrochloric acid, conc.**

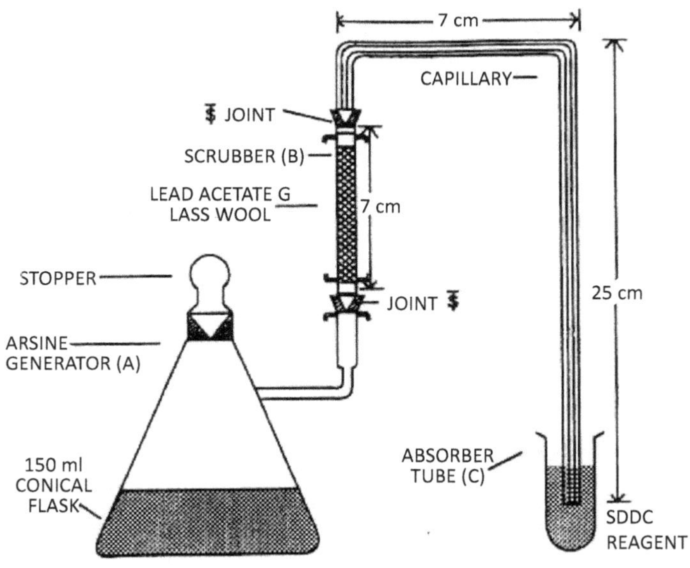

Fig 44-1-1 Modified Arsenic Estimation Apparatus

6. **Stannous chloride solution:** Dissolve 40g. arsenic-free $SnCl_2.2H_2O$ in 100 ml. conc. HCl.

7. **Arsenic stock solution:** Dissolve 1.320g. arsenic-trioxide, As_2O_3 in 10 ml. distilled water containing 4g.

NaOH and make up to 1000 ml. with distilled water or alternatively dissolve 1.734g. sodium meta arsenite $NaAsO_2$ in water and make up to 1000 ml.

1.00 ml. = 1.00 mg. As

7. **Arsenic intermediate solution:** Dilute 10.0 ml. arsenic stock solution to 1000 ml. with distilled water.

   1.00 ml. = 10.0 µg As

8. **Arsenic standard solution:** Dilute 10.00 ml. intermediate solution to 100 ml. with distilled water.

   1.00 ml. = 1.00 µg As

Procedure

Standards

1. In to a series of arsenic generators, Pipet out 0.0, 1.0, 2.0, ... 10.0 ml. of arsenic standard solution. Dilute to 50 ml. with distilled water.
2. Add 5 ml conc. HCl, and 2 ml. potassium iodide solution. Add 8 drops of stannous chloride solution, and allow 15 minutes for all arsenic to be reduced to trivalent state.
3. Meantime, impregnate the glass wool in the scrubber with lead acetate solution. (Do not make too wet, since water will be carried over into the absorber solution).
4. Take 4.0 ml silver diethyl dithiocarbamate reagent in the absorber tube. Connect the generator - scrubber-absorber assembly. Make sure that all the connections are tightly fitted.

5. Add 3g. zinc to the generator by opening the stopper, and immediately replace the stopper. Allow 30 minutes for complete evolution of arsine.

6. Pour the solution from the absorber directly into a 1 cm. cell and measure the absorbance of the solution at 535 nm using a spectrophotometer. Prepare a calibration curve.

Samples

Place 50 ml. of the well mixed sample (acid digested sample in the case of effluents) in the arsine generator. If the sample is polluted water containing organic matter suitable aliquot (100 or 200 ml) of the sample has to be concentrated and organic matter must be destroyed by acid digestion.

Proceed from step No. 2 to 6 of the procedure given above. Find out the microgram equivalent of arsenic to the net optical density.

Express the result as mg/l Arsenic (as As).

44-2 BORON

Boron is an essential element for the nutrition of higher plants, but concentrations exceeding 0.5 mg/l may be deleterious to certain crops.

Boron is best determined by the following curcumin method.

CURCUMIN COLORIMETRIC METHOD

Curcumin method is applicable to samples containing boron in 0.10 - 1.0 mg/l range.

Principle

When a water or waste water containing boron is acidified and evaporated with curcumin, a red colored product rosocyanin is formed which is dissolved in alcohol and the red color is measured colorimetrically.

Interferences

Organic matter and color interfere seriously but are eliminated by ashing the sample. Nitrate nitrogen interferes above 20 mg/l. If the total hardness (as $CaCO_3$) exceeds 100 mg/l, opalescence is produced in the alcoholic solution of rosocyanin which affects the color measurement. However, this interference can be eliminated by filtration of the final solution or by passing the sample through a column of strongly acidic cation exchange resin (in the hydrogen form).

Reagents

1. **Stock boron solution:** Dissolve 571.6 mg. anhydrous boric acid H_3BO_3 in distilled water and make upto 1000 ml. in a volumetric flask.

   $$1.00 \text{ ml} = 100\ \mu g\ B$$

2. **Standard boron solution:** Dilute 10.00 ml. of stock boron solution to 100 ml. with distilled water in a volumetric flask.

   $$1.00 \text{ ml} = 1\ \mu g\ B$$

3. **Ethyl alcohol, 95 %.**

4. **Curcumin reagent:** Dissolve 40 mg. finely ground curcumin and 5.0 g. oxalic acid in 80 ml. of 95% ethyl alcohol in a beaker. Add 4.2 ml. conc. HCl and make up

the solution to 100 ml. with ethyl alcohol in a volumetric flask. This reagent is stable for several days if kept in a refrigerator.

5. **Sodium hydroxide, 1N (app):** Dissolve 4g. NaOH. in 100 ml. distilled water.

6. **Hydrochloric acid 1+11:** Pour slowly 10 ml. conc. HCl into 110 ml. distilled water and mix well.

Procedure

Standards

1. Pipet 0.1, 0.20,0.40, 0.60, 0.80, 1.00 ml of standard boron solution into evaporating dishes (of 20 - 25 ml capacity made up to Vycor glass) and make upto 1.0 ml, with distilled water. Include a dish containing 1.0 ml distilled water as the blank.

2. Add 4.0 ml. curcumin reagent to each dish and swirl gently.

3. Float the dishes on a water bath set at 55 ± 2°C until the contents are completely evaporated to dryness and the odor of HCl is gone.

4. Remove the dishes from the water bath and cool to room temperature. Add 10.0 ml. 95% ethyl alcohol to each dish stirring gently with a polyethylene rod to insure complete dissolution of the red colored product.

5. Transfer the contents of the dish to a 25 ml. volumetric flask (or Nessler tube) using 95% ethyl alcohol. Make upto the mark with alcohol and mix thoroughly.

6. Read the absorbance or transmittance of the standards at a wave length of 540 nm after setting the blank at 100 % transmittance using 20 mm. cells.

7. Prepare a calibration curve relating optical density to µg boron.

Effluent Samples

Place a suitable aliquot of the sample in an evaporating dish (silica or porcelain) and make it alkaline using 1 N. sodium- hydroxide solution. Add 1 to 2 ml. in excess. Evaporate to dryness by keeping it on a water bath and then ignite at 500 to 550°C. Cool and add 2.5 ml, 1+11 HCl. Ensure that the solution is acidic. Bring the residue into solution by trituration with a police man. Centrifuge the solution and take the clear solution in a 25 ml. or 50 ml. volumetric flask and make upto to the mark with distilled water. Use 1 ml. of this made up solution for estimation. Start from step No. 2 of the procedure for standards. Conduct a blank simultaneously by taking distilled water in place of sample starting from the addition of alkali. From the calibration graph find out the microgram equivalent of boron to the observed optical density.

Express the result as mg/l Boron (as B).

If the sample is known to have a total hardness greater than 100 mg/l (as $CaCO_3$), remove the hardness by passing the prepared sample (i.e. ashed, leached with acid and made up sample) through a column of strongly acidic cation exchange resin (in the hydrogen form). The column is prepared as follows:

Soak a quantity of ion exchange resin "Zeokarb- 225" or equivalent grade in 1+5 HCl. Transfer the slurry to a column of approximately 50 cm. long and 1.2 cm. dia. fitted with a tap. Back wash the column with distilled water to remove air bubbles and to ensure even packing. Pass 50 ml 1+5 HCl through the column and then wash with distilled water until free from chloride. Keep the resin remains covered with liquid at all times.

Pipet 25 ml. prepared sample or a smaller aliquot if the sample contains excess boron, in to the resin column. Adjust the delivery tap so that the liquid flows out at a rate of 2 drops per second. Collect the eluate in a 50 ml volumetric flask. Wash the column with several small portions of distilled water until the flask is full to the mark. Stopper the flask and mix well. Care should be taken to keep liquid remains over the resin at all times.

Pipet 1.0 ml. of the made up sample into an evaporating dish and proceed from step No.2 of the procedure for standards.

Procedure for polluted waters

Follow the procedure described for effluent samples. If the total hardness of the sample exceeds 100 mg/l. (as $CaCO_3$) remove it as described above and use the softened sample for estimation.

Procedure for unpolluted waters

Place 1.0 ml. of the sample in a Vycor glass dish and proceed from step No.2 of the procedure for standards.

44-3 CHLORINE (RESIDUAL)

Effluents from bleaching operations in textile mills and

paper mills contain significant concentration of chlorine. Some other organic effluents may also contain chlorine, as they are chlorinated for disinfection before disposal.

Note: The residual chlorine content of water and effluent samples decrease after collection. This effect is more with rising temperatures and in the presence of organic matter. Hence the residual chlorine test should be conducted without delay, and preferably in the absence of sunlight.

IODOMETRIC METHOD

Both free available chlorine and combined available chlorine liberates iodine on reaction with potassium iodide and the liberated iodine is titrated with standard sodium thiosulfate solution using starch indicator. The reaction is preferably carried out at a pH of about 3 to 4.

Interferences

Nitrite ions interfere seriously. Ferric iron in excess of 2 mg/l interferes. Manganic ions are also reported to interfere. These interferences however are tolerable if the pH of the sample is adjusted between 4.5 and 8.0 (See modified procedure).

Reagents

1. **Acetic acid, conc. (glacial).**
2. **Potassium iodide, crystals.**
3. **Starch indicator solution:** See "Dissolved Oxygen" (Chapter 41-1, Page 328).
4. **Sodium thiosulfate solution, 0.1N.:** Dissolve 24.82g.

sodium thiosulfate, $Na_2S_2O_3.5H_2O$ in boiled and cooled distilled water and make upto 1000 ml. in a volumetric flask. Standardize it against standard potassium -dichromate solution (0.1N) as described in Chapter 41-1 "Dissolved Oxygen", (Page 328). Preserve it by adding a few drops of chloroform.

5. **Sodium thiosulfate solution, 0.0025N:** Dilute appropriate volume of 0.1 N. sodium thiosulfate to give 0.0025N. solution.

   1.0 ml. 0.0025N sodium thiosulfate = 0.08865 mg.Cl_2

Procedure

Place 500 ml. or a suitable aliquot of the sample in a conical flask. Add 0.5 g. to 1.0g. potassium iodide crystals and 5 ml. acetic acid. Mix and immediately titrate with 0.0025N. thiosulfate until the color of the liberated iodine is nearly discharged. Add 1 ml. starch solution and titrate until the blue color is discharged.

Modified procedure when nitrite, manganese, and ferric iron (more than 2 mg/l) are present.

Adjust the pH of the sample to be between 4.5 and 8.0 by the addition of acetic acid or sodium hydroxide. Add 0.5 to 1.0 g. potassium iodide crystals. Titrate immediately with 0.0025 N. thiosulfate solution as described above.

Calculation

mg/l chlorine (as Cl_2)

$$= \frac{\text{ml. 0.0025N. sodium thiosulfate} \times 0.0887 \times 1000}{\text{ml. sample taken for estimation}}$$

44-4 CYANIDE (TOTAL)

The major industrial effluents containing cyanide in significant concentrations are from the process of electroplating, gold mining, heat treatment of steel from gas works, coke ovens and from the scrubbing of gases at steel plants.

The samples have to be preserved by increasing the sample pH to 12 or more by the addition of sodium hydroxide. Cool to 4°C.

Cyanide determination suffers from many interferences but are eliminated by preliminary screening and distillation.

PRELIMINARY SCREENING - DISTILLATION-PHOTOMETRIC METHOD USING BIS - PYRAZOLONE

The method is applicable to all types of industrial effluents, domestic waste waters, polluted waters, raw and treated waters.

Principle

Cyanide in the sample is distilled and absorbed in alkali. CN^- in the alkaline distillate on treatment with bromine water is converted to cyanogen bromide (CNBr) which reacts with pyridine–pyrazolone reagent to form a blue dye. Most of the simple cyanides M(CN) are readily converted to HCN by acid distillation. The complex cyanides such as alkali ferri and ferrocyanides are not converted to HCN during distillation. Acid cuprous chloride and/or magnesium chloride are added to the sample which converts the complex cyanide to simple cyanides which are then converted to HCN by distillation.

Interferences

Sulfide in the sample will convert CN^- to SCN^- rapidly especially at high pH. Further, sulfide will also distill over with the cyanide and affect both colorimetric and titrimetric determinations.

Oxidizing agents destroy most of the cyanides during storage and manipulation, and must be removed prior to distillation.

Fatty acids distill along with cyanide and form soaps under alkaline conditions.

Thiocyanates at concentrations higher than 0.5 mg/l may interfere if the distillation is carried out with cuprous-chloride. These samples may be distilled with magnesium-chloride.

Aldehydes convert cyanide to nitrites under distillation conditions. Prior to distillation they are removed by treatment with silver nitrate solution (1.7%)

Removal of sulfides: Treat the alkaline sample at pH 11.0 with small increments of lead carbonate powder until no more precipitate is formed.

Removal of oxidizing agents: If the sample contains any oxidizing agents, determine the quantity of sodium sulfite required to neutralize it by titration using starch as indicator on a separate aliquot, and add this amount of sodium sulfite to the sample aliquot prior to distillation.

Removal of fatty acids: Acidify the sample to pH about 6.5 with acetic acid and extract with chloroform or preferably by hexane.

Apparatus

Distillation apparatus consisting of a distillation flask of 500 ml. capacity, fitted with a splash head and water cooled condenser. (For an efficient and special type of distillation apparatus, refer "Standard Methods for the Examination of Water and Waste Water", 14th edition (1976). For the general distillation assembly see Figure 41-5-1. (Page 365)

Reagents

(A) For distillation

1. **Acid cuprous chloride solution:** Dissolve 2 g. cuprous-chloride, Cu_2Cl_2 in 100 ml. 5N.HCl. Prepare freshly. (5N.HCl is prepared by diluting 43 ml. conc. HCl to 100 ml. with distilled water).

2. **Magnesium chloride solution:** Dissolve 51 g. $MgCl_2.6H_2O$ in 100 ml. distilled water.

3. **Sodium hydroxide solution, 1 N:** Dissolve 40 g. NaOH in distilled water and make upto 1000 ml. with distilled water.

4. **Sodium hydroxide solution. app. 0.5 N:** Dilute 100 ml. 1N. sodium hydroxide to 200 ml. with distilled water.

(B) For photometric measurement

5. **Sodium hydroxide solution, app. 0.1N:** Dilute 20 ml. of 1N. sodium hydroxide solution to 200 ml. with distilled water.

6. **Cyanide stock solution (1 ml = 1.0 mg CN):** Dissolve 2.5g. potassium cyanide, KCN, in distilled water. Add

100 ml. 1 N. sodium hydroxide, and make upto 1000 ml. in a volumetric flask. Standardize the solution against 0.1 N. silver nitrate solution as described below:

Standardization*: Place 25.0 ml cyanide stock solution into a 250 ml. conical flask. Dilute to 100 ml. Titrate with 0.1N. silver-nitrate solution (16.9895 g. $AgNO_3$/1000ml) using rhodanine indicator solution. (0.02% in acetone). The color change is from canary yellow to salmon pink.

Calculate the cyanide in the whole of the solution as follows:

$$\text{Total cyanide as CN mg/l} = \frac{5.2 \times \text{ml.} AgNO_3 \times 1000}{\text{ml. cyanide solution taken}}$$

(Strength of this stock solution decreases on standing. Hence it should be checked before use).

7. **Cyanide intermediate stock solution:** Dilute appropriate volume of cyanide stock solution to 100 times with distilled water, such that 1 ml of the resulting solution is 10 µg CN. During dilution, add 1 ml. of 1 N. sodium hydroxide solution for each 10 ml. of the final volume. Prepare the solution freshly.

8. **Cyanide working solution:** Dilute appropriate volume of cyanide intermediate stock solution 10 times so as to give a solution of strength 1 ml = 1.0 µg CN. During dilution add 1 ml. of 1N. sodium hydroxide solution for each 10 ml of the final volume. Prepare the solution freshly.

9. **Bromine water, saturated solution.**

---

* Standardization itself constitutes one method for the determination of cyanide. Higher concentration of cyanide can be determined by this method after distillation.

10. **Arsenious acid solution:** Dissolve 2 g. arsenious oxide, in 100 ml. distilled water by boiling under reflux.

11. **1-phenyl - 3methyl - 5pyrazolone solution:** Prepare a saturated aqueous solution (approximately 1g. in 200 ml) by adding pyrazolone to water at 75°C. Agitate occasionally as the solution cools to room temperature. Prepare daily.

12. **Mixed pyridine - pyrazolone reagent:** Mix 125 ml of the filtered saturated aqueous pyrazolone (above solution) with a solution containing 25mg. bis-pyrazolone in 25 ml pyridine. Mix well prepare daily.

Procedure

Distillation

If the sample contains sulfide, oxidizing agents and fatty acids, remove them first (see interferences). Adjust the pH to be about 7 with HCl.

Place in the distillation flask appropriate volume of the pretreated sample (not exceeding 500 ml) containing not more than 50 μg. cyanide.

Caution: Do not use mouth pipets for dispensing volumes of cyanide containing samples and solution.

Dilute to 200ml. with distilled water, if the aliquot is less than 200 ml. Add 10 ml. acid cuprous chloride (or 20 ml. magnesium chloride solution if thiocyanate interference is suspected) and a few anti-bumping granules. Assemble the apparatus immediately. Place 20 ml. 0.5N sodium hydroxide solution in the receiving flask. Make the tip of the condenser dipped in sodium hydroxide solution. Start the distillation and collect a total volume of about 90 ml. in the receiving flask. Transfer the content of the receiving flask to a 100 ml. volumetric flask and make upto the mark with distilled water.

Photometric Measurement

1. Place 20 ml. or a suitable aliquot of the distillate (dilute to 20 ml. if necessary) in a 50 ml. volumetric flask.

2. Place 20 ml. 0.1 N. NaOH in a 50 ml. volumetric flask which is the blank.

3. Into a series of 50 ml. volumetric flasks, place 1.0 ml. 2.0 ml, 3.0 ml.....10.0 ml of cyanide working solution and dilute to 20 ml. with distilled water.

4. To each flask, add 0.5 ml. conc. HCl and 0.5 ml. bromine water. Dispense bromine water by means of an automatic pipet. Stopper the flask immediately, mix well and allow to stand for about 2 minutes.

5. Add 0.5 ml arsenious acid solution to each flask to remove excess bromine. Mix well and add 5 ml. mixed pyridine - pyrazolone reagent. Stopper and mix well. Allow 20 minutes for (blue) color development. Make upto the mark with distilled water.

6. Measure the optical densities of the blank, standards, and sample using either a spectrophotometer at 620 nm. or filter photometer with a suitable red filter taking water as the reference. Prepare a calibration curve and find out the mg. cyanide equivalent to the optical

density. Visual comparison can also be made by transferring to 50 ml Nessler tubes.

Express the result as mg/l Total Cyanide (as CN).

44-5 OILS & GREASE (SOLVENT EXTRACTABLE MATTER)

Determination of oils and grease is useful to determine treatment plant efficiencies and to overcome the difficulties during treatment.

Presence of oils and grease in trade effluents is quite common. The source for oils and grease are natural raw materials used in the process and/or from the lubricants applied for machineries. The oil and grease fraction is reported to contain hydrocarbons, lipids, fatty acids, soaps, fats & waxes and oils. It should be stated that the result obtained by the solvent extraction method described below indicates not only the oil and grease content but also the quantity of all extractable matter by the solvent. It does not indicate the quantity of volatile oils and grease. However, the results are useful for practical considerations. When the other constituents of organic matter are more, the results should be reviewed and interpreted carefully.

SOLVENT EXTRACTION METHOD

Principle

Oils, grease and other extractable matter are dissolved in a suitable solvent and extracted from the aqueous phase. The solvent layer is then evaporated and the residue is weighed as oil and grease.

Reagents

1. **Flocculating agent:** Use any one of the suitable combination of flocculating agents, as required by the sample.
   a) Magnesium sulfate ($MgSO_4.7H_2O$), 1% aqueous solution, and
   b) Milk of lime (2% aqueous solution)

   or

   a) Aluminium sulfate, (1% aqueous solution), and
   b) Milk of lime (2% aqueous solution)

   or

   a) Ferric chloride, (1% aqueous solution), and
   b) Ammonium hydroxide.

   or

   a) Zinc acetate, (10 % aqueous solution) and
   b) Sodium carbonate, (5% aqueous solution)

2. **Dilute Hydrochloric acid, 1+3:** Add 25 ml. conc. HCl. to 75 ml distilled water and mix well.

3. **Petroleum ether:** Boiling range 40° to 60°C

4. **Sodium sulfate, anhydrous.**

Procedure

It is always better to collect separate samples for oils and grease determination.

1. Place 500 ml. or 250 ml. of the well mixed sample in a beaker. (If any floating layer of oil to considerable thickness is observed in the sampling bottle, it may be

decanted carefully to the separating funnel. The water layer settles down to the bottom of the funnel may be drawn and transferred to the sample beaker. The oil layer remaining in the funnel may be weighed).

2. Add 5 ml. of magnesium sulfate solution to the sample in the beaker and add continuously with stirring small amounts of milk of lime until flocculation occurs. Continue stirring for 2 minutes; allow to settle for 5 minutes. If a clear layer does not begin to form within 5 minutes, carryout this preliminary flocculation with other flocculating agents with a fresh aliquot of sample.

3. When the precipitate has completely settled, siphon off the clear top layer leaving about 1.5 cm clear layer above the precipitate level.

4. Dissolve the precipitate in dil. hydrochloric acid and transfer the liquid to the separating funnel (Avoid transferring large adventitious materials such as leaves, twigs, and sticks). Wash the beaker with 50 ml. of petroleum ether and add this to the separating funnel.

5. Shake the funnel continuously and gently for one minute. Draw off the aqueous layer into another separating funnel and again extract with 50 ml. of petroleum ether.

6. Combine the two ether extracts in a 200 ml. beaker. Add 2g. sodium sulfate (anhydrous) and mix well with a glass rod. Cover the beaker with a watch glass. Mix the contents of the beaker at frequent intervals for about 30

minutes, wash the glass rods into the beaker with the solvent.

7. Filter the contents through a filter paper (Whatman No.42) containing sodium sulfate in its cone and moistened with the solvent, into a tared evaporating dish. Wash the beaker with two 20 ml. portions of petroleum ether and filter them one by one through the same filter.

8. Evaporate the petroleum ether by keeping it on a water bath. Dry the outside, cool, and weigh. (After evaporation, if droplets of water are seen in the residue, add 2 ml. acetone and evaporate on a water bath, repeat the addition and evaporation until the residue shows no visible water). The difference in weight is the amount of oil present in the aliquot of the sample.

Calculation

mg/l l Oil and grease

$$= \frac{\textit{mg. residue in the evaporating dish}}{\textit{ml. sample taken for determination}} \times 1000$$

While expressing the result, specify the solvent used.

44-6 PHENOLS

Phenolic compounds, collectively referred as phenols are hydroxy derivatives of benzene or its condensed nuclei. They are present in varying concentrations in coal-tar and gas liquor, and in the waste water of synthetic resin, plastics and polymer manufacturing, dye manufacturing, and rubber proofing. Disinfectant wastes and cutting oil wastes also contain significant quantities of phenol.

Phenol impart a characteristic objectionable odor to receiving waters. They are well known disinfectants and hence they resist biological activity and are toxic to fish.

The presence of varied type of inorganic substances present in effluents and polluted waters interfere with the colorimetric procedure, and samples should be subjected to preliminary treatment, distillation and extraction. The phenol concentration is then determined by 4-amino antipyrine colorimetric method.

Collection & Preservation of samples

Collect 1000 ml. of sample in a glass bottle. Acidify to pH below 4 using H_3PO_4 (1+3) (to eliminate interferences due to alkaline conditions) and add 1g. copper sulfate (per litre) to precipitate sulfides as cupric sulfide and to prevent biological growths etc., Keep the sample at 5° to 10°C. Start the analysis within 24 hrs.

Removal of oxidizing agents

If oxidizing agents (such as chlorine) are present (detected by acidification and addition of Kl and starch) remove them by adding sodium arsenite.

Removal of sulfur compounds

Bring the pH of the sample below 5 with H_3PO_4 (1+9) and aerate briefly by stirring prior to the addition of copper sulfate.

DISTILLATION

The phenols are separated from the other non-volatile impurities by distillation. As the rate of volatilization of phenols is gradual, the volume of the distillate must be equal to that of sample being distilled.

Procedure

Use an all glass distillation apparatus (See Chapter 41-5, Ammonia Nitrogen, Page-365)

1. Place 500 ml. sample in the distillation flask (If the sample is known to contain higher quantities of phenols, smaller aliquots may be taken and volume of distillate may be reduced accordingly).

2. Add 5 ml. copper sulfate solution (10%) and bring the pH below 4 using H_3PO_4 (1+9). (Omit these additions if the sample is already preserved).

3. Start distillation and collect 450 ml. distillate. Cool the flask and add 50 ml. distilled water and continue the distillation until the total volume of the distillate is 500 ml.

   (If distillate is turbid, acidify with 1+9 H_3PO_4 to pH below 4 and add 5 ml. $CuSO_4$ solution (10%) and distill as described above. If the second distillate is also turbid, acidify 500 ml. of original sample using 1N. H_2SO_4 and methyl orange indicator and add 150 g. NaCl. Extract with 40 ml. chloroform. Extract further with 25 ml. chloroform for four times. Combine all the chloroform extracts in a second separating funnel and extract with 2.5 N. sodium hydroxide solution (100g. NaOH/litre) first with 4.0 ml. and then with 3.0 ml. volumes two times. Combine the alkaline extracts, heat on a water bath until chloroform is removed. Cool and dilute to 500 ml. and proceed distillation as described above starting from step No. 2.

Note 1: Appropriate smaller volumes of reagents may be used in cases where smaller volumes of samples are taken for distillation.

Note 2: Diethyl ether is recommended in place of chloroform as it eliminates emulsion problem and the use of NaCl. Chloroform is preferred because of the hazards in handling ether.

4-AMINO ANTIPYRINE COLORIMETRIC METHOD

This method is applicable to most of the phenolic compounds excepting para cresol and similar para substituted phenols.

Principle

Phenols react with 4-amino antipyrine at a pH of 10.0±0.2 in the presence of potassium ferricyanide to form a red antipyrine dye.

Interferences

All the interferences are eliminated or reduced to a minimum level by preliminary treatment and distillation.

Reagents

1. **Phenol stock solution:** Dissolve 1.0 g. phenol in distilled water and make up to 1000 ml. in a volumetric flask. Standardize* this solution every time as described below:

---

* Standardization itself constitutes one method for the determination of phenols if present in higher concentrations.

Pipet 25.0 ml phenol stock solution into an iodine flask and dilute to 100 ml. Pipet 20.0 ml. bromate-bromide solution (0.5 N) (Prepared by dissolving 6.96 g. potassium bromate and 37.5 g. potassium bromide in 500 ml. distilled water.) into the flask. Add 10 ml. 1+3 sulfuric acid solution and stopper the flask immediately. Seal the neck of the flask with 10 ml. 10% potassium iodide solution. Keep the flask in the dark for one hour. Wash the sealing solution into the flask and add a further 10 ml. of 10% potassium iodide solution. Titrate the liberated iodine with 0.25N. sodium thiosulfate using starch indicator. Conduct a blank simultaneously using 100 ml. distilled water and 20 ml bromate-bromide solution.

Calculate the amount of phenol present in one litre as follows:

mg/l Phenol

$$= \frac{(\text{ml. thio used for blank} - \text{ml. thio used for back titration}) \times 3.921 \times 1000}{25}$$

2. **Phenol intermediate solution:** Dilute appropriate volume of phenol stock solution to 1000 ml. such that 1.0 ml of the diluted solution contains 10 μg. phenol.

3. **Phenol working solution:** Pipet 50.0 ml. of phenol intermediate solution into a 500 ml volumetric flask and make upto the mark with distilled water.

   1.00 ml = 1 μg Phenol

4. **Ammonia-ammonium chloride buffer:** Dissolve 16.9 g. ammonium chloride, NH_4Cl, in 143 ml. conc. ammonia solution and dilute to 250 ml. with distilled water. Place it in a tightly stoppered bottle.

5. **4-amino antipyrine solution:** Dissolve 2g. of 4-aminoantipyrine in distilled water and dilute to 100 ml. with distilled water. Prepare daily.

6. **Potassium ferricyanide solution:** Dissolve 8 g. potassium ferricyanide $K_3Fe(CN)_6$ in distilled water and dilute to 100 ml. This solution may be kept for one week.

7. **Chloroform**

8. **Sodium sulfate, anhydrous, granular.**

Procedure

1. Into a series of 1000ml separating funnels, place 0.0, 2.0, 5.0, 20.0 ..... 50.0 ml. of phenol working solution and dilute each to 500 ml. with distilled water.

2. Place 500 ml. distillate in 1 litre separating funnel. (Lower aliquots may also be used if phenol concentration is more, but dilute to 500 ml, with distilled water).

3. To the blank, standards and sample, and 10 ml. ammonia-ammonium chloride buffer solution (Addition of 1 ml. of this buffer to 50 ml. distillate brings the pH of the sample to 10±0.2).

4. Add 3.0 ml of 4-amino antipyrine solution. Mix immediately and add 3.0 ml. potassium ferricyanide and again mix immediately.

5. Allow to stand for 3 mts. and extract with 25 ml. chloroform if 5 cm. cell is used or with 50 ml. if 10 cm cell is used.

6. Pass through fritted-glass funnel containing 5 g. layer of

sodium sulfate and collect the dried extract in 50 ml Nessler tubes.

7. Measure the optical densities of the sample and standards at 460 nm setting blank at 100% transmittance and prepare a calibration curve. Find out the mg. equivalent of phenol from the curve. Visual comparison can also be made.

Express the result as mg/l Phenolic substances (as Phenol (C_6H_5OH).

44-7 SURFACTANTS

Effluent from textile wet processing contains excess quantities of surfactants. Household wastes also contain significant amounts of surfactants.

DETERMINATION OF ANIONIC-SURFACTANTS BY LONGWELL AND MANIECE METHYLENE BLUE COLORIMETRIC METHOD

This method is applicable to industrial effluents, polluted waters, raw and treated waters.

Principle

Methylene blue reacts with anionic surfactants to form a blue colored salt. The intensity of blue color depends on the concentration of the anionic surfactants.

Interferences

Sulfide interferes by reducing the methylene blue. However, it can be oxidized by hydrogen peroxide.

Reagents

1. **Alkaline phosphate solution 1%:** Dissolve 5 g. anhydrous disodium hydrogen phosphate, Na_2HPO_4 in water. Adjust the pH to 10 using sodium hydroxide and make upto 500 ml. in a volumetric flask.

2. **Neutral methylene blue solution:** Dissolve 350 mg methylene blue in distilled water and make upto 1000 ml. in a volumetric flask.

3. **Chloroform.**

4. **Acid methylene blue solution:** Dissolve 35mg. methylene blue in 500 ml. distilled water in a 1000 ml. volumetric flask and add 6.5 ml. conc. H_2SO_4 and make upto the mark with distilled water.

5. **LAS (Linear alkyl sulphonate) stock solution:** Dissolve 0.100 g. Manoxol OT (Sodium dioctyl sulfosuccinate) or any other reference LAS material in water and make upto 1 litre in a volumetric flask.

   1.0 ml = 100 μg Manoxol OT (or reference LAS material)

6. **LAS standard solution:** Pipet 10.0 ml. stock solution in to a 100 ml. volumetric flask and make upto the mark with distilled water.

   1.0 ml = 10 μg Manoxol OT (or reference LAS material)

7. **Hydrogen peroxide (20-Volume).**

Procedure

1. Place a suitable volume of the sample containing 20 - 200μg. surfactant in a separating funnel. If the volume is below 100 ml. make upto 100 ml. with distilled water. Add 10 ml. alkaline phosphate solution, 5 ml. neutral methylene blue solution and 15 ml. chloroform.
2. Shake gently for 30 seconds. Excessive agitation may cause emulsion troubles. Allow the layers to separate and draw the chloroform layer into another separating funnel containing 110 ml. distilled water and 5 ml. acid methylene blue solution. Rinse the first separating funnel with 2.0 ml. chloroform and add the rinsings to the second separating funnel.
3. Shake the second separating funnel for 30 seconds and allow the layers to separate. Pass the chloroform layer through a small funnel plugged with cotton wool (moistened with chloroform) into a 50 ml. volumetric flask. Rinse the separating funnel with 2.0 ml. chloroform and add the rinsings to the volumetric flask.
4. Repeat the extraction with two further 10 ml portions of chloroform and collect the extracts in the 50 ml volumetric flask and make upto the mark with chloroform. Measure the optical density at 650 nm.
5. Prepare standards simultaneously as described above covering the range from 20 to 200μg. surfactant. Conduct a blank also. Measure the absorbance at 650 nm (or with a suitable orange filter) using a light path of 1

cm. Plot a calibration curve. Read the number of micrograms of surfactant in the sample equivalent to the observed optical density.

Express the result in mg/l Surfactant (as Manoxol OT) or (as the reference material used).

For Samples containing sulfide

Place a suitable volume of the sample in a separating funnel and add 10 ml. alkaline phosphate solution followed by 2 ml. hydrogen peroxide (20 volume). Stand for five minutes and dilute to 110 ml. with water. Add 15 ml. chloroform and 5 ml. neutral methylene blue solution and proceed from step No.2 of the procedure.

44-8 TANNIN AND LIGNIN

The effluent from pulp & paper industries and tanneries contribute tannin and lignin to water courses. Tannin is also used in the internal treatment of boiler waters to prevent scale formation. Hence the blowdown waters contain appreciable quantities of tannin.

Tannin and Lignin compounds are best determined by the following reduction method.

REDUCTION METHOD (COLORIMETRIC)

Principle

Determination of tannin and lignin is based upon the capacity of them ot reduce tungstophosphoric acid and molybdophosphoric acids to produce a blue color.

Reagents

1. **Tannin-lignin reagent:** Dissolve 25g. sodium tungstate-dihydrate, $Na_2WO_4.2H_2O$ and 5g. molybdophosphoric acid (also known as phospho molybdic acid) $20MoO_3$.-$2H_3PO_4$. $48H_2O$, and 12.5 ml. 85% phosphoric acid, H_3PO_4 in 200 ml. distilled water. Reflux the mixture for 2 hours. Cool well and make upto 250 ml. with distilled water.

2. **Sodium carbonate solution, (saturated):** Dissolve 100 g. sodium carbonate. Na_2CO_3 in 250 ml. warm distilled water and dilute to 500 ml. Keep it in a rubber stoppered bottle.

3. **Stock tannin (or lignin) solution:** The sample under investigation may contain either tannin or lignin compounds. According to the substance present in the sample, a standard solution has to be prepared. Dissolve 1.000 g. tannic acid or lignin compound (which is known to be present in the sample) in distilled water and make upto 1000 ml. in a volumetric flask.

   1 ml= 1.0 mg tannin (or lignin) compound

4. **Standard tannin (or lignin) solution:** Pipet 10.0 ml. of the above stock solution, and make upto 1 litre in a volumetric flask.

   1.00 ml. = 0.01 mg. tannin (or lignin) compound

Procedure

Place 50.0 ml. filtered sample in a 100 ml. Nessler tube. Add 2 ml. tannin-lignin reducing agent and mix well. After 5 minutes add 10 ml. sodium carbonate solution and stir well.

Allow 10 minutes for maximum color development. Simultaneously prepare standards by using 1.0, 2.0, 4.0, 6.0, 8.0 and 10.0 ml. of standard tannin (or lignin) solution and adding the reagents in the order as mentioned above. Prepare a blank also. Compare the color of the sample with that of the standards visually. To make photometric comparison, use a light path of 1 cm. or longer at a wavelength of 700nm.

Report the result as "Soluble Tannin like compounds (as Tannic acid) mg/l "or "Soluble Lignin like compounds mg/l" or simply as "Soluble hydroxylated aromatic compounds (as the standard taken) mg/l".

REFERENCES

44-1. ARSENIC

1) G. Stratten and H.C. Writehead, "JAWWA", 1962, 54, 861.
2) National Environmental Engineering Research Institute, "Technical Digest" No. 54, April - 1977.

44-2. BORON

1) Indian Standards Institution, IS: 2488 (Part III) 1968, "Methods of Sampling & Test for Industrial Effluents".
2) Bunton, N.G. & B.H. Tait, "JAWWA" 61,357, 1969
3) N. Manivasakam, "Physico-chemical Examination of Water, Sewage & Industrial Effluents" - 1984, Pragati-Prakashan, Meerut.

44-4. CYANIDE (TOTAL)

1) American Public Health Association, "Standard Methods for the Examination of Water and Waster water" - 14th ed, 1976, APHA, Washington.

2) Department of Environment “Analysis of Raw, Potable and Waste waters” - 1972, Her majesty's Stationery Office, London.

44-5. OILS AND GREASE

1) Indian Standards Institution, 1S: 2488 (Part - 1) - 1966. “Methods of Sampling and Test for Industrial Effluents” Part-I, ISI, New Delhi.

44-6. PHENOLS

1) Ettinger, M.B. etal, “Sensitive 4-amino antipyrine method for phenolic compounds” Anal. Chem 23. 1783, 1951.

2) American Public Health Association, “Standard Methods for the Examination of Water and Waste water” - 14th ed, APHA, Washington.

3) A.I. Vogel, “A Text book of Quantitative Inorganic Analysis including Elementary Instrumental Analysis” - 3rd ed, 1964, ELBS & Longmans Green & Co London.

44-7. SURFACTANTS

1) J.Longwell and W.O. Maniece, “Analyst”. 1955, 80, 167.

2) D.C. Abbott, “Analyst,” 1962, 87, 286.

44-8. TANNIN AND LIGNIN

1) American Public Health Association, “Standard Methods for the Examination of Water and Waste Water” - 14th ed,1976 APHA, Washington.

PART -V

TREATMENT OF INDUSTRIAL EFFLUENTS

TREATMENT OF INDUSTRIAL WASTES-GENERAL

Treatment of industrial waste is carried out in three stages. They are:

I. Primary Treatment

II. Secondary Treatment

III. Tertiary Treatment

For some wastes, primary treatment only is sufficient. Most of the wastes require secondary treatment also. Still other effluents require tertiary treatment also.

Primary treatment is employed to remove suspended solids, and to certain extent color and to bring the pH in the neutral range. It includes screening, equalization, neutralization, coagulation, sedimentation etc., Specific toxic substances such as cyanide are given special treatment.

Secondary treatment is a biological process applied to reduce BOD and COD. Trickling filtration, activated sludge process, oxidation ditch and oxidation ponds are some of the common secondary treatments.

Tertiary treatment aims further purification of waste water. (i.e.) final polishing, reduction of dissolved solids content, disinfection of effluent etc.,

Chapter - 45

PRIMARY TREATMENT

45-1 SCREENING

It is the process by which the larger suspended and floating materials such as fibres, paper, rags, wooden pieces, fruit skins, string substances and other materials are removed. The conventional screening system is comprised of two units: The first one known as coarse screen consists of metal bars or heavy wires spaced 25 to 50 mm apart. They are set in the flow channel, either at an angle of about 60 to horizontal axis of the flow or curved to the radius of a circle. The second unit fine screens, as the name implies are used to remove the finer particles. The size of opening ranges from 0.8mm to 6mm depending upon the of type of effluent to be treated.

The accumulated materials at screens must be removed at frequent intervals. This may be done manually or automatically (screenings are usually incinerated)

The modern mechanical screens (rather filters) are of two types: (i) rotary, self-cleaning, gravity type units and (ii) circular overhead fed vibratory units. Both these are costlier when compared to conventional bar screens but they are more effective in reducing the suspended solids and BOD. The degree of purification is so high such that some wastes may be discharged directly after screening without further treatment.

Some industries use micro-strainers also. The major

component of these micro-strainers is the filter-fabric the apertures of which vary from 25 to 35 microns. (The principal use of micro strainer is for the removal of fine suspended solids after secondary biological treatment)

45-2 SEDIMENTATION

Sedimentation is the settlement of solids by gravity and does not require any mechanical accessories. It is not used in most of the industrial waste treatment systems as the other stages involve the chemical coagulation and clarification. It is employed as a primary treatment (after screening), of wastes containing high percent of settleable solids or when the waste is given a combined treatment with sewage.

45-3 EQUALIZATION

Different types of wastes may be generated and discharged at different intervals in an industry. Hence the wastes exhibit different characteristics from time to time and a proper treatment is not possible, unless the character is uniform. For this, different streams of effluents are allowed to held in a holding tank for a specified period such that the character will be uniform. For example, if an industry produces a material, through a series of operations for 10 hrs, than the equalization tank should have a capacity to hold 10 hrs wastes. Equalization is the best means of stabilization of pH and BOD. During equalization, most of the solids also settle along with heavy metals. Further, it reduces the cost of treatment by way of mixing acid and alkaline effluents during the holding period.

Some times equalization may produce an effluent which requires no further treatment.

The size and shape as well as dimension of equalization tanks vary with the quantity of waste and mode of discharge. Generally the basins are rectangular and square. As already indicated, the equalization tank should have a capacity to hold the volume of waste produced upto the completion of a cycle of operations. Even if the cycle exceeds 24 hours, it should be large enough to hold the entire volume of waste generated.

The mere holding of waste will not produce an equalized effluent. Each unit volume of waste must be mixed thoroughly with already discharged other unit volume of wastes to render a homogeneous effluent. Proper mixing is brought about by (i) appropriate distribution and baffling, (ii) mechanical agitation, and (iii) aeration. For some wastes, a combination of these three is necessary.

Proper distribution involves the use of several inlet pipes spaced at regular intervals across the width of the tank. For more efficient horizontal and vertical distribution, over and under baffles are provided. Mechanical agitation with the use of paddles results in better mixing than baffling. The agitators are normally placed along the center line of the length of the tank spaced equidistantly. About 10 rpm to 20 rpm is sufficient. When limited space is available this is the effective and economical method. Aeration is the most effective but expensive method, but is especially applicable to wastes containing sulfur compounds and reducing substances. It also prevents accumulation of settleable solids.

45-4 NEUTRALIZATION

Wastes with extreme pH values should not be discharged without treatment. They are highly dangerous especially during peak loading and hence should be neutralized properly. Depending on the type and volume of waste, the neutralizing material, and equipment are chosen.

45-4-1 NEUTRALIZATION OF ACID WASTES

1. Mixing of wastes

Mixing of acid wastes with alkaline wastes generated within the same plant or nearby industry is the cheapest method. These may be stored in separate holding tanks and mixed in appropriate proportions.

2. Limestone Treatment

Acid waste is passed through the beds of lime stone either from the top or from bottom depending on the head available. They react with the lime stone to form $CaSO_4$ or $CaCl_2$ or $CaNO_3$ according to the acid present. Sulfuric acid is the major component of acid wastes. It forms $CaSO_4$ which coats the limestone and prevents further reaction. To counteract this, waste containing sulfuric acid is diluted to about 5% or less and applied at a rate of 5 gpm/ft^2. The limestone beds must be replaced at periodic intervals depending on the type and quantity of waste neutralized. Disposal of used up limestone beds, however is a serious problem.

3. Lime-slurry Treatment

The advantages using lime slurry for neutralization are:

(i) it has a high neutralizing power and high reactivity (ii) full utilization of lime-slurry and (iii) calcium sulfate formed during neutralization does not interfere in the further reaction, as it is carried out along with the neutralized waste and (iv) relatively inexpensive.

Dolomitic lime slurries are also more effective for neutralizing acid wastes (upto concentration 1.5%) but expensive. The major advantage of dolomitic lime is the formation of magnesium sulfate which is highly soluble.

4. Caustic-soda Treatment

It is the most effective method but expensive. The products formed during neutralization with caustic soda are highly soluble in water and unlike other treatments, it does not increase the hardness. Only smaller volumes of caustic solutions are required.

When small volumes of wastes are neutralized, caustic soda solution is normally led into the suction side of the pump discharging acid wastes. But for treating larger volumes, special proportioning equipments are necessary. In addition, storage tanks and variable speed pumps for direct addition of these caustic solutions into the flow of acid wastes are also required.

45-4-2 NEUTRALIZATION OF ALKALINE WASTES

1. Sulfuric acid Treatment

This is the most efficient method but expensive. Though there is some difficulty in handling due to its high corrosivity, it is compensated by the minimum requirements of storage and feeding equipments.

2. *Carbon dioxide Treatment*

Bottled carbondioxide may be employed to neutralize alkaline waste but is more expensive.

3. *Utilizing waste Boiler-flue gas*

This method is reported to be economical. Carbondioxide is present in significant quantities (about 14%) in stack gases. It reacts with the alkalinity of waste water to form carbonates. The equipment consists of a blower placed in the stack, pipeline to carry the gas, filter to remove sulfur and unburnt carbon particles, and a gas diffuser for diffusing CO_2 in to waste water. It is to be noted that H_2S may be released during neutralization and it must be vented to the upper atmosphere.

45-5 COAGULATION

Coagulation is a process by which the colloidal particles are destabilized by the addition of certain chemicals and consequently aggregated and bound to form a floc which settles rapidly. The formed flocs absorb and entrap the colloidal and other suspended matter and bring them down. The size of colloidal particles range from 1μ (10^{-6} mm) to 100μ (10^{-4} mm). These particles do not settle out on standing due to the repulsive forces because of the same charge they possess and cannot be removed by conventional physical treatment. Coagulation is the only effective and economical means for their removal.

COAGULANTS

The commonly employed coagulants are i) Alum (Aluminium sulfate-$Al_2(SO_4)_3$.. $14H_2O$) ii) Ferric sulfate $Fe_2(SO_4)_3$ iii) Copperas, $FeSO_4.7H_2O$ iv) Chlorinated copperas (a mixture of ferric sulfate and chloride, $FeSO_4.Cl$) and v) Ferric chloride $FeCl_3$. Among these, alum is the most popular coagulant used both in water and waste water treatment. It is reported that sulfate coagulants are more effective than chloride coagulants.

Neutralization of colloid charge is the aim of coagulation. Most of the colloids present in effluents coagulate at a pH value less than 7 but the hydroxides of coagulants are insoluble only at pH values above 7.0 and over 9.0 Hence lime is added to raise the pH and also to aid in precipitation of colloids. The most favorable pH range for aluminium coagulants is 5.5 to 6.8; with iron coagulants the usual range is 3.5 to 5.5 and above 9.0. (However there are exceptions for these ranges with some wastes). Ferrous iron coagulates only at a pH of above 9.5. Copperas is useful only for highly alkaline wastes. Lime itself is a coagulant added with aluminium and iron salts to raise the pH to isoelectric point of the coagulant. At this point, the colloid carry only a minimum charge and is least stable.

COAGULANT-AIDS

In addition to the coagulants, some other chemicals termed as coagulant aids are also added in minor quantities to enhance coagulation by promoting the growth to large and rapidly settling flocs. Activated silica and polyelectrolytes (high molecular wt. polymers) are typical coagulant-aids. They are unaffected by pH and can serve as coagulant by reducing the effective charge on a colloid. The usual dose of polyelectrolytes is about 0.10-0.25 mg/l. The dosage of coagulants should be minimum otherwise reverse effect may take place. Jar test and final plant trials will indicate the appropriate dose.

DETERMINATION OF COAGULANT DOSAGE

Determine the dosage of the coagulant by the jar test procedure described below.

To 500 ml. sample in a 1 litre beaker add small increments of coagulant at pH of 6.0 to 6.5 and mix rapidly after each addition followed by a 3 minutes slow mix. Continue addition until visible floc is formed. Note this amount and add to each of 2000 ml. beakers containing effluent sample. Adjust the pH of the sample to be 5.0, 6.0, 7.0, 8.0 and 9.0 with lime solution. Mix rapidly and stir with slow speed for about 15 minutes. Measure the effluent concentration of settled sample from each beaker. Plot a curve relating percent removal of characteristic Vs pH and find out the optimum pH. Repeat the experiment with different doses of coagulant with the same optimum pH and find out the appropriate dose. (If the dosage of polyelectrolytes are to be determined, repeat the above procedure, adding different doses of polyelectrolyte at the end of rapid mix).

Note: The presence of anionic surfactants increases the coagulant dosage.

Sequence of addition of chemicals

If there is no sufficient alkalinty, lime must be added first. Coagulants are added next and given a rapid mix for 4 to 6 minutes. Coagulant-aids may be added at the end of this rapid mix. Then slow stirring (or gentle agitation) for flocculation has to be given for 30 to 40 minutes.

Chapter - 46

SECONDARY TREATMENT

Secondary treatment removes the dissolved organic matter and thereby reduces BOD. It also removes oil and phenol to a considerable extent. But commissioning and maintenance of secondary treatment systems require huge capital and recurring expenditures. All the secondary treatments are biological processes and bacteria are the major organisms that carryout the oxidation. They consume organic matter as food and thus positively remove dissolved organics.

The most common biological treatments are:

1. Aerated Lagoon
2. Trickling Filtration
3. Activated Sludge process
4. Oxidation Ditch
5. Oxidation Pond, and
6. Anaerobic Digestion.

Other treatments not much practised are,

1. Dispersed-growth aeration,
2. Contact stabilization,
3. Bio-Disc system, and
4. Spray irrigation.

Mention has to be made that the effectiveness of the treatment is dependent on the bacterial slime, pH value, temperature, and nutrients. It is highly affected by the presence of extreme pH values of the effluents and toxic constituents such as chromium, lead, nickel and hydrogen sulfide etc.,

46-1 AERATED LAGOONS

These are large holding tanks or ponds lined with cement or preferably by butyl rubber or polythene. Their depth varies from 3-5 m. These lagoons are used for oxidation of dissolved organics. The effluent after primary treatment (particularly after clarification) is passed on to these tanks and aerated mechanically. Floating aerators are the most commonly employed to supply the necessary oxygen and mixing power. (These aerators are mounted on a circular float which rests on the surface of the liquid and rises or falls according to the change in liquid level).

Recommended detention period ranges from 2 to 6 days, during which time a healthy flocculent sludge is formed which carryout oxidation of organic matter. It is reported that 90% of BOD removal is achieved with aerated lagoons. Another advantage is that the operation and maintenance are much simpler. The effluent from lagoons may contain bacteria and need further biological purification which is achieved by passing the lagoon effluent to maturation ponds or to secondary sedimentation and sludge digestion.

46-2 TRICKLING FILTRATION

This is an effective aerobic oxidation and used widely. In this method, waste water after primary treatment is sprinkled over a bed of broken stones. The bacterial slimes formed on the stones oxidise the waste water during its passage. The effluent from trickling filter is allowed to settle and then discharged. (See Fig. 46-2-1).

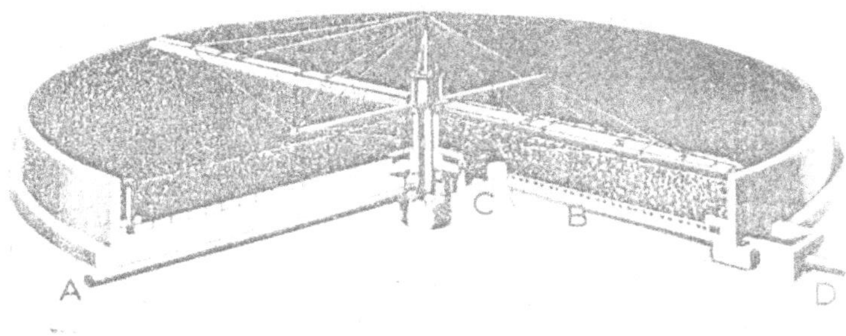

Fig 46-2-1. *Sectional Perspective View of a Circular Trickling Filter Showing Rotatory Distributors and Filter Media

A- inlet pipe, B-under drain blocks,
C-effluent channel, D-outlet pipe

The trickling filters are usually circular. (Rectangular, hexagon and octagon are not uncommon). Instead of stones, plastic and PVC media are also used now a days due to their ability to afford maximum surface area for microbial film formation, and light weight which helps in the reduction of construction costs especially in laying under drain of the filter. When these materials are used, the depth of the bed may be increased more than 3 m. Waste water is sprinkled by moving (or fixed) sprinklers The gelatinous film of aerobic organisms

* (Courtesy of Dorr-oliver Co., Ltd and ELBS & John Wiley & Sons, Chichester)

formed on the whole surface of the filter medium (stone, plastic or PVC) absorbs both dissolved and colloidal organic matter which are then oxidized by the metabolic activities of the bacteria in the slime. Oxygen requirement is satisfied as air flows through the filter by convection.

The removal of organics depends on the organic concentration and temperature of the waste, size of the stone (or other media), microbial surface area, depth of the filter and retention time and oxygen transfer. The organic concentration of the waste is suitably reduced by recirculation of the filtered waste.

The microbial layer is very sensitive to temperature, and the metabolic activities are directly proportional to the temperature of the waste passing through the filter. It should be remembered that during winter seasons, efficiency of these filters decreases.

The shape of the stone is spherical, the size of which ranges from 2 to 10 cm. (Flat stones cannot be used as they tend to compact too tightly). The size requirement of filter stone depends on the type of waste to be treated. It is advantageous to fill the filter with stones or other media as small as possible to yield the maximum surface area. But the problem with the smaller stones is the decrease in size of void space which is as important as media for the flow of liquid and air. For maximum efficiency, the microbial mass should be kept as thin as possible. Plastic and PVC media serves this purpose and are economical also.

The depth of the filter bed varies from 1 m to 3 m.

Though the depth of low-rate and high rate filters is almost the same, they differ by the hydraulic and organic loading. With low-rate trickling filters, a hydraulic loading of 2 to 4 millions gallons per acre per day (mgad) and an organic loading of 4 to 8 kg. BOD per 27 m^3 (1000 ft^3) of stone, are applied. High rate trickling filters operate satisfactorily at a rate of 10 to 40 mgad and 35 to 40 kg. BOD per 27m^3 (1000 ft^3) BOD removal is 85% with the former and 65 to 75% with the latter.

The base underdrain system is normally vitrified-clay-block construction. Vitrified clay is preferred because of their high structural strength and they can withstand the corrosive and abrasive action of wastes. The underdrain system not only collects the filtered waste and passes it to the collection channel (which runs along the middle (bottom) of the filter) but it also allows sufficient air for circulation through the filter. If the filter is devoid of oxygen, anaerobic organisms grow and aerobic oxidation will not take place.

The final sedimentation tank is an integral part of the filter. It removes the large masses of biological growth (which embeds a certain percent of organic matter of the waste) that have been dropped from the filter media. It also aids to discharge a clarified effluent.

Though the trickling filtration is classed as an aerobic oxidation, truly it is a facultative system. Aerobic bacteria (spore forming bacteria and bacillus are predominant) present in the upper layer of filter, the anaerobic (such as Desulfovibrio) in the interfaces of stones. The facultative bacteria include the species of the genera, Pseudomonas, Alcaligens, Micrococcus,

Flavo bacterium and Enterobactericeae. Fungi and algae are also present in trickling filters. Protozoa, ciliates, worms, snails and insect larvae (which feed on micro organisms) are also present.

A few advantages and disadvantages of trickling filtration over the other methods are summarized below:

Advantages

1. Produce effluents of better and consistent quality.
2. Recover quickly from shock loads.
3. Constant manual attention is not necessary.

Disadvantages

1. Loss of head is necessary to operate revolving arms (distributors).
2. Need of ventilation ducts for the under drain system.
3. Cost of construction is a limiting factor.

The effluents that are treated effectively with trickling filters are from beet sugar mfr, brewery, cannery, dairy, distillery, electroplating, fermentation and food processing industry, pulp and paper mill, petrochemical industry, pharmaceuticals, slaughter house and poultry processing.

46-3 ACTIVATED SLUDGE PROCESS

The most versatile biological oxidation method for treating industrial wastes is activated sludge process. In this process, industrial waste or sewage is continuously exposed and subjected to biological degradation carried out by the microbial floc suspended in the reaction tank into which oxygen is introduced by mechanical means. The effluent from the reaction tank is allowed to settle and a portion of the sludge is recycled to the tank itself for effective microbial population. Fig. 46-3-1 illustrates the principle of activated sludge process.

The flocs formed in activated sludge process are zoogleal masses of living organisms, embedded with their food and slime material and the active centres of biological mass include bacteria, fungi, protozoa, rotifers and sometimes nematodes. Among these, bacteria are the most important as they are the workhorses degrading the organic matter. Variety of bacteria and other microorganism are present in the slimes but the nature of organic matter in the waste is the determining factor of predominancy of the exact type and species of bacteria and other micro-organisms required. The desired concentration of the floc is maintained in the system by recirculation of the settled sludge (about 20%). Higher volumes of returned sludge though removes a greater percent of BOD, will induce a situation requiring more air and food. A young light sludge is preferable over an old heavy sludge as the latter would be mineralized and devoid of oxygen.

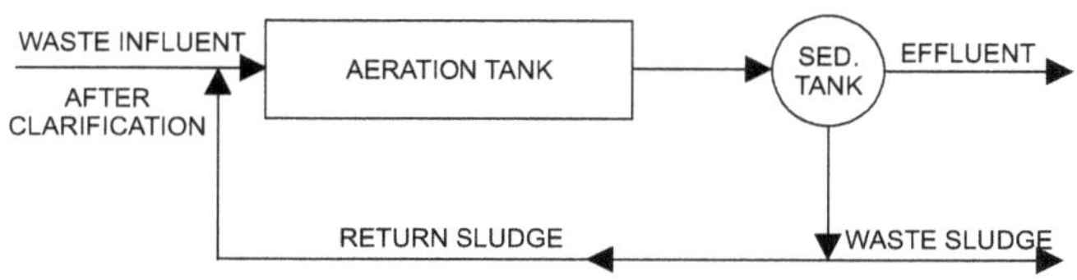

Fig 46-3-1. Schematic Diagram of Activated Sludge Process

Oxygen is the primary material supplied to the activated sludge. If the system has to be operated properly, atleast a residual of 0.5 mg/l should present at all times. Oxygen is supplied either by mechanical or diffused aeration systems. The average period of aeration for industrial waste ranges from 6 to 24 hrs, and for sewage about 6 hrs. If the oxygen supplied is less than the amount required (by the bacteria and

Fig 46-3-2. Activated Sludge Tank with 14 Surface Aerators

(Courtesy of Ames Crosta Ltd and ELBS & John Wiley sons-Chichester)

other microbes for their metabolism), anaerobic conditions develop resulting in turbid effluents due to the growth of filamentous bacteria than the other required bacteria and protozoa. The filamentous bacterial growths retard floc compaction and settling.

The micro-organisms should also be supplied with essential nutrients. Among them the most important are nitrogen and phosphorus. If these are deficient, the waste will stimulate the growth of fungi over bacteria. Fungi are filamentous in nature and consequently prevent settling. Usually nitrogen and phosphorus are supplied by the addition of either urea, or mono ammonium or diammonium hydrogen-phosphate. The nitrogen and phosphorus requirements are estimated to be about 10% and 2% respectively of volatile solids of the sludge weight (on dry basis) produced per day. The amount to be added daily can be calculated from the total quantity required and quantity present in the sludge. Other nutrient elements such as potassium, calcium, magnesium are usually present in the waste. Iron, molybdenum and cobalt are also required in trace quantities.

Other important factors that determine the efficiency of activated sludge are pH value, temperature and oxidation-reduction potential. The optimum pH range is 6.5 to 9.0. Below 6.5 fungi will compete with the bacteria and above 9.0 rate of metabolism will be decreased. An activated sludge system must not only have the proper pH, but sufficient buffer to resist pH changes brought out by microbes. The neutral materials such as aldehydes, ketones, alcohols and carbohydrates are converted into acids and if sufficient buffer is not maintained, pH drops, slowing down the metabolic rate. (In most biological systems, bicarbonate buffer system predominates). It is to be mentioned that measurement of pH at start and at the end of aeration period do not show the lack of sufficient buffer. Hence, it is essential that pH measurement must be made at regular intervals along the aeration tank or aeration period to ascertain deviation from normal condition. The effect of temperature need not be emphasized again and again. Low temperature results in a slow rate of metabolism. High temperature increases metabolic activity and oxygen is used up immediately, leading to anaerobic conditions.

Some of the modifications over conventional activated sludge process are: (1) Tapered aeration (2) Dispersed aeration (3) Extended aeration (4) Contact stabilization and (5) High rate aerobic treatment.

In conventional system, oxygen is supplied at a uniform rate throughout the aeration tank. In fact, the oxygen demand decreases along the length of the tank as the waste stabilizes. Hence an appreciable economy is achieved by reducing the oxygen supply progressively which is termed as "Tapered Aeration". Instead of reducing oxygen supply, the influent may be passed at several places of the aeration tank and than the process is termed as "Stepped Aeration". In all of these systems, the usual hydraulic retention time is 8-12 hrs at mean flow. If the aeration period is extended to 18-48 hrs, the process is known as "Extended Aeration" in which the rate of sludge autolysis increases resulting in less sludge production.

In the "Contact Stabilization" (or biosorption) process BOD removal is carried out in two stages, waste is aerated for a short time (0.5 to 1 hr) only in the aeration tank for entanglement and absorption. The solids are settled out in secondary clarifier (sedimentation tank) and retained there for about 2 hrs for clarification. The effluent from sedimentation tank is discharged as treated waste. One part of the sludge is also discharged as waste while the other portion is passed on to a stabilization tank where it is aerated for about 2-4 hrs for solubilization, oxidation and reactivation of the sludge which is mixed with the incoming waste for treatment. Fig. 46-3-3.

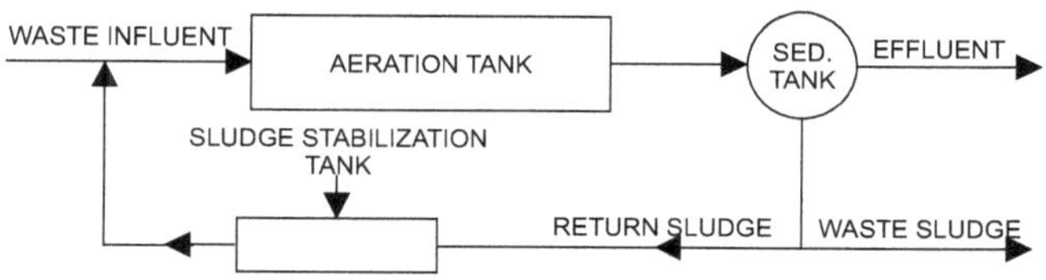

Fig 46-3-3. Schematic Diagram of Contact Stabilization Process

This modification is highly useful for wastes with BOD associated with suspended and colloidal solids. Considerable reduction in construction cost (only small tanks are enough), and aeration volume and period are the major advantages of this process.

For oxidizing organic wastes, a modified activated sludge process named as 'High-rate aerobic treatment' is applied. This process involves comminution of the waste, aeration for longer periods (24 to 72 hrs), settling and returning of settled sludge to aeration tank. The major advantage with this method of treatment is the elimination of primary clarification and sludge digestion and requires only a little maintenance. However, the aeration tank must be large enough to hold the waste for the complete aeration period.

MICROBIAL INDICATORS TO ASSESS PERFOAMANCE

It is possible to determine the operating characteristics of the various systems by regular microscopical examination of activated sludges. A good activated sludge does not contain dispersed bacteria. It contains relatively good population of free swimming (eg. stylonichia) and stalked ciliates (eg. vorticella). A few rotifers would also present. If a very high population of dispersed bacteria is observed, it is an evidence of a poor activated sludge system. Spirillum is an typical example of dispersed bacteria. The presence of filamentous micro organisms indicates the deficiency of nitrogen and/or phosphorus, low pH and low oxygen levels. If the system goes completely anaerobic, bacteria begin to proliferate.

ADVANTAGES AND DISADVANTAGES OF ACTIVATED SLUDGE PROCESS

This process produces a high quality effluent with relatively small areas and with loss of head than required by trickling filters.

The disadvantages are:

1. Sensitive to shock loads of toxic and organic substances.
2. Careful attention for control of volume and concentration of return sludge.
3. High cost of operation and maintenance.

The effluents that are treated satisfactorily with the activated sludge process are antibiotic, food processing, gelatin manufacture, phenol manufacture, sugar manufacture, textile processing (such as cotton kiering, bleaching and dyeing) etc.,

46-4 OXIDATION DITCH

The oxidation ditch is a modification of the conventional activated sludge process. The major advantages of oxidation ditch are that no primary clarification and sludge treatment are necessary. The oxidation ditch is a long continuous channel, (lined with butyl rubber, plastic or tar) usually oval in shape and 1.0-2.0 m. deep. Waste, after being subjected to preliminary treatment (screening or comminution) is passed into oxidation ditch where longer retention times are provided. The usual hydraulic retention time is 12 to 24 hrs and for solids 20-30 days. This much of efficiency is achieved by recycling about 95% of the sludge, such that overall sludge production is reduced, and even the sludge handling and treatment are almost negligible as it can be dewatered and dried readily (without odor) on sand drying beds.

The major difference between oxidation ditch and conventional activated sludge process is the shape and the aerator. The aerator used in oxidation ditch is cage rotor and is placed across the channel. See Fig 46-4-1.

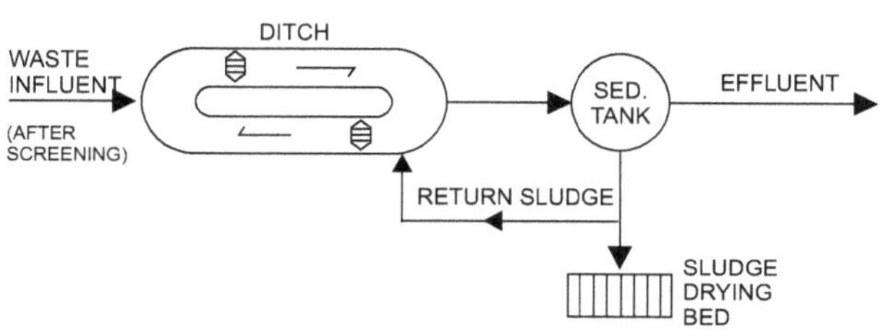

Fig 46-4-1. Schematic Diagram of Oxidation Ditch

The solids are maintained in suspension by the velocity (0.3 to 0.4m/sec) induced by the rotors. The usual range of concentration of total suspended solids in the ditch is 3000 to 5000 mg/l. Whenever the concentration exceeds this range, the return sludge flow is diverted to the sludge drying bed for a short period.

The volume of an oxidation ditch is about 2.5 m^3 per kg. of BOD. The oxygen requirement per gm. of BOD is about 1.5 to 2.0g (This rate is essential to maintain aerobic conditions along the entire length of the ditch). Cage rotors (of 70 cm.dia) are reported to provide 3 kg. of O_2 per metre length per hour with the usual immersion depth of 15 cm. and 70 rpm. It is reported that 80 to 95% BOD is removed.

The major advantages of oxidation ditch are,

1. Low-cost of construction, operation and maintenance.
2. Extreme simplicity of operation.
3. Overall efficiency and flexibility in the degree of treatment.

The wastes treated effectively with oxidation ditch are from (1) dairy (2) beet-sugar mfr (3) edible oil refinery (4) slaughter house and meat packing and (5) vegetable and fruit canning. Phenolic wastes are also reported to be treated effectively by oxidation ditch.

46-5 OXIDATION PONDS

(Waste Stabilization Ponds)

An oxidation pond is a large shallow pond in which wastes are added at one end and effluent removed at another end. Stabilization of organic matter is brought about by bacteria (and to certain extent by flagellated protozoa). Oxygen required for its metabolism is supplied by algae in the pond. In turn it utilizes the carbondioxide released by bacteria for its photosynthesis. This mutual symbiotic action is the characteristic of oxidation ponds. Hence for effective treatment of wastes, maximum (sun) light penetration for photosynthesis is necessary. To satisfy this requirement, oxidation ponds are constructed shallow (0.5m to 1.5m depth) and not so deep.

Though the oxidation pond on an overview is considered to be aerobic, the purification is carried out by a combination of aerobic, facultative and anaerobic processes. Solids in the waste settle at the bottom and this solid layer acts as anaerobic phase and anaerobes convert organic matter to methane, carbondioxide and ammonia. The top liquid layer (major portion) is the place where aerobic oxidation takes place and the end products are carbondioxide and water. The facultative zone is near the anaerobic phase. The more the algae in the pond more will be the amount of oxygen released and consequently increase in rate of purification. The growth of algae is enhanced by the nutrients present in the waste. If the waste is devoid of nutrients they have to be added. During day time, algae indulge in photosynthesis and oxygen requirement by them is much less, but during night time they also require oxygen which is taken out from the pond itself. This oxygen must be supplemented by surface aeration. If the pond is devoid of oxygen, then anaerobic conditions set up leading to putrefaction of wastes and release of foul odors.

Micro-organisms

Bacteria and algae are the predominant organisms,

while protozoa and rotifers are also present under certain loading conditions. The microbial population of a low organic load pond will be different from that of high organic load pond. The prominent bacteria are pseudomonas, flavo-bacterium and alcaligenes. The algal predominancy depends on the type and quantity of nutrients available. When the nutrient level is high, phytoflagellates such as euglena and chlorella predominate. When the nutrient level decreases, filamentous algae such as spirogyra, ulothrix and vaucheria appear. Higher organic concentration lead to the growth of flagellates such as chilamonas but quickly they give way to the free swimming ciliates such as colpidium, paramecium, glaucoma and euplotes. Stalked ciliates such as vorticella and epistylis grow when the bacterial population decreases. When the oxygen level is high and organic load is less, higher animals such as daphnia and rotaria appear. They are useful in clarification of the waste and not in purification.

Design of Oxidation Ponds

Pond size (area) and detention period is dependent on the type and quantity of the waste to be treated. The pond depth should be more than 1 m. to prevent weed emergence. Weeds reduce the effective area and make the pond a breeding ground for mosquitoes and midges. Depths greater than 1.5 m. make the pond anaerobic rather than aerobic. But in arid climates, evaporation rates are high and the depth has to be increased to 2 m. The hot climate is ideal for pond operation. But for the satisfactory working of the pond at all times, the mean temperature of the coldest month is commonly taken as the design temperature. The bottom of the pond should be impermeable to prevent seepage though the sludge solids clog and seal the pores. Polythene sheeting, linings of bitumen and asphalt are recommended. An embankment slope of 1 in 3 is usually satisfactory. Placement of concrete slabs at the sides to prevent vegetation and erosion is also recommended. It is also advisable to deepen the pond near the inlet to create an anaerobic zone for solids deposition and digestion. The pond should have arrangements to measure the inflow and outflow.

Advantages of Oxidation Ponds

1. Operation and maintenance are much simple and carried out at low cost. Skilled operators are not needed.
2. Applicable to all types of industrial wastes and any degree of purification can be achieved. (Anaerobic ponds are particularly useful for strong wastes). They withstand organic and hydraulic shock loads.
3. The (high) pH of the pond aids the precipitation of heavy metals as hydroxides which settle as sludge.

Disadvantages of Oxidation Ponds

1. They require much larger areas of land.
2. Effluent from the pond may have to be disinfected or has to be treated in a separate maturation pond before final discharge.

46-6 ANAEROBIC DIGESTION

This treatment is mainly employed for the digestion of sludges. However a few liquid wastes containing soluble organics such as from dairy, slaughter house, cotton and paper mill are effectively and economically treated by this method. As the name implies, oxidative digestion is carried out in the absence of air in a closed vessel and methane and ammonia are liberated as the end products. It is to be mentioned that pH must be controlled to near neutral point in the digesters.

Species of pseudomonas, flavobacterium. alcaligenes, escherichia and aerobacter are responsible for the production of acids, and bacteria belong to the genera, methano bacterium, methanosarcina and methanococcus are responsible for the generation of methane. The bacteria desulfo vibrio reduce sulfates to sulfides with the release of hydrogen sulfide.

Though anaerobic treatment is slow it is highly useful for wastes of small volume containing readily oxidizable dissolved organic solids in finely divided or in liquid form. Operation and maintenance costs are much less with this treatment. Needless to say that anaerobic digestion should be followed by aeration/ aerobic treatments before discharge.

Chapter - 47

TERTIARY TREATMENT

Tertiary treatment is applied to the waste for the removal of dissolved inorganic solids and bacteria and some times for the removal of final traces of organics.

Removal of bacteria especially of fecal origin is achieved by keeping the effluents from secondary biological treatment units in maturation ponds for specified periods. If the final effluent still contains greater number of bacteria, it is chlorinated. Chlorination of waste water is similar to that of water excepting for the requirement of higher doses of chlorine and a contact period of 1 to 2 hrs.

Removal of inorganic dissolved solids is a major problem with tannery, mercerizing and other textile processing, electroplating and fertilizer wastes. The proved treatment methods include 1) Evaporation 2) Reverse Osmosis 3) Dialysis 4) Ion exchange 5) Chemical Precipitation and 6) Removal by algae. Mention has to be made that all these treatments are not economical. If enough water is available, dilution is the best method for reducing the concentration of these solids.

47-1 EVAPORATION

It is a simple process of boiling the waste and vaporizing the water phase. This method is especially useful for the concentration of radioactive waste and other wastes of smaller volumes. It is suitable only when the recovered solids or concentrated solutions are reused. (For example, chrome, nickel and copper acid-type plating wastes may be reclaimed and returned to the plating system).

The major disadvantage of the method is the cost incurred for fuels. Only when the recovered materials are reused, this method is advantageous. Further, evaporating vessels are also attacked by the waste. Scale formation on the heating surface is another problem. Removal of these hard scales from the surface is very difficult. Because of all these limitations, this method is not applied generally.

In tropical countries, if space is available, wastes may be solar evaporated, which leads to a greater reduction in economy and operations.

47-2 REVERSE OSMOSIS

In recent times, Reverse Osmosis is gaining prominence as a method of tertiary treatment. This method is especially suitable for the removal of dissolved solids from waste water. When the liquid waste containing much dissolved solids is passed over the surface of a semipermeable membrane at a pressure in excess of the osmotic pressure of the feed waste (usually 40 to 50 atms) the water in the effluent passes through the membrane leaving a concentrated liquor on the surface.

Cellulose acetate and polyamide-hydrazide are the most effective membranes. The former is used in flat or tubular forms while the latter is used in the form of hollow fine fibers. Though the method is highly effective, cost of the membranes is a limiting factor. Much attention is also required for the pretreatment of effluents.

47-3 DIALYSIS

It is a method of separation of solutes from the solution, based on the difference in the rates of diffusion. This is practiced in metal industries (copper, stainless steel) for the recovery of sulfuric acid. It is reported that about 70% of the acid is recovered and just a recovery of 20% justifies the cost. In textile processing house, caustic soda is recovered. The waste containing moderate concentration of caustic soda is passed upward countercurrent to a down stream of water supply - both the phases are separated by a semipermeable membrane. The sodium hydroxide permeates the membrane at a faster rate than the other impurities and combines with the water yielding a dilute solution of caustic soda, while the dissolved solids content of the waste is also reduced. Also the water permeating to the waste phase dilutes it further. Another advantage of dialysis is the requirement of little attention. In electro dialysis permeation is enhanced by the application of electrical energy. The membranes commonly employed are cellulose nitrate, cellophane and parchment.

Dialysis is applicable only when the wastes contain materials that could be reused. It is not applicable to voluminous wastes.

47-4 ION EXCHANGE

Application of ion exchange as a method of waste treatment is a development over the traditional method of water softening. It is economical only when exchanged salts are reused in the process. In electroplating industry, most of the metals are recovered, simultaneously the toxicity of the wastes due to these metals is also reduced. If removal of dissolved solids is the aim, this method may not be economically feasible, though the operations are simple.

47-5 CHEMICAL PRECIPITATION

If the dissolved solids present in the waste water are precipitable, they can easily be removed. Most of the heavy metals are precipitated as hydroxides at elevated pH. Lime is the cheapest and effective precipitating agent. For example, chromates present in electroplating waste water and cooling waters are reduced by treatment with ferrous sulfate and then precipitated as chromium hydroxide by reaction with lime.

47-6 REMOVAL BY ALGAE

Removal of dissolved solids by algae is still in the experimental stage. Algae require carbon, nitrogen, phosphorus sulfur, potassium, calcium and magnesium for their growth. In addition they also need iron, manganese, zinc, boron, cobalt, copper, molybdenum and silicon in trace quantities. Algal uptake of these elements reduces the dissolved solids content of waste water. It is to be mentioned that algae must be removed prior to final disposal in to a water course used as a public water supply.

47-7 ACTIVATED CARBON TREATMENT

(For the removal of organics)

Final traces of organics not treatable by other methods are removed by adsorption on activated carbon. This method is particularly useful for the removal of pesticides such as DDT,

2,4-D, and other carbamate insecticides. However adoption of this method is limited for economical reasons.

REFERENCES

1) Mckinney R.E., "Microbiology for Sanitary Engineers"-1962, McGraw Hill Book Co. Inc, New York.

2) Arceivala, S.J.etal, "Waste Stabilization ponds : Design, Construction and Operation in India, "Central Public Health Engineering Research Institute, Nagpur-20.

3) W.J.Muller, "Disposal of Sewage and other Water borne Wastes"-1971, Butterworths, London.

4) Duncan Mara "Sewage Treatment in Hot Climates", - 1976, ELBS & John Wiley & Sons, Chichester.

5) R.C.Culp & G.L.Culp "Hand Book of Advanced Waste water Treatment", 2nd ed, 1978, Van Nostrand Reinhold Co., Newyork.

6) Nemerow N.L. "Industrial Water Pollution, Origins, Characteristics and Treatment" - 1978, Addison Wesley Publishing Co., Massachuesetts.

7) Parker H.W. "Waste water Systems Engineering" - 1978, Prentice Hall of India Pvt. Ltd, New Delhi.

APPENDIX

TABLE - I

LENGTH

1 Kilometer	= 1000 meters	1 micron	= 0.000001 meter (10^{-6} m)
1 meter	= 100 centimeters	1 millimicron	= 0.000000001 meter (10^{-9} m)
1 meter	= 1000 millimeters	1 Angstrom	= 0.0000000001 meter (10^{-10} m)
1 millimeter	= 0.03937 inch	1 inch	= 2.540 centimeters
1 centimeter	= 0.3937 inch	1 foot	= 30.480 centimeters
1 meter	= 39.37 inches	1 foot	= 0.3048 meter
1 meter	= 3.2808 feet	1 yard	= 91.440 centimeters
1 meter	= 1.09361 yards	1 yard	= 0.9144 meter
1 kilometer	= 0.6214 mile	1 mile	=.1.6093 kilometers

TABLE - II

AREA

1 square centimeter	= 0.155 square inch	1 sq. inch	= 6.4516 sq. cms
1 square meter	= 1555.0 sq. inch	1 sq. foot	= 929.0341 sq.cms
1 sq. meter	= 10.764 sq. feet	1 sq. foot	= 0.0929 sq. meter
1 sq. meter	= 1.196 sq. yards	1 sq. yard	= 0.8361 sq. meter
1 sq. km	= 0.3861 sq. mile	1 acre	= 43560 sq. ft
1 sq. mile	= 2.590 sq. km	1 acre	= 0.4047 hectare
		1 hectare	= 2.471 acres

TABLE - III

VOLUME

1 kilo litre	= 1000 litres	1 cubic meter	= 1000 litres
1 deci litre	= 0.1 litre	1 cubic meter	= 35.145 cubic feet
1 centi litre	= 0.01 litre	1 cubic foot	= 1.3079 cubic yard
1 milli litre	= 0.001 litre	1 cubic centimer	= 0.06102 cubic inch
1 micro litre	= 0.000001 litre	1 cubic inch	= 16.3872 cubic centimeter
1 litre	= 1000 milli litres	1 cubic foot	= 0.02832 cubic meter
		1 cubic yard	= 0.7646 cubic meter

TABLE - IV
WEIGHT

1 ton	= 1000 Kilograms	1 microgram	= 0.001 mg (10^{-3} mg or 10^{-6}g)
1 kilogram	= 1000 grams	1 microgram	= 1000 nanograms (10^{-6} mg or 10^{-9}g)
1 gram	= 1000 milligrams	1 kilogram	= 2.205 pounds
1 milligram	= 1000 micrograms	1 pound	= 453.6 grams

TABLE - V
APPLICATION OF METRIC UNITS

Description	Unit	Symbol
Precipitation run-off	milli meter (1 mm of rain = 1 litre/m^2)	mm
River flow	cubic meter per second	m^3/S
Flow in pipes	cubic meter per second or litre per second	m^3/S or l/s
Discharges	cubic meter per day	m^3/d
Usage of water	litre per person per day	l/person/day
Density	Kilogram per cubic meter	Kg/m^3
Concentration	milligram per litre or gram per cubic meter	mg/l or g/m^3
B.O.D. loading	Kilogram per cubic meter per day	Kg/m^3/d
Hydraulic load per unit area	cubic metre per square-meter per day	m^3/m^2/d
Hydraulic load per unit volume	cubic metre per cubic metre per day	m^3/m^3/d
Velocity	metre per second	m/s

TABLE - VI

WATER ANALYSIS UNITS INTER-RELATIONS AND EQUIVALENTS

1 part per million (ppm)	1 milligram per litre
	0.1 part per hundred thousand
	0.07 grain per imperial gallon
	0.0583 grain per U.S. gallon
1 part per hundred thousand (pts/100,000)	10 part per million
	10 milligrams per litre
	0.7 grain per imperial gallon
	0.583 grain per U.S. gallon
1 grain per imperial gallon	14.3 parts per million
	14.3 milligram per litre
	1.43 parts per hundred thousand
	0.833 grain per U.S. gallon
1 grain per U.S. gallon	17.1 parts per million
	17.1 milligrams per litre
	1.71 parts per hundred thousand
	1.2 grains per imperial gallon

TABLE - VII

HARDNESS CONVERSION FACTORS

Units of Measure	mg/l $CaCO_3$	British gr/gal (Imperial) $CaCO_3$	American gr/gal (US) $CaCO_3$	French Degree Parts/100000 $CaCO_3$	German Degree Parts/100000 $CaCO_3$	meq/l	g/l CaO
mg/l $CaCO_3$	1.0	0.07	0.058	0.1	0.056	0.02	5.6×10^{-4}
British gr/gal (Imperial) $CaCO_3$	14.3	1.0	0.83	1.43	0.83	0.286	8.0×10^{-3}
American gr/gal (US) $CaCO_3$	17.1	1.2	1.0	1.72	0.96	0.343	9.66×10^{-3}
French Degree Parts/100000 $CaCO_3$	10.0	0.7	0.58	1.0	0.56	0.2	5.6×10^{-3}
German Degree Parts/100000 $CaCO_3$	17.9	1.25	1.04	1.79	1.0	0.358	1.0×10^{-2}
meq/l	50.0	3.5	2.9	5.0	2.8	1.0	2.8×10^{-2}
g/l CaO	1790.0	125.0	104.2	179.0	100.0	35.8	1.0

TABLE - VIII

FACTORS FOR INTERCONVERSIONS BETWEEN MILLIGRAM PER LITRE (mg/l) AND MILLIEQUIVALENT PER LITRE (meq/l)

Ion	Multiplying Factors	
	mg/l to meq/l	meq/l to mg/l
Na^+	0.04349	22.90
K^+	0.02557	39.10
Ca^{2+}	0.04990	20.04
Mg^+	0.08226	12.16
Ba^{2+}	0.01456	68.67
Sr^{2+}	0.02283	43.81
Zn^{2+}	0.03060	32.68
Cu^{2+}	0.03148	31.77
Pb^{2+}	0.009653	103.6
Ni^{2+}	0.03047	29.36
Fe^{2+}	0.03581	27.92
Mn^{2+}	0.03640	27.47
Al^{3+}	0.1112	8.994
Fe^{3+}	0.05372	18.62
Cr^{3+}	0.05770	17.33
Mn^{4+}	0.07281	13.73
Cr^{6+}	0.1154	8.666
F^-	0.05264	19.00
Cl^-	0.02821	35.45
Br^-	0.01251	79.91
I^-	0.007880	126.9
NO_2^-	0.02174	46.01
NO_3^-	0.01613	62.00
HCO_3^-	0.01639	61.01
OH^-	0.05860	17.01
CO_3^{2-}	0.03333	30.00
SO_3^{2-}	0.02498	40.03
SO_4^{2-}	0.02082	48.03
S^{2-}	0.06238	16.03
CrO_4^{2-}	0.01724	58.00
PO_4^{3-}	0.03159	31.66

TABLE - IX
VOLUME CONVERSION FACTORS

From to	cm^3	liters	m^3	in^3	ft^3	yd^3	gal (U.S)	gal (U.K)
cm^3	1	0.001	1×10^{-6}	0.06102	353×10^{-5}	1.3×10^{-6}	2.64×10^{-4}	2.2×10^{-4}
litre	1000	1	0.001	61.02	0.03532	0.00131	0.2642	0.220
m^3	1×10^{-6}	1000	1	6.1×10^4	35.31	1.308	264.2	220
in^3	16.39	0.01639	1.64×10^{-5}	1	5.79×10^{-4}	2.14×10^{-5}	0.00433	0.00360
ft^3	283×10^4	28.32	0.02832	1728	1	0.03704	7.481	6.229
yd^3	7.65×10^5	764.5	0.7646	4.67×10^4	27	1	202.0	168.2
gal (U.S)	3785	3.785	0.00379	231.0	0.1337	0.00495	1	0.8327
gas (U.K)	4546	4.546	0.00455	277.4	0.1605	0.00595	1.201	1

TABLE - X

FLOW RATE CONVERSION FACTORS

From / to →	litre/s	gal/min	ft^3/s	ft^3/min	m^3/s	m^3/min
litre/s	1	15.85	0.03532	2.119	0.001	0.06
gal/min	0.06309	1	0.00223	0.1337	0.000063	0.00379
ft^3/s	28.32	448.8	1	60	0.02832	1.699
ft^3/min	0.4719	7.481	0.01667	1	0.000472	0.02832
m^3/s	1000	15852	35.31	2118.6	1	60
m^3/min	16.667	264.2	0.5885	35.31	0.01667	1

TABLE -XI

CONVERSION FACTORS FOR TEMPERATURE SCALES

The following formula may be used to convert temperatures from one scale to another

	Temperature wanted in			
Temp. given in	^0C	^0K	^0F	^0Rank
^0C	C	C + 273.16	1.86 C + 32	1.8C + 491.4
^0K	K – 273.16	K	1.8K – 459.4	1.8K
^0F	0.556F – 17.8	0.556F + 255.3	F	F + 459.4
^0Rank	0.556 Rank – 273.1	0.556 Rank	Rank – 459.4	Rank

TABLE - XII
CONVERSION FACTORS FOR GASES

Multiply	By	To obtain
p. p. m. of oxygen	0.698	ml/l
ml/l of oxygen	1.43	p.p.m
p.p.m of carbon dioxide	0.506	ml/l
ml/l of carbon dioxide	1.98	p.p.m

TABLE - XIII

SIEVE MESH CONVERSION TABLE - SCREEN EQUIVALENTS

Sieve No.	U.S. Standard		Meshes per Inch	Tyler Standard		Meshes per Inch	British Standard	
	Opening			Opening			Opening	
	mm	Inches		mm	inches		mm	inches
12	1.68	0.0661	10	1.65	0.065	10	1.68	0.0660
14	1.41	0.0555	12	1.40	0.055	12	1.40	0.0553
16	1.19	0.0469	14	1.17	0.046	14	1.20	0.0474
18	1.00	0.0394	16	0.991	0.039	16	1.00	0.0395
20	0.84	0.0331	20	0.833	0.0328	18	0.853	0.0336
25	0.71	0.0280	24	0.701	0.0276	22	0.669	0.0275
30	0.59	0.0232	28	0.589	0.0232	25	0.599	0.0236
35	0.50	0.0197	32	0.495	0.0195	30	0.500	0.0197
40	0.42	0.0165	35	0.417	0.0164	36	0.422	0.0166
45	0.35	0.0138	42	0.351	0.0138	44	0.353	0.0139
50	0.297	0.0117	48	0.295	0.0116	52	0.295	0.0116
60	0.250	0.0098	60	0.246	0.0097	60	0.251	0.0099
70	0.210	0.0083	65	0.208	0.0082	72	0.211	0.0083
80	0.177	0.0070	80	0.175	0.0069	85	0.178	0.007
100	0.149	0.0059	100	0.147	0.0058	100	0.152	0.006
200	0.074	0.0029	200	0.074	0.0029	200	0.076	0.003
325	0.044	0.0017	325	0.043	0.0017	240	0.066	0.0026

TABLE - XIV

CHEMICAL VOLUMETRIC FACTORS FOR USE IN TITRATIONS

1 ml. 0.01N (N/100) potassium dichromate	= 0.558 mg iron
1 ml. 0.0125N (N/80) potassium permanganate	= 0.25 mg Ca = 0.1 mg oxygen
1 ml. 0.01 N (N/100) potassium permanganate	= 0.2 mg Ca
1 ml. 0.1250N (N/8) sodium thiosulfate	= 1.0 mg oxygen
1 ml. 0.1N (N/10) sodium thiosulfate	= 3.546 mg chlorine = 1.568 mg phenol = 1.734 mg Cr = 3.8697 mg CrO_4
1 ml. 0.0250N (N/40) sodium thiosulfate	= 0.2 mg oxygen = 0.887 mg chlorine = 0.426 mg H_2S = 0.4 mg S = 0.434 mg chromium = 1.589 mg copper = 0.392 mg phenol
1 ml. 0.01250N (N/80) sodium thiosulfate	= 0.1 mg oxygen = 0.443 mg chlorine
1 ml. 0.02N (N/50) H_2SO_4 or HCl	= 1.0 mg $CaCO_3$
1 ml. 0.1N (N/10) NaOH	= 4.8 mg SO_4 = 4.0 mg SO_2
1 ml. 0.02N (N/50) EDTA	= 1.0mg $CaCO_3$
1 ml. 0.01N (N/100) silver nitrate	= 0.54 mg HCN = 0.52 mg CN = 0.3546 mg chloride
1 ml. 0.01428N (N/70) sodium sulfite	= 0.5 mg chlorine
1 ml 1.0N ferrous ammonium sulfate	= 17.425 mg Cr

1 ml conc. H_2SO_4 (Sp. Gr. 1.84/36N) neutralizes 1800 mg alkalinity as $CaCO_3$
1 ml. conc. HCl (Sp. Gr. 1.18/11.6N) neutralizes 580 mg alkalinity as $CaCO_3$
1 g lime (100% CaO) neutralizes 1785 mg acidity as $CaCO_3$
1 g lime (90% CaO) neutralizes 1606 mg acidity as $CaCO_3$

TABLE - XV

SOLUBILITY OF OXYGEN IN FRESH WATER

Temperature ^0C	*^0F*	*Dissolved oxygen p.p.m*
0	32.0	14.62
1	33.8	14.23
2	35.6	13.84
3	37.4	13.48
4	39.2	13.13
5	41.0	12.80
6	42.8	12.48
7	44.6	12.17
8	46.4	11.87
9	48.2	11.59
10	50.0	11.33
11	51.8	11.08
12	53.6	10.83
13	55.4	10.60
14	57.2	10.37
15	59.0	10.15
16	60.8	9.95
17	62.6	9.74
18	64.4	9.54
19	66.2	9.35
20	68.0	9.17
21	69.8	8.99
22	71.6	8.83
23	73.4	8.68
24	75.2	8.53
25	77.0	8.38
26	78.8	8.22
27	80.6	8.07
28	82.4	7.92
29	84.2	7.77
30	86.0	7.63

TABLE - XVI
SOLUBILITIES OF BICARBONATES, CARBONATES, CHLORIDES, HYDROXIDES AND SULFATES OF CALCIUM, MAGNESIUM AND SODIUM

Name	Formula	Solubility, mg/l	
		At 0^0 C	at 100^0 C
Calcium bicarbonate	$Ca(HCO_3)_2$	1620	Decompose
Calcium carbonate	$CaCO_3$	15	13
Calcium chloride	$CaCl_2$	336000	554000
Calcium hydroxide	$Ca(OH_2)$	2390	888
Calcium sulfate	$CaSO_4$	1290	1250
Magnesium-bicarbonate	$Mg(HCO_3)_2$	37100	Decompose
Magnesium carbonate	$MgCO_3$	101	75
Magnesium chloride	MgCl	362000	443000
Magnesium hydroxide	$Mg(OH)_2$	17	8
Magnesium sulfate	$MgSO_4$	170000	356000
Sodium bicarbonate	$NaHCO_3$	38700	Decompose
Sodium carbonate	Na_2CO_3	61400	290000
Sodium chloride	NaCl	225000	243000
Sodium hydroxide	NaOH	370000	970000
Sodium sulfate	Na_2SO_4	33600	210000

TABLE - XVII
EQUIVALENT WEIGHTS OF COMMONLY USED ACIDS, BASES AND SALTS

Hydrochloric acid	36.46
Sulphuric acid	49
Nitric acid	63
Oxalic acid	63
Acetic acid	60
Sodium hydroxide	40
Sodium carbonate	53
Sodium bicarbonate	84
Potassium hydroxide	56
Ammonia solution (sp. 0.90)	35
Potassium dichromate	49.03
Potassium permanganate	31.6
Sodium thiosulphate	248.17
Ferrous sulphate	278
Ferrous ammonium sulphate (Mohrs' salt)	392

TABLE - XVIII

CONCENTRATION OF ACIDS AND BASES
Common Commercial Strengths

	Molecular weight	*Moles per litre*	*Grams per litre*	*Percent by wt.*	*Sp. Gravity*
Acetic acid glacial	60.05	17.4	1045	99.5	1.05
Acetic acid	60.05	6.27	376	36	1.045
		5.75	264	25	1.06
Hydriodic acid	127.9	7.57	969	57	1.70
		5.51	705	47	1.50
		.86	110	10	1.10
Hydrobromic acid	80.92	8.89	720	48	1.50
		6.82	552	40	1.38
Hydrochloric acid	36.5	11.6	424	36	1.18
		2.9	105	10	1.05
Hydrofluoric acid	20.01	32.1	642	55	1.167
		28.8	578	50	1.155
Nitric acid	63.02	15.99	1008	71	1.42
		14.90	938	67	1.40
		13.3	837	61	1.37
Perchloric acid	100.5	11.65	1172	70	1.67
		9.20	923	60	1.54
Phosphoric acid	98.0	14.8	1445	85	1.70
		9.20	923	60	1.54
Sulfuric acid	98.1	18.0	1766	96	1.84
Ammonia solution	17.0	14.8	252	28	0.898
Potassium hydroxide	56.1	13.5	757	50	1.52
		1.94	109	10	1.09
Sodium carbonate	106.00	1.04	110	10	1.10
Sodium hydroxide	40.0	10.1	769	50	1.53

TABLE - XIX

AMMONIA SOLUTION
Specific Gravity and Percentage

Sp. Gravity	*Percentage Ammonia*
0.88	35%
0.885	33%
0.90	28 – 29%
0.91	25%

TABLE - XX

PERIODIC TABLE OF THE ELEMENTS

1 H 1.008	Atomic Number Symbol Atomic Mass

1 IA	2 IIA	3 IIIB	4 IVB	5 VI	6 VIB	7 VIIB	8	9 VIII	10	11 IB	12 IIB	13 IIIA	14 IVA	15 VA	16 VIA	17 VIIA	18 O
1 H 1.008																	2 He 4.003
3 Li 6.939	4 Be 9.012											5 B 10.81	6 C 12.01	7 N 14.01	8 O 16.00	9 F 19.00	10 Ne 20.18
11 Na 22.99	12 Mg 24.31											13 Al 26.98	14 Si 28.09	15 P 30.97	16 S 32.06	17 Cl 35.45	18 Ar 39.95
19 K 39.10	20 Ca 40.08	21 Sc 44.96	22 Ti 47.90	23 V 50.94	24 Cr 52.00	25 Mn 54.94	26 Fe 55.85	27 Co 58.93	28 Ni 58.71	29 Cu 63.54	30 Zn 65.39	31 Ga 69.72	32 Ge 72.59	33 As 74.92	34 Se 78.96	35 Br 79.91	36 Kr 83.80
37 Rb 85.47	38 Sr 87.62	39 Y 88.91	40 Zr 91.22	41 Nb 92.91	42 Mo 95.94	43 Tc (99)	44 Ru 101.1	45 Rh 102.9	46 Pd 106.4	47 Ag 107.9	48 Cd 112.4	49 In 114.8	50 Sn 118.7	51 Sb 121.8	52 Te 127.6	53 I 126.9	54 Xe 131.3
55 Cs 132.9	56 Ba 137.3	57 La 138.9	72 Hf 178.5	73 Ta 181.0	74 W 183.9	75 Re 186.2	76 Os 190.2	77 Ir 192.2	78 Pt 195.1	79 Au 197.0	80 Hg 200.6	81 Ti 204.4	82 Pb 207.2	83 Bi 209.0	84 Po (210)	85 At (210)	86 Rn (222)
87 Fr (223)	88 Ra (226)	89 Ac (237)	104 Unq	105 Unp	106 Unh	107 Uns	108 Uno	109 Une									

58 Ce 140.1	59 Pr 140.9	60 Nd 144.2	61 Pm (147)	62 Sm 150.4	63 Eu 152.0	64 Gd 157.3	65 Tb 158.9	66 Dy 162.5	67 Ho 164.9	68 Er 167.3	69 Tm 168.9	70 Yb 173.0	71 Lu 175.0
90 Th 232.0	91 Pa (231)	92 U 238.0	93 Np (237)	94 Pu (242)	95 Am (243)	96 Cm (247)	97 Bk (247)	98 cF (249)	99 Es (254)	100 Fm (253)	101 Md (256)	102 No (253)	103 Lr (257)

TABLE - XXI

INTERNATIONAL ATOMIC WEIGHTS 1973

Element	*Symbol*	*Atomic No.*	*Atomic Weight*
Actinium	Ac	89	(227)
Aluminium	Al	13	26.9815
Americium	Am	95	(243)
Antimony	Sb	51	121.75
Argon	Ar	18	39.948
Arsenic	As	33	74.9216
Astatine	At	85	(210)
Barium	Ba	56	137.34
Berkeylium	Bk	97	(247)
Beryllium	Be	4	9.0122
Bismuth	Bi	83	208.9804
Boron	B	5	10.81
Bromine	Br	35	79.904
Cadmium	Cd	48	112.40
Calcium	Ca	20	40.08
Californium	Cf	98	(251)
Carbon	C	6	12.011
Cerium	Ce	58	140.12
Cesium	Cs	55	132.9054
Chlorine	Cl	17	35.453
Chromium	Cr	24	51.996
Cobalt	Co	27	58.9332
Copper	Cu	29	63.542
Curium	Cm	96	(247)
Dysprosium	Dy	66	162.50
Einsteinium	Es	99	(254)
Erbium	Er	68	167.26
Europium	Eu	63	151.96
Fermium	Fm	100	(257)
Fluorine	F	9	18.9984
Francium	Fr	87	(223)
Gadolinium	Gd	64	157.25
Gallium	Ga	31	69.72
Germanium	Ge	32	72.59
Gold	Au	79	196.9665
Hafnium	Hf	72	178.49
Helium	He	2	4.0026
Holmium	Ho	67	164.9304
Hydrogen	H	1	1.0079
Indium	In	49	114.82
Iodine	I	53	126.9045
Iridium	Ir	77	192.22
Iron	Fe	26	55.847

Element	*Symbol*	*Atomic No.*	*Atomic Weight*
Krypton	Kr	36	83.80
Lanthanum	La	57	138.9066
Lawrencium	Lr	103	(260)
Lead	Pb	82	207.0
Lithium	Li	3	6.941
Lutetium	Lu	71	174.97
Magnesium	Mg	12	24.305
Manganese	Mn	25	54.9380
Mendelevium	Md	101	(258)
Mercury	Hg	80	200.59
Molybdenum	Mo	42	95.94
Neodymium	Nd	60	144.24
Neon	Ne	10	20.179
Neptunium	Np	93	237.0482
Nickel	Ni	28	58.70
Niobium	Nb	41	92.9064
Nitrogen	N	7	14.0067
Nobelium	No	102	(255)
Osmium	Os	76	190.2
Oxygen	O	8	15.9994
Palladium	Pd	46	106.4
Phosphorus	P	15	30.9738
Platinum	Pt	78	195.09
Plutonium	Pu	94	(244)
Polonium	Po	84	(209)
Potassium	K	19	39.098
Praseodymium	Pr	59	140.9077
Promethium	Pm	61	(145)
Protactinium	Pa	91	231.0359
Radium	Ra	88	226.0254
Radon	Rn	86	(222)
Rhenium	Re	75	186.207
Rhodium	Rh	45	102.9055
Rubidium	Rb	37	85.4678
Ruthenium	Ru	44	101.07
Samarium	Sm	62	150.4
Scandium	Sc	21	44.9559
Selenium	Se	34	78.96
Silicon	Si	14	28.086
Silver	Ag	47	107.868
Sodium	Na	11	22.9898
Strontium	Sr	38	87.62
Sulphur	S	16	32.06
Tantalum	Ta	73	180.9479
Technetium	Tc	43	(97)
Tellurium	Te	52	127.60
Teribium	Tb	65	158.9254
Thallium	Tl	81	204.37
Thorium	Th	90	232.0381
Thulium	Tm	69	168.9342
Tin	Sn	50	118.69

Element	Symbol	Atomic No.	Atomic Weight
Titanium	Ti	22	47.90
Tungsten	W	74	183.85
Uranium	U	92	238.029
Vanadium	V	23	50.9414
Xenon	Xe	54	131.30
Ytterbium	Yb	70	173.04
Yttrium	Y	39	88.9059
Zinc	Zn	30	65.38
Zirconium	Zr	40	91.22

TABLE - XXII

CHEMICAL FORMULA, FORMULA WEIGHTS AND EQUIVALENT WEIGHTS

Name	Chemical Formula	Formula Weight	Equivalent Weight
Alum, ammonia	see Ammonium alum		
Alum, filter	see Aluminium sulphate		
Alum, potash	See Potassium oxide		
Aluminium	Al	27.0	9.00
Aluminium chloride	$AlCl_3$	133	44.4
Aluminium chloride	$AlCl_3.6H_2O$	241	80.5
Aluminium hydroxide	$Al(OH)_3$	78.0	26.0
Aluminium oxide	Al_2O_3	102	17.0
Aluminium sulphate	$Al_2(SO_4)_3.I8H_2O$	666	111
Ammonia	NH_3	17.0	17.0
Ammonium ion	NH_4^+	18.0	18.0
Ammonium alum	$Al_2(SO_4)_3.(NH_4)_2SO_4.24H_2O$	906	151
Ammonium chloride	NH_4Cl	53.5	53.5
Ammonium - hydroxide	NH_4OH	35.1	35.1
Ammonium sulphate	$(NH_4)_2SO_4$	132	66.1
Barium	Ba	137	68.7
Barium carbonate	$BaCO_3$	197	98.7
Barium chloride	$BaCl_2.2H_2O$	244	122
Barium sulphate	$BaSO_4$	233	117
Bicarbonate ion	HCO_3^-	61.0	61.0
Bicarbonate of soda	see Sodium bicarbonate		
Blue vitriol	see Copper sulphate		
Calcite	see Calcium carbonate		
Calcium	Ca	40.1	20.05
Calcium bicarbonate	$Ca(HCO_3)_2$	162	81.0
Calcium carbonate	$CaCO_3$	100	50.0
Calcium chloride	$CaCl_2$	111	55.5
Calcium hydroxide	$Ca(OH)_2$	74.1	37.1
Calcium nitrate	$Ca(NO_3)_2$	164	82.1
Calcium phosphate (tri-)	$Ca_3(PO_4)_2$	310	51.7

Name	Chemical Formula	Formula Weight	Equivalent Weight
Calcium sulphate	$CaSO_4$	136	68.1
Calcium sulphate		145	72.6
Calcium sulphate	$CaSO_4.(1/2)H_2O$ (Plaster of Paris) $CaSO_4.2H_2O$ (gypsum)	172	86.1 3.00
Carbon	C	12.0	
Carbon dioxide (gas)	CO	44.0	22.0
Carbonate ion	CO_3^{2-}	60.0	38.0
Carbonic acid	H_2CO_3	62.0	31.0
Caustic potash	see Potassium hydroxide		
Caustic soda	see Sodium hydroxide		
Chalk	see Calcium carbonate		
Chemical lime	see Calcium oxide		
Chile saltpeter	see Sodium nitrate		
Chloride ion	Cl^-	35.5	35.5
Chlorine (gas)	Cl_2	70.9	35.5
Copper	Cu	63.6	31.8
Copper sulphate	$CuSO_4.5H_2O$	250	125
Copperas	See Ferrous sulphate	184	46
Dolomite	$CaCO_3.MgCO_3$		
Dolomitic lime	CaO + various amounts MgO		
Dolomitic lime, hydrated	$Ca(OH_2)$ + various amounts MgO		
Dolomitic limestone	$CaCO_3$ + various amounts $MgCO_3$		
Epsom Salt	See Magnesium sulphate		
Ferric chloride	$FeCl_3$	162	54.1
Ferric chloride	$FeCl_3.6H_2O$	270	90.1
Ferric hydroxide	$Fe(OH)_3$	107	35.6
Ferric oxide	Fe_2O_3	160	26.6
Ferric sulphate	$Fe_2(SO_4)_3$	400	66.6
Ferroso-Ferric hydroxide	$Fe_3(OH)_8$	304	
Ferroso-ferric oxide	Fe_3O_4	232	
Ferrous bicarbonate	$Fe(HCO_3)_2$	178	89.0
Ferrous carbonate	$FeCO_3$	116	57.9
Ferrous hydroxide	$Fe(OH)_2$	89.9	44.9
Ferrous oxide	FeO	71.8	35.9
Ferrous sulphate	$FeSO_4.7H_2O$	278	139
Fluoride ion	F^-	19.0	19.0
Fluorine gas	F_2	38.0	19.0
Fluosilicic acid	H_2SiF_6	144	72.0
Green vitriol	see Ferrous sulphate		
Gypsum	see $CaSO_4.2H_2O$		
Hydrochloric acid	HCl	36.5	36.5
Hydrofluoric acid	HF	20.0	20.0
Hydrofluosilicic acid	See Fluosilicic acid		
Hydrogen ion	H^+	1.01	1.01

Name	Chemical Formula	Formula Weight	Equivalent Weight
Hydrogen gas	H_2	2.02	1.01
Hydrogen sulphide (gas)	H_2S	34.1	17.0
Iodide ion	I^-	127.0	127
Iodine	I_2	254	127
Iron (ferric ion)	Fe^{+++}	55.8	18.6
Iron (ferrous ion)	Fe^{++}	55.8	27.9
Lead	Pb	207	104
Lime	see Calcium oxide		
Lime, chemical	see Calcium oxide		
Lime, dolomitic	see Dolomitic lime		
Lime, hydrated	see Calcium hydroxide		
Lime, slaked	see Calcium hydroxide		
Lime, quick	see Calcium oxide		
Magnesia	see Magnesium oxide		
Magnesite	see Magnesium carbonate		
Magnesium - bicarbonate	$Mg(HCO_3)_2$	146	73.2
Magnesium - carbonate	$MgCO_3$	84.3	42.2
Magnesium chloride	$MgCl_2$	95.2	47.6
Magnesium - hydroxide	$Mg(OH)_2$	58.3	29.2
Magnesium nitrate	$Mg(NO_3)_2$	148	74.2
Magnesium oxide	MgO	40.3	20.2
Magnesium phosphate(tri-)	$Mg_3(PO_4)_2$	263	43.8
Magnesium sulphate	$MgSO_4$	120	60.2
Magnesium sulphate	$MgSO_4.7H_2O$ (Epsom salt)	246	123
Manganese (manganic ion)	Mn^{+++}	54.9	18.3
Manganese (manganous ion)	Mn^{++}	54.9	27.5
Manganese dioxide	MnO_2	86.9	21.7
Manganic hydroxide	$Mn(OH)_3$	106	35.3
Manganic oxide	Mn_2O_3	158	26.3
Manganous - bicarbonate	$Mn(HCO_3)_2$	177	88.5
Manganous carbonate	$MnCO_3$	115	57.5
Manganous hydroxide	$Mn(OH)_2$	89.0	44.5
Manganous oxide	MnO	70.9	35.5
Muriatic acid	see Hydrochloric acid		
Nitrate ion	NO_3^-	62.0	62.0

Name	Chemical Formula	Formula Weight	Equivalent Weight
Nitric acid	HNO_3	63.0	63.0
Nitrogen	N (valencies 3 and 5)	14.0	
Nitrogen (gas)	N_2	28.0	
Oil of vitriol	see Sulphuric acid		
Oxygen	O	16.0	8.0
Oxygen (gas)	O_2	32.0	8.0
Pearl ash	see Potassium carbonate		
Permanganate of potash	see Potassium permanganate		
Phosphate (ortho ion)	PO_4^{3-}	95.0	31.7
Phosphorus	P (valencies 3 and 5)	31.0	
Phosphorus pentoxide	P_2O_5	142	23.7
Plaster of paris	see $CaSO_4.1/2H_2O$		
Potash	see Potassium carbonate		
Potash alum	see Potassium alum		
Potassium	K	39.1	39.1
Potassium alum	$Al_2(SO_4)_3.K_2SO_4.24H_2O$	949	158
Potassium bicarbonate	$KHCO_3$	100	100
Potassium carbonate	K_2CO_3	138	69.1
Potassium chloride	KCl	74.6	74.6
Potassium hydroxide	KOH	56.1	56.1
Potassium iodide	KI	166	166
Potassium nitrate	KNO_3	101	101
Potassium oxide	K_2O	94.2	47.1
Potassium permanganate	$KMnO_4$	158	31.6
Potassium sulphate	K_2SO_4	174	87.1
Quartz	see Silicon dioxide		
Quicklime	see Calcium oxide		
Sal ammoniac	see Ammonium chloride		
Sal soda	see $Na_2CO_3.10H_2O$		
Salt (common)	see Sodium chloride		
Salt cake	see Sodium sulphate		
Saltpeter	see Potassium nitrate		
Saltpeter, Chile	see Sodium nitrate		
Silica	see Silicon dioxide		
Silicon	Si	28.1	7.03
Silicon dioxide	SiO_2	60.1	
Silver	Ag	108	108
Silver chloride	AgCl	143	143
Silver nitrate	$AgNO_3$	170	170
Soda	see Sodium carbonate		
Soda ash	see Na_2CO_3		

Name	Chemical Formula	Formula Weight	Equivalent Weight
Sodium	Na	23.0	23.0
Sodium bicarbonate	$NaHCO_3$	84.0	84.0
Sodium bisulphate	$NaHSO_4$	120	
Sodium bisulphite	$NaHSO_3$	104	
Sodium carbonate	Na_2CO_3 (Soda ash)	106	53.0
Sodium carbonate	$Na_2CO_3.10H_2O$ (Sal soda)	286	143
Sodium chloride	NaCl	58.5	58.5
Sodium chromate	Na_2CrO_4	162	
Sodium dichromate	$Na_2Cr_2O_7.2H_2O$	298	
Sodium hydroxide	NaOH	40.0	40.0
Sodium nitrate	$NaNO_3$	85.0	85.0
Sodium oxide	Na_2O	62.0	31.0
Sodium phosphates:			
meta (comm. Grade 69.0% P_2O_5)	$NaPO_3$	102	34.0
mono (comm. Grade 58.0% P_2O_5)	NaH_2PO_4	120	40.0
mono (comm. Grade 52.0% P_2O_5)	$NaH_2PO_4.H_2O$	138	46.0
di (comm. Grade 49.0% P_2O_5)	Na_2HPO_4	142	47.3
di (comm. Grade 19.0% P_2O_5)	$Na_2HPO_4.12H_2O$	358	119
tri (comm. Grade 39.5% P_2O_5)	$Na_3PO_4.H_2O$	182	60.7
tri (comm. Grade 18.9% P_2O_5)	$Na_3PO_4.12H_2O$	380	127
Sodium sulphate	Na_2SO_4 (Salt cake)	142	71.0
Sodium sulphate	$Na_2SO_4.10H_2O$ (Glauber's salt)	322	161
Sodium sulphite	Na_2SO_3	126	63.0
Sodium tetrathionate	$Na_2S_4O_6$	270	
Sodium thiosulphate	$Na_2S_2O_3$	158	
Sulphate ion	SO_4^{2-}	96.1	48.0
Sulphide ion	S^{2-}	32.1	16.0
Sulphite ion	SO_3^{2-}	80.1	40.0
Sulphur	S (valencies 2, 4 and 6)	32.1	
Sulphur dioxide	SO_2	64.1	32.0
Sulphur trioxide	SO_3	80.1	40.0
Sulphuric acid	H_2SO_4	98.1	49.0
Sulphurous acid	H_2SO_3	82.1	41.0
Tin	Sn (valencies 2 and 4)	119	
Vitriol	see Sulphuric acid		
Water	H_2O	18	9.00
Zinc	Zn	65.4	32.7

TABLE - XXIII

DATA ON CHEMICALS USED IN WATER AND WASTE WATER TREATMENT

Chemical Name, Formula and Formula Weight	Common Name/Trade Name	Available form	Commercial strength	Charact–eristic	Solubility gm/100 ml water				Weight lb/cu.ft	Storage space cu.ft / ton	Shipping containers	Suitable handling materials
					0°C	10°C	20°C	30°C				
Activated Carbon – See Carbon activated												
Aluminium sulfate $Al_2(SO_4)_3.14H_2O$ 594	Alum Filter Alum Sulfate of Alumina	Ivory/Brown colored slab, lump, granule, powder, Also available in liquid form	14.5% to 17.5% Al_2O_3	pH of 1% solution 3.4	60.8	65.3	71.0	78.8	Slab lump (or) ground 57-67 powdered 38-45	30-35 45-53	Bags, barrels and in bulk; liquid is transported by Tankers	Dry : Iron, steel Liquid: Lead lined rubber, silicon asphalt
Alum, ammonium $Al_2(SO_4)_3.(NH_4)_2.SO_4.2H_2O$ 906	Ammonia Alum (Sometimes called as crystal alum)	Lump, nut, powder	11% Al_2O_3	pH of 1% solution 3.5	3.9	9.5	15.1	20.0	64-68	30-31	Bags, barrels, boxes, Cartons, Kegs and bulk	Lead lined rubber, silicon, Iron, stone-ware.
Alum potassium $Al_2(SO_4)_3.K_2SO_4.24H_2O$ 949	Potash alum	Lump, crystal	10.7% Al_2O_3	pH 1% solution 3.5	5.7	7.6	11.4	16.6	64-68	30-31	Bags, barrels, boxes, cartons, and kegs	Lead lined rubber, stone-ware.
Bentonite	Colloidal clay, Volclay, Wilkinite	Powder, pellet, mixed sizes	-	-	-	-	-	-	60	-	Bags bulk	Iron steel

Chemical Name, Formula and Formula Weight	Common Name/Trade Name	Available form	Commercial strength	Charact-eristic	Solubility gm/100 ml water				Weight lb/cu.ft	Storage space cu.ft / ton	Shipping containers	Suitable handling materials
					0°C	10°C	20°C	30°C				
Bleaching powder See "chlorinated lime"												
Calcium Carbonate $CaCO_3$ 100	Lime stone, Calcite, Chalk, Whiting	Crushed and graded granules; also available in powder form	–	–	–	–	13	–	Granule 100-115 powder 48-71	18-20 28-42	Bags, barrels bulk	
Calcium chloride $CaCl_2$ 111	–	Flake: 77-78% $CaCl_2$ Solid: 73-75% $CaCl_2$ Liquid: 40% $CaCl_2$	–	–	59.5	65.0	74.5	102	60-66	29-31	Flakes: moisture proof bags or drums Solids: drums	–
Calcium hydroxide $Ca(OH)_2$ 74.1	Hydrated lime, slaked lime	Powder (light and dense)	85-99% $Ca(OH)_2$	Hopper agitation required for dry feed of light form.	0.18	0.17	0.16	0.15	25-50	40-80	Bags, barrels, bulk	Rubber, Iron, Steel, cement and asphalt
Calcium hypochlorite $CaOCl_2.4H_2O$ 199	"ATH" "Perchloron" "Pittchlor"	Powder	70% available chlorine	Generally 1-3% available Cl_2 solution used.	Highly soluble				–	–	Cans and drums	Glass, rubber, stoneware and wood

Chemical Name, Formula and Formula Weight	Common Name/Trade Name	Available form	Commercial strength	Charact-eristic	Solubility gm/100 ml water				Weight lb/cu.ft	Storage space cu.ft / ton	Shipping containers	Suitable handling materials
					0°C	10°C	20°C	30°C				
Calcium oxide CaO 56.1	Lime, burnt lime, chemical lime, quick lime, unslaked lime	Lump, pebble ground and powder.	75 to 99% CaO	Slakes with water to form hydrated lime $Ca(OH)_2$ For slaking 1 Kg. lime, 10 litres of water is required	–	–	–	–	Lump 50-65 Ground 50-70 Pulv. 39-71	31-40 29-40 28-52	Bags, barrels and in bulk	Rubber, Iron, steel, cement and asphalt
Calcium sulfate $CaSO_4.2H_2O$ -172	Gypsum	Ground	–	–	0.18	0.19	0.20	0.21	50-60	33-40	Bags, barrels and in bulk	–
Carbon activated C 12	–	Black granules and powder	–	–	Insoluble				Granular 12-30 Powder 15-28	67-167 72-134	Bags, bulk	Dry : iron, steel, Wet :rubber, stainless steel.
Caustic soda See Sodium hydroxide												
Chlorinated lime $CaO.2CaOCl_2.3H_2O$ 364	Bleaching Powder, chloride of lime	White powder	25-37% available chlorine	Deteriorates on exposure to air	Very soluble				48	–	Drums	Glass, rubber, stone–ware, wood
Chlorine Cl_2 70.9	–	Chlorine gas, Liquid chlorine (liquified gas under pressure in steel cylinders)	99.8% Cl_2	–	1.46	0.980	0.716	0.562	–	–	Cylinders Tankers, Tank trucks	Dry: Iron Copper, steel, wetgas: glass, hardrubber.
Chlorine dioxide ClO_2 102.9	Chlorine–dioxide	Yellow red gas	26.3% available chlorine	–	–	–	–	–	–	–	Generated at the site itself	Plastic, soft rubber (Hard rubber to be avoided)

Chemical Name, Formula and Formula Weight	Common Name/Trade Name	Available form	Commercial strength	Charact-eristic	Solubility gm/100 ml water				Weight lb/cu.ft	Storage space cu.ft / ton	Shipping containers	Suitable handling materials
					0°C	10°C	20°C	30°C				
Copper sulfate $CuSO_4.5H_2O$ 249.6	Blue vitriol Blue stone	Crystal, lump powder	99% $CuSO_4$	–	19.2	–	26.4	31.2	Crystal 75-90 lump 73-80 Powder 60-64	–	Bags, barrels, drums	Iron, Stainless steel, Silicon, asphalt
Disodium phosphate (see Sodium phosphate)												
Dolomitic lime CaO + MgO (Content of MgO varies)	Dolomitic – lime, High magnesium lime	Lump, pebble, ground or powder	58% CaO and 40% MgO	Typical for Silica removal Slakes with water forming $Ca(OH)_2$+ MgO the later Slakes very slowly	–	–	–	–	Lump 50-65 Pebble 60-65 Ground 50-75 Pulv 37-63	31-40 31-34 27-40 32-54	Bags, barrels and in bulk	Wood, Iron steel
Dolomitic lime, hydrated $Ca(OH)_2$ + MgO content of MgO varies	Dolomitic – lime, hydrated	Powder	62% $Ca(OH)_2$ and 32% MgO	Typical for silica removal	–	–	–	–	28-52	39-72	Bags, barrels and in bulk	Wood, Iron Steel
Ferric chloride (i) $FeCl_3$ (ii) $FeCl_3.6H2O$ (iii) $FeCl_3$ (35045%) solution	Anhydrous ferric chloride Crystal ferric Chloride "Ferrichlor" Chloride of Iron	Green black powder Yellow brown lump Dark brown Syrupy liquid	98% $FeCl_3$ 34% Fe 59-61% $FeCl_3$ 20-21% Fe 37-47% $FeCl_3$ 20-21% Fe	Highly hygroscopic; lumps and powder should be stored in air tight containers	Completely soluble				–	–	Casks, kegs barrels car-boys, trucks, tankers.	Rubber, glass, stoneware synthetic-resins

Chemical Name, Formula and Formula Weight	Common Name/Trade Name	Available form	Commercial strength	Charact-eristic	Solubility gm/100 ml water				Weight lb/cu.ft	Storage space cu.ft / ton	Shipping containers	Suitable handling materials
					0°C	10°C	20°C	30°C				
Ferric Sulfate $Fe_2(SO_4)_3$ 400	Ferrisul Ferrifloc	Granules Reddish brown powder	Composition from different sources varies 70%-90% $Fe_2(SO_4)_3$ some grades contain $Al_2(SO_4)_3$ also	Mildly hygroscopic. coagulated at pH 8.5 – 11.0	Highly soluble, to dissolve in cold water, add 2 parts water to 1 part ferric sulfate with stirring.				60-70	29-34	Bags, barrels, kegs, drum and in bulk	Rubber, plastic, lead ceramic, stainless-steel
Ferrous sulfate $FeSO_4.7H_2O$ 278	Copperas green vitriol	Crystals, granules	55% $FeSO_4$ 20% Fe	Hygro-scopic: cakes on storage. Optimum pH 8.5 – 11.0	28.7	37.5	48.5	60.2	63-66	30-32	Bags, barrels kegs and in bulk	Wood, lead asphalt, concrete
Hydrated lime (See Calcium hydroxide)												
Lime (See Calcium oxide)												
Hydrazine N_2H_4 32.05	–	Colorless oily liquid	–	Powerful reducing agent fuming in air; penetrating odor resembling that of ammonia	Miscible with water				–	–	Glass	Glass
Hydrochloric acid HCl 36.45	Muriatic acid	sometimes colored by traces of iron chlorine and organic matter	38% HCl	Fumes in air; pH of 1N HCl is 0.1 0.1N HCl = 1.1 0.01N HCl = 2.02	–	–	–	–	–	–	Bulk in Tankers	Plastic rubber

Chemical Name, Formula and Formula Weight	Common Name/Trade Name	Available form	Commercial strength	Charact-eristic	Solubility gm/100 ml water				Weight lb/cu.ft	Storage space cu.ft / ton	Shipping containers	Suitable handling materials
					0°C	10°C	20°C	30°C				
Meta Sodium-phosphate (See sodium phosphate meta)	–	–	–									
Soda ash (See Sodium carbonate)												
Sodium aluminate $Na_2O.Al_2O_3$ 164	Soda alum	Brown powder, liquid (27°Be)	70-80% $Na_2Al_2O_4$ Min. 32% $Na_2.Al_2O_4$	For dry feed hopper agitation required	Complete				50-60	–	Bags, drums tankers	Iron, plastic, rubber, steel
Sodium carbonate Na_2CO_3 106	Soda ash	White powder Extra light, dense	Light soda ash is used in water treatment. 99.4% Na_2CO_3 58% Na_2O	pH of 1% solution 11.3 Hopper agitation required for dry feeding	7.0	12.5	4.5	38.8	23 35 65	–	Bags, barrels bulk [Carloads, trucks]	Iron, rubber, steel
Sodium chloride NaCl 58.5	Salt, common salt, rock salt evaporated salt	Crystal, powder	98% NaCl	–	35.7	35.8	36.0	36.3	50-70	29-40	Bags, barrels, bulk	Rubber, cement, plastic, bronze
Sodium hydroxide NaOH 40	Caustic soda	Flakes, solid, ground, liquid	98.06 NaOH (76% Na_2O) 50-70% solution	Solid Highly hygoscopic pH of 1% solution 12.9	42.0	51.5	109	119	–	–	Flake : drums Solid : drums Solution: in tankers	Cast iron steel, rubber.
Sodium hypochlorite NaOCl 74.5	Sodium – hypochlorite	Light yellow liquid	12-15% available chlorine	–	–	–	–	–	–	–	Carboys, Bulk in tankers	Glass plastics, rubber, ceramics
Sodium hexameta – phosphate $(NaPO_3)_6$ 750	'Calgon', glassy phosphate, vitreous phosphate	Crystal, flakes, powder	66% P_2O_5 (unadjusted)	pH of 0.25% solution 6.0-8.3	–	–	–	–	47	–	Bags	Plastics, hard rubber, stainless steel

Chemical Name, Formula and Formula Weight	Common Name/Trade Name	Available form	Commercial strength	Charact-eristic	Solubility gm/100 ml water				Weight lb/cu.ft	Storage space cu.ft / ton	Shipping containers	Suitable handling matcrials
					0°C	10°C	20°C	30°C				
Sodium phosphate, mono NaH_2PO_4 (anhydrous) 120 $NaH_2PO_4.H_2O$ 138	–				Monohydrate				Anhydrous 55-70 Mono hydrate 58-79	29-37 26-35	Bags, barrels, boxes, kegs drums	Cast iron, steel
					66.6	80.4	98.0	12.3				
Sodium phosphate, Disodium – phosphate, Disodium hydrogen – phosphate Na_2HPO_4 – 142 (anhydrous) $Na_2HPO_4.12H_2O$ 358	Basic sodium phosphate , DSP, secondary sodium phosphate	Crystal	19.5% P_2O_5	Precipitates calcium and magnesium pH of 1% solution 9	Anhydrous				Anhydrous 53-62 Dodeca hydrate 46-53	32-39 38-44	Bags, barrels, kegs, drum	Cast Iron, steel
					1.7	3.6	7.7	20.8				
					Dodeca hydrate							
					4.2	8.9	19.3	52.4				
Sodium phosphate Trisodium – phosphate $Na_3PO_4.H_2O$ 182 $Na_3PO_4.12H_2O$ 380	Normal sodium phosphate Tertiary sodium phosphate	Crystal coarse Medium standard	19% P_2O_5	pH of 1% solution 11.9	Monohydrate				83-90 56-*60	22-24 32-36	Bags, barrels boxes, kegs, drum	Cast Iron, steel
					1.7	4.6	12.3	22.2				
					Dodeca hydrate							
					3.5	9.5	25.5	46.4				
Sodium phosphate, meta $NaPO_3$ 102	–	Flake Glass	–	Soluble in water with alkaline reaction	Highly soluble				Flake 17–19 Glass 39–41	105–118 49–51	–	–
Sodium tripoly – phosphate (STPP) polypenta sodium triphosphate $Na_5P_3O_{10}$ – 367.9	–	Slightly hygroscopic granules	–	pH of 1% solution 9.7 to 9.8	20 gm at 25°C and 86.5 gm at 100°C				–	–	–	–
Sodium Silicate $Na_2O.SiO_2$ (variable ratio of Na_2O to SiO_2)	Water glass, Silicate of soda	Opaque viscous liquid	38° – 40° Be solution	pH of 1% solution 12.3	Miscible in water in all proportions				86.1	–	Drums Tankers	Cast iron rubber, steel
Sodium sulfite Na_2SO_3 126	Sulfite	White crystals, powder	23% SO_2	Sulfurous taste and odor	14.0	20.0	27.0	36.0	80-91	23-26	Bags, barrels kegs and drums	Iron, steel wood

Chemical Name, Formula and Formula Weight	Common Name/Trade Name	Available form	Commercial strength	Characteristic	Solubility gm/100 ml water				Weight lb/cu.ft	Storage space cu.ft / ton	Shipping containers	Suitable handling materials
					0°C	10°C	20°C	30°C				
Sulfur dioxide SO_2 64.1	Sulfurous acid anhydride	Colorless gas (liquefied gas under pressure in steel cylinders)	99% SO_2	Highly corrosive and irritating gas	22.8	16.2	11.3	7.8	-	-	Steel cylinders, Tankers	Aluminium, brass, stainless steel
Sulfuric acid H_2SO_4 98.1	Oil of vitriol	Liquid	98% 93.2% (66° Be) 77.7% (60° Be)	Approximate pH value of 0.5% solution 1.2	Miscible with water in all proportions				-	-	Bottles, carboys, drums, tankers	Concentrated: Iron, steel, Dilute: glass, lead, porcelain rubber.
Tetra Sodium phosphate Pyro $Na_4P_2O_7.10H_2O$ 445.8	Alkaline sodium pyrophosphate	White powder	53% P_2O_5	pH of 1% solution 10.5	-	-	-	0.6	68	-	Bags barrels, kegs	Cast Iron, steel.

INDEX

A

B

C

D

E

F

G

H

I

K

L

M

N

O

P

R

S

U

V

W

Y

Z

www.ingramcontent.com/pod-product-compliance
Lightning Source LLC
LaVergne TN
LVHW021002300926
849160LV00007B/229